22 W

Introduction
to
Forest Biology

Harold W. Hocker, Jr.
University of New Hampshire

Introduction to Forest Biology

John Wiley & Sons
New York Chichester Brisbane Toronto

Library of Congress Cataloging in Publication Data

Hocker, Harold W. 1926-
  Introduction to forest biology.

  Includes index.
  1. Forests and forestry. 2. Trees. 3. Forest
ecology. I. Title.

SD395.H62   582'.15   78-26878
ISBN 0-471-01978-X

Printed in the United States of America

10 9 8 7 6 5 4 3 2 1

DEDICATED TO DOTTIE AND BRAD

Preface

Expansion in knowledge requires that students, before they undertake a course in silviculture, have a basic understanding of forest genetics, physiology, soils, hydrology, and ecology. A background in the subjects of forest entomology, pathology, and wildlife is necessary, as well as an understanding of forest stand structure, development, and reproduction. Most students cannot obtain an adequate background in biology before they enter silviculture if they must take each subject as a single course. Therefore, it seemed appropriate to include an introduction to biology in a single introductory course, offered in one semester with a recognition that it cannot substitute for a thorough study of the subjects involved. This course provides a background for silviculture, and at the same time serves as a foundation for students who plan advanced study.

I did not use the title *Forest Biology* without reservation; it was derived in part from one B.E. Fernow proposed in 1905: biological dendrology. Dendrology has since been adopted for use as a title for the subject of forest tree identification and taxonomy. Forest biology seemed an apt description of the collection of topics found in a course that followed dendrology but preceded silviculture.

I have presented each topic so that students with limited backgrounds will understand principles, yet I have also tried to develop each sufficiently. Chapters are arranged so they may be studied in any order. References to research and choice of illustrations have been limited to those most appropriate.

I wish to thank those who have reviewed the manuscript in various stages of development; their suggestions have helped immeasurably. I am also grateful to Phyllis Groves, Jean Gilman, and Cindy Grimard, who each typed and re-typed the drafts of the manuscript.

Harold W. Hocker, Jr.

Contents

Introduction

The forest can be envisioned as an aggregate of plant and animal life subsisting within an abiotic environment.* Large trees form the main canopy of the forest, with smaller trees sharing space in the understory along with shrubs and herbs, and grasses may occur in openings. Mosses and lichens grow on the ground, the rocks, and on the boles and branches of some trees. Fungal fruiting bodies may be found during certain seasons, or perennially, on dead or diseased trees. Interspersed within the plant community—but an integral part of the system—are animals, both large and small, insects, and other smaller microorganisms. Each component makes its contribution to the flow of energy and minerals through the system.

The natural forest is a combination of relationships where no single phase can describe it clearly; it can only be understood in terms of the

*Terms that may be unfamiliar to some readers are defined in the Glossary.

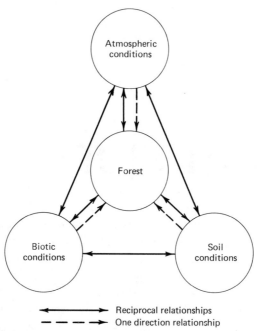

Reciprocal relationships
One direction relationship

FIGURE I.-1. Schematic diagram of the biotic and abiotic components of a forest and their functional relationships.

reciprocal and direct dependence of all the biotic and abiotic components (Figure I.1). A forest community must also be considered a dynamic structure that responds to the dynamic laws of cause and effect where all organisms intertwine their existence within an environment to form a harmonic whole. Disturbances are normal, occurring as fire, wind, insect infestations, or disease; they result in destruction of small or large segments of the whole or, when extreme, contribute to the destruction of the whole forest. The effect of disturbance is to produce sites where new communities of trees, plants, and animals can exist, and which differ from those of the mature forest. The annual leaf fall triggers a recycling of energy through the systems that augments the breakdown and decay of whole plants and trees.

The various aspects of the forest system can be divided into a number of special areas of study, but the forest can only be considered as an intact biological system, and it is this point of view which will be taken in this book.

Forest biology also provides an understanding of the principles that govern the practice of silviculture. Silviculture, as the name implies, deals with the tending and reproducing of forest stands, the culture of the forest. Forest biology is sometimes referred to as forest ecology or silvics. However, if either ecology or silvics is taken in too narrow a sense, it will not encompass all that is forest biology. Forest biology needs to consider the genetic and environmental factors that separately and in interaction determine the form and functioning of life systems and the growth and development of forest trees and stands. In addition, it encompasses consideration of the influence of the trees and stands upon other biological communities and systems. It is important that foresters in practicing their profession have an understanding of forest stand structure and development and the elements related to the distribution of forest stands. They must have, as well, an understanding of the interaction of influencing factors and of the immediate and long-range effects change in stand structure may have on other systems. The material that follows is designed to define the circumstances which explain the occurrence and growth of forest trees, and at the same time to provide background on the relationship between a number of the biological subsystems that occur in forest stands.

The location in which a particular tree grows can be defined as a *niche*. A niche has biotic and abiotic dimensions. Tree species have evolved to take advantage of the various niches that exist in natural environments. The niche of a forest tree species can be defined by the site factors that identify its particular needs; these factors are the soil, climate, physiography—the elements of site. To complete the definition of niche requires that the biotic, the living phase of the habitat be defined. Here other plants and the animals that inhabit the site and exist with trees need to be studied. There is competition within a niche because the requirements of one species may overlap those of another. Theoretically, no two species can occupy the same niche at the same time. If one species is more aggressive than another with which it shares the same niche requirements, the more aggressive species will prevail. Competition among trees takes the form of one species' ability to utilize the site more effectively, for example, where one species may be more resistant to damage from a biotic agent or to stress imposed by a climatic or soil factor or factors.

The approach to forest biology in this text will be through the exploration of four general topics: (1) individual trees, (2) forest stands, (3) forest site, and (4) forest biotic populations and influences.

Individual Trees

Trees in a stand differ because of genetics and environment and from the interaction between genetic and environmental factors. The taxonomic ordering of forest trees does not always provide a complete understanding of the genetic effects that are involved in the development of trees as individuals within a taxonomic group. It is necessary then not only to consider the factors that influence species formation, but also to examine the effects of hybridization, provenance, and inheritance patterns as they influence different tree traits. The mode of reproduction of species of trees must be included here because of the effect sexual and asexual reproduction have upon changing or maintaining genetic likeness.

The tree-environmental interaction may be expressed as

$$\text{Environment} \rightarrow \text{Genotype} \rightarrow \text{Phenotype}$$

The environment applies pressure upon the organism, the genotype, and in response certain traits may be accentuated or, more specifically, certain gene combinations will be perpetuated with greater frequency in a population in a given environment thus bringing about a predominance of a particular phenotype. A phenotype can be either an individual specimen—a tree—or a family of trees. The important thing to consider is that although trees in a stand may have similar genotypes, these genotypes will vary according to their immediate environment within that stand. Further, genotype differences are the rule rather than the exception, thereby making phenotypic differences all the more evident. Tree characteristics such as bole form, branch size, branch angle, and crown width are all apparently under sufficient genetic control so that members of a family will be similar over a wide range of environments. Gross morphological species differences are quite apparent, but individual genotypic differences are less apparent and more difficult to discern among a group of trees of the same species.

Differences in morphological characteristics can be traced to differences in the character of individual genes or to the characters of several genes at different loci. Genes are located on the chromosomes in the nuclei of cells; they originate in the germ cell nuclei of the individual female and male gametes. When the gametes pair upon fertilization of the female gamete by the male, they produce an embryo that, in turn, results in a new individual miniature tree contained within the seed. Sexual reproduction thus provides a means whereby new combinations of genetic material can be produced and is the grist for the evolutionary mill. The fact that two genes which affect character are paired at homologous loci (alleles) results in a variety of phenotypic expressions.

Adaptation of individuals to the environment occurs as a result of individual trees and races within species being able to survive and reproduce with greater facility in a particular environment than non-adapted types; thus adapted genetic types will eventually prevail in a community. The genetic effects that concern foresters involve the different genetic combinations which influence the morphological, anatomical, and physiological traits of trees which affect their use as a raw material for industry, and the environmental adaptability of trees to various locations in a region.

Forest Stands

The stand is the primary working unit of silviculture since it is by reproducing and tending them that the forest is made to produce the variety of goods and services needed for mankind. This book defines ways in which separate stands can be identified and measured to permit classifying them on the basis of their growth potential.

A forest is made up of a number of stands, some of which are similar and can be placed into composite groups on the basis of their growth potential. Other stands represent different stages in successional change that was initiated as a result of some type of climatic or biotic disturbance. However, these successional stands can be arranged into categories that represent the particular stage of succession.

The consideration of forest stands will provide the means by which diverse species of trees growing in different locations and at different stages of development can be arranged into a logical order that can be used in management planning. Needed to classify forest stands are a designation of a species structure, measurements of tree density, and estimates of the growth potential of sites on which stands grow.

Forest Site

The components of the forest site are its atmosphere, soil, and physiographic location.

Atmosphere includes solar radiation and its effects upon air temperature, soil temperature, atmospheric moisture and as a source of energy for production of carbohydrates. Precipitation may be viewed in terms of a source of the soil water that supplies the physiological needs of the tree; or it may be viewed as ice and snow which damage trees and hinder their development. Wind is a diurnally recurring phenomenon in one sense, affecting water loss through transpiration, or it can be an occasional hazard that razes whole stands.

The soil that underlays a forest stand is a microcosm in itself with its living phase subsisting within the mineral, gas, and liquid nonliving phases.

Physiographic location can reflect a modification of the climate of a forest stand. Difference in physiography can present different soil or biotic complexes. Physiography must be taken into account in evaluating the forest environment. One may find similar environments occurring at the same or at different physiographic locations.

Forest Biotic Communities and Influences
The environment of an open field changes with the influx of trees. The environment of a western hemlock stand that follows the Douglas-fir is not the same as with the fir. Forest stands alter their environment. These alterations affect air and soil temperature, soil moisture, and soil physical properties as well as water run-off, snow pack, and other phenomenon associated with stream flow. Forests exhibit different biotic facets because different plants and animals are associated with different forest situations. Animals that live in the forest adapt to the conditions found there so that their needs for food and shelter can be met. Insects, fungi, and man modify stand and forest composition. Forest conditions determine which animals are present and how many; in turn, the animals will influence the condition of the forests.

The preceding topics embody the consideration of the elements of forest biology. It is necessary for teaching purposes to separate the components, or elements, so that they may be examined in some detail. It should be remembered, however, that a tree is not influenced by single elements one at a time; it is affected by all elements, not at the same time necessarily, but over time all elements will have played their part in shaping the phenotype. Trees will, at the same time, have an influence on the environment. These relationships can be expressed as a factorial display of linear interactions. For the analytically minded, there are no doubt nonlinear relationships that exist, and there are other analytical approaches, but consideration of these would simply compound what is already a complex function, and the factorial approach is reasonably straightforward. Functionally arranged the components are:

$$P = G + E + GE$$

where P = phenotype
$\quad G$ = genotype—the genetic component
$\quad E$ = environment—the environmental component

Partially expanding each of the components, the following can be developed:

$$G = f(aa, dd, ad, \cdot \cdot \cdot)$$
$$E = f(c, s, b, p, cs, cb, cp + sb, sp, bp, csb, cbp, spb, csbp)$$

where for one allele
$\quad aa$ = additive by additive
$\quad ad$ = additive by dominance
$\quad dd$ = dominance by dominance

$\quad c$ = atmosphere
$\quad s$ = soil
$\quad b$ = biotic
$\quad p$ = physiographic

Where the phenotypic function is expanded to include the expansion of the genotype and the environment functions, we have

$$P = (aa + dd + \cdot \cdot \cdot) + (c + s + b + p + cs + \cdot \cdot \cdot)$$
$$+ (aac + aas + aab + aap + aacs + \cdot \cdot \cdot + ddst + \cdot \cdot \cdot)$$

The statement above helps to explain why it is difficult to tell how one tree differs from another; or why trees of the same species in the same stand are different; or why two forest stands growing in similar environments are not exactly alike. The statement also offers a partial definition of niche as related to forest trees, but it does not take into account the time factor, an important biological factor. The function as stated can only define a static system that does not exist in nature.

Perhaps there will be developed, in time, a single model for the forest ecosystem. Until then, it will be necessary to work with simple or incomplete models that define only one part of the ecosystem. Models have been developed by population geneticists to explain the distribution of genotypes in different environmental locations, for example, the reaction of genotypes to environmental stress. Other models have been developed by physiological ecologists and tree physiologists to define the physiological response of trees to various environmental situations. Still other models have been proposed by forest hydrologists to help them develop an understanding of the interrelationship between the forest and the yield and quality of water from forested watersheds. Since the beginning of forestry mensurationists have been developing growth models of forest stands that are perhaps the best developed models available to forestry. Each of the models developed by the different groups impinges on models of the other groups to a certain extent. This is true because there is a definite interrelationship between each segment of the ecosystem with all of the other segments, and there is no way to make the system simple.

GENERAL REFERENCES

Major, J. 1961. A functional, factorial approach to plant ecology. *Ecology* 32:392–412.

Daubenmire, R. 1968. *Plant Communities*. Harper and Row, New York.

Köstler, J. 1956. *Silviculture*. Oliver and Boyd, London and Edinburgh.

Stern, K., and L. Roche. 1974. *Genetics of Forest Ecosystems*, vol. 6 of *Ecological Studies*. Springer-Verlag, New York.

Part One
Forest Trees

Chapter 1

RACES, CLINES, AND ECOTYPES

Evolution within forest tree populations is no different from that of other plant populations.* Species evolution begins when populations within a species begin to diverge from the center of geographic origin. As these groups migrate and extend the species range, they encounter, at different locations, different environmental conditions. One group may experience dry climate, another wet climate, others hot climate, others cold climate or various combinations of climatic factors. In each instance, the degree of hot, cold, wet, or dry is relative to the environmental conditions at the center of origin. Each environmental factor will exert a selection pressure on individuals making up the species. A tree species has the ability to adapt to a range of temper-

Evolution of Species and Genetics of Forest Trees

*See L. E. Mettler, and T. G. Gregg [*Population Genetics and Evolution*, in Foundations of Modern Genetics Series, Prentice-Hall, Englewood Cliffs, N.J. (1969)] for a more detailed presentation of this complex subject.

ature, moisture, soil, and biotic conditions; however, in the process of adapting to a selection pressure, the frequency of different gene combinations is changed. As a hypothetical example, if a conifer species had a geographic range that extended into a climatic zone where heavy snow occurred, a phenotype might arise in which the trees might exhibit short branches with short needles; this would permit the trees to withstand heavy snow pressure that would cause the branches to break if the needles and branches were longer. Trees not having the genetic ability to produce progeny with short needles and branches would disappear since their progeny would not be able to survive in large enough numbers to maintain the genotype in the population. On the other hand, another population within the species that best survived in warm climates could be one in which long needles and wide spreading crowns predominated. These traits would adapt individuals to grow faster and to survive in greater numbers in competition with other plants.

In these two instances the species would have responded to a need to adapt to different environmental conditions, and the exploitation of the genetic diversity that was basic to the genotype of the species occurred. A species can be viewed as a composite of all the genotypes of the individuals representing it; therefore, most tree species can adapt to a number of conditions of climate and soil. These climatic and edaphic provenances are not entirely uniform because climate and soil factors vary throughout a provenance. The hypothetical conifer species used in the example above probably started as a rather homogeneous morphological type; however, it exhibited different morphological types as different populations within it became adapted to the environments within which each of them grew. Initially, different populations, as they develop within the species, can be recognized as representative of different races. Individual races maintain the ability to interbreed with other races within the species, but each may exhibit differences in morphological or physiological characteristics, making it possible to distinguish between separate races within the species. In the binomial system of taxonomic names, a race may be designated as a variety or, if sufficiently evolved, a subspecies.

Selection of adapted genotypes in trees is a continuing process. A genotype that does not adapt to a particular environment will die a genetic death. Genetic death may mean that an unadapted individual can die either as a result of its inability to survive, or of its failure to produce progeny. In either case, the relative frequency of nonadapted genotypes in the population is reduced, and the relative frequency of the adapted genotypes increased.

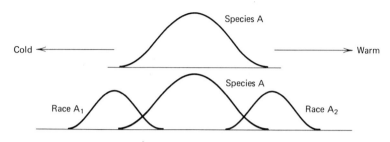

In the illustration above, species A originated because it became isolated from the base population of the species and formed a base population from which subpopulations A_1 and A_2 evolved. These subpopulations may be designated as races. For the sake of argument, assume race A_1, to the left, is better adapted to cold climate, and race A_2, to the right, is better adapted to warm climate. These two new races could evolve only after many generations of interbreeding of surviving genotypes, resulting in increasing number of adapted individuals within each group. Additional subgroups may extend the range of the species even farther into new regions where the climates are colder or warmer than those of A_1 or A_2 or at the center of origin. It can be expected that as a result of continuing migration additional new races would develop. In some instances the physiological changes occurring in race formation carry an associated morphological change, and as a result a taxonomic distinction can be made among subspecies.

There are several well-documented cases where forest species have formed races. Provenance trials with loblolly pine have demonstrated that it is difficult to grow Texas loblolly pine in Maryland or to grow Maryland loblolly pine in Texas (Wells and Wakeley, 1966). Although the species is well adapted to these two locations, the Maryland race cannot survive the dry climate of Texas, and the Texas race cannot survive the cold climate of Maryland. Comprehensive studies of provenance are being carried out with shortleaf pine (Wells and Wakely, 1970). Reports of these studies indicate areas where seed should be collected and the areas in which the resulting trees should be planted. With shortleaf pine, because of their latitudual nature, the seed zones appear to follow a thermal clinal pattern of adaptation with the zones following isothermal change (Figure 1.1). Seed can be collected any-

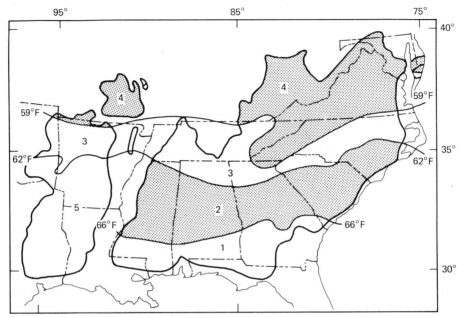

FIGURE 1.1. Suggested seed-collection and planting zones for shortleaf pine, with isotherms of average annual temperature. Natural range is outlined by heavy black line (Wells and Wakely, 1970).

where within a thermal zone and the trees which result will be better adapted to that zone than trees which are grown from seed collected in another zone but planted outside the zone of collection. Loblolly pine appears to respond to differences in rainfall as well as in temperature (Hocker, 1956). The seed zones for loblolly pine recommended by Wells exhibit a thermal clinal difference, but also show an apparent response to rainfall in that zones 1 and 2 represent two rainfall regimens and are specific locations recommended from which seed should be collected (Figure 1.2). This would seem to indicate that in loblolly pine there is a response to moisture as well as to temperature. However, there is the possibility of a loblolly x shortleaf pine hybrid in Texas; this will be discussed in a later section. The recommendation that loblolly seed for zone 2 be collected from a restricted area stems from the fact that tests show there is an apparent resistance to fusiform rust (*Cronartium fusiforme*) inherent in trees growing in this area. Fusiform rust is a disease that can severely limit growth of loblolly pine.

Where the terrain is more mountainous and there is a marked dif-

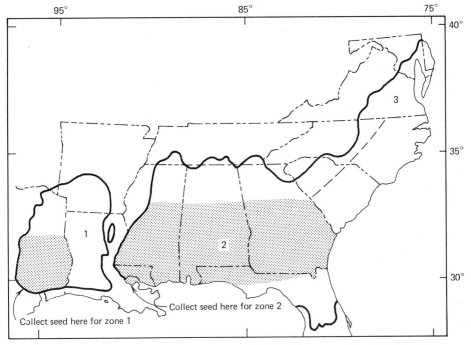

FIGURE 1.2.   Suggested seed-collection and planting zones for loblolly pine. Natural range is outlined by solid black line (Wells, 1969).

ference in the temperature-precipitation pattern, seed zones may be more restricted in size (Figure 1.3). The zones in California represent geographic provenances that were established on physiographic as well as climatic differences. These zones conform to provenance studies carried out with ponderosa pine. The numbers on the map denote physiographic-climatic regions (first digit), physiographic and climate subregions (second digit), and the seed collecting zones (third digit). Seedlings produced from seed collected in a zone are returned to the zone with the expectation that they will grow as well as or better than seedlings produced from seed from other zones that might be used to regenerate stands in the zone. It has been assumed that locally adapted races have developed within most of the local species.

Silen (1970) has demonstrated the existence of geographic races of ponderosa pine that apparently developed as a result of tolerance to drought and temperature differences. The west coast race of Douglas-fir is quite distinct from the Rocky Mountain race. There is sufficient evidence also to indicate that other forest tree species have evolved

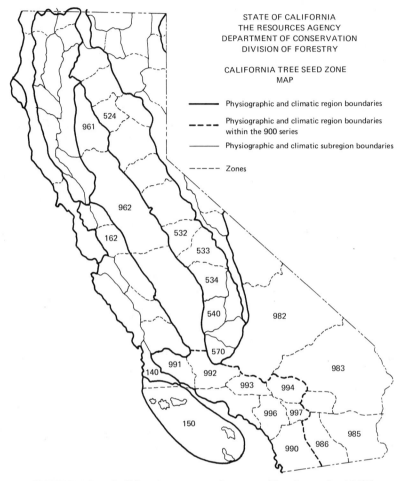

FIGURE 1.3. California tree seed zones (Buck et al., 1970).

races that are adapted to particular climatic, edaphic, or geographic situations, and that the geographic source of seed must be taken into account when artificial reproduction of a species is planned. If seed from one provenance is used to produce seedlings that are to be planted in a different provenance, it can be expected that the trees will not grow as well as trees which are grown from the seed of the adapted local race. Race differences must be taken into account when individual plus trees are to be selected to produce genetically improved varieties. Before an individual tree selection program can be started, it is first necessary to define the geographic boundaries of adapted races. Selection of plus trees then is made using trees from within the geographic area of each race.

Critchfield (1957) proposed that lodgepole pine be divided into four subspecies: Pacific Coastal region, *Pinus contorta* ssp. *contorta*; Medocino plains, *Pinus contorta* ssp. *bolanderi*; Rocky Mountains, *Pinus contorta* ssp. *latifolia;* and Sierra Nevada, *Pinus contorta* ssp. *murrayana.* In this instance he determined that there was sufficient morphological and physiological difference among the groups for them to evolve beyond a simple racial designation. The groups are also rather well separated geographically so there is little chance for them to interbreed.

Variation of traits in morphology, physiology, and anatomy that permit separation of races can take either of two forms. Variation can be *clinal* in that a trait exhibits change in character in a continuous pattern with no distinct breaks in the pattern of variation. Clinal patterns have been shown to occur in cone size, needle length, stomate number, wood fiber length, cell wall thickness, and seed size. These patterns are generally associated with gradient changes in temperature or precipitation, or with an interaction between temperature and precipitation.

Lowe (1974) found that the flushing date (the time when apical buds began to show growth in the spring) of balsam fir exhibited a clinal trend related to latitude, mean annual maximum daily temperature, and July mean daily temperature. However, the flushing date could be best related to July mean daily maximum temperature and growing degree days (Figure 1.4). Late flushing races of balsam fir are desired because they are not so susceptible to damage from late spring frosts, or to damage from the balsam twig aphid (*Mindraus abientinus*). Ledig (1974) showed that at increasing elevation balsam fir trees exhibited a decrease in the temperature at which they attained an optimum photosynthesis rate (Figure 1.5).

Ledig and Fryer (1971) showed that pitch pine exhibited a clinal pattern for serotinous cone character (Figure 1.6). Trees in New Jersey had the highest frequency of closed cones when there had been numerous fires, and as the distance from this center increased, the number of trees with closed cones decreased. It is interesting to note that toward the southern extent of pitch pine, pond pine occurs; at one time pond pine was considered a taxonomic variety of pitch pine.

In other situations a race may become adapted to a specific edaphic, climatic, or biotic condition, and there evolves a difference in one

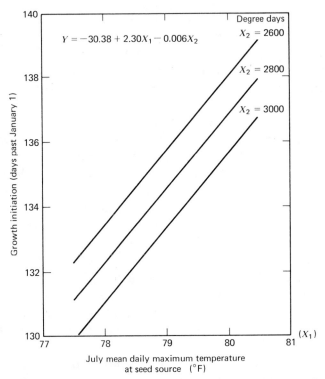

$$Y = -30.38 + 2.30X_1 - 0.006X_2$$

FIGURE 1.4. Prediction equation for time of growth initiation of balsam fir (Lowe, 1974). X_2 is the growing degree days at seed source (days temperature average exceeded 40°F).

trait, or differences in several traits, that make it possible for it to be recognized as a race. *Ecotypic* races occur within a species as a result of adaptation to environmental pressure resulting in a distinct break in a pattern of a particular trait. For example, Fuentes (1971) reported that loblolly pine trees grown from seeds collected from trees growing on (a) a wet site and (b) a dry site produced trees that grew better on a site similar to that from which the seed was collected (Table 1.1). Knauf and Bilan (1974) found that two-year old loblolly pine seedlings from xeric sites had fewer stomates than seedlings from mesic sites. This trait changed with age and there were no differences at 15 years. Silen (1970) also implied that there is adaptability of one race of ponderosa pine to warm-moist sites and another race to hot-dry sites; however, the patterns of growth traversed between warm-moist and hot-dry produced a clinal pattern of growth. Ecotypes appear to exist

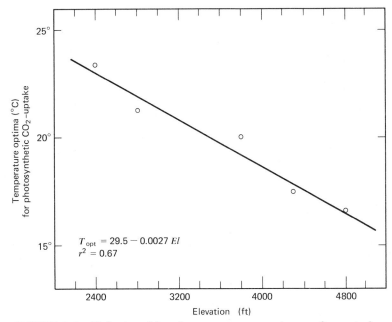

$$T_{opt} = 29.5 - 0.0027\ El$$
$$r^2 = 0.67$$

FIGURE 1.5. Relationship of temperature optimum for net photosynthetic CO_2 uptake with elevational origin of balsam fir seedlings (from Ledig, 1974).

within a number of species: see Squillace and Bingham (1958) for western white pine, Youngman (1967) for loblolly pine in Texas, Hermann and Lavender (1968) for Douglas-fir, and Musselman, Lester, and Adams (1975) for northern white-cedar. In some instances it is quite difficult to distinguish between clinal and ecotypic patterns because of incomplete knowledge of the physiology of most tree species.

Continuing research on the source of seed most likely will demon-

TABLE 1.1 Mean dbh and Height of Loblolly Pine Six Years after Outplanting One-year-old Seedlings (From Fuentes, 1971)

| SEED SOURCE | MEAN dbh (in.) | | MEAN HEIGHT (ft) | |
| | PLANTING SITE[a] | | | |
	Dry	Wet	Dry	Wet
Dry	4.5	4.2	22.7	21.0
Wet	4.1	4.8	21.7	23.1

[a]There was a significant interaction between sources and sites indicating there could be an adaptability to soil moisture supply.

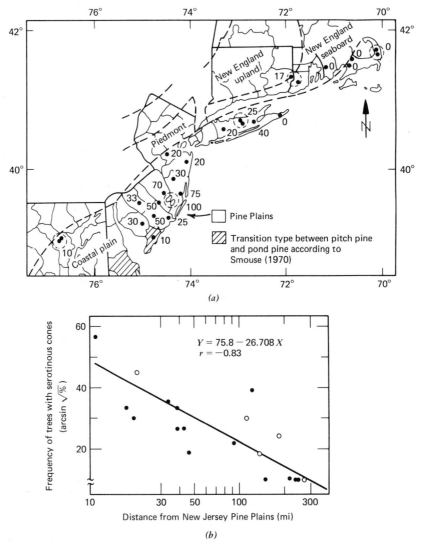

FIGURE 1.6. Frequencies of trees with serotinous cones in pitch pine (from Ledig, and Fryer, 1971). (a) Trees bearing 100% serotinous cones are found in the pine plains of New Jersey. From this center frequencies decrease in all directions (b)

strate racial differences within a number of species. There is danger, however, that a complete fragmenting of a species could occur if an attempt were made to separate it on the basis of all the ecotypic and clinal differences that can be distinguished. However, it will no doubt be necessary to recognize ecotypes as well as clines within a number of our important commercial species. By so doing, trees can be grown in those locations to which they are best adapted.

Contrary to being fragmented into two or more racial groups there is one rather stable forest tree species that exhibits only minor differences from place to place over its range. Red pine (Fowler, 1963) is a species that apparently shows only minor differences in character. Sand pine, a species with a restricted range, needs to be separated either into two ecotypes or clinally based on an open-coned versus closed-cone character, and there appear to be clinal differences in seed size, seed color, and seedling shoot weight, particularly as the latter relates to fertilizer treatment (Morris, 1967). It would appear that it is not necessary for a species to have a wide geographic range for it to have been affected by the selection pressures of climate; however, not all species are necessarily changed.

HYBRIDS

Isolation of populations within species may result in the development of races having different morphological, anatomical, and physiological traits as a consequence of their becoming adapted to different environmental conditions; but more important, from an evolutionary standpoint the races may develop reproductive incompatibility with other races within the species. In time, sexual barriers to reproduction become so sufficiently fixed that races, even though they are morphologically similar, must be placed in separate species or subspecies categories because they do not continue to interbreed. They are recognized as sibling species—species that developed from a common ancestral species, but which have lost the ability to interbreed.

Interspecies hybridization, crossing between taxonomically distinct species, is considered to be evidence of a relic stage in speciation, and a lack of ability to hybridize shows advanced evolution. In a taxonomic sense, there have been few intergeneric crosses accomplished (Wright, 1976). It can be argued that if two species in separate genera were to hybridize, then a reclassification within the genera should be considered. Taxonomic classification is a product of humans, and as a result may not always represent the actual condition of nature. Forest trees present a number of examples where natural hybridization occurs. Grant (1971) identifies these syngameons as communities of species and shows them to occur in pine, oak, birch, and willow.

There are several ecological situations where two pine species produce interspecific hybrids and the hybrids in turn produce viable seed. Oaks appear to be notorious for their ability to produce hybrids and do not always fit the taxonomic categories established by plant taxonomists. However, oaks are confusing in their taxonomy, one species varying in morphology from one location to another and from one growth stage to another; thus it may be that putative oak hybrids are the result of confusing morphological characters rather than active hybridization occurring among the species.

Where hybridization between species occurs and is a continuing process, it is termed *introgressive hybridization.* If a hybrid is better adapted to a site than either parent, it can be assumed that the hybrid will persist in the breeding population, and through recombinations occurring in subsequent generations, a genetic change in population will take place. Manley (1972) found that on gentle slopes and flat uplands red and black spruce show a continuum between the sites on which each occurs, black spruce on the low lands and red spruce on the uplands. In zones of contact hybrid populations have become established. Introgression of black spruce predominates in the hybrid population where fire, logging, and damage to red spruce by the spruce budworm (*Choristoneura fumiferana*) has occurred.

The following is a partial list of known coniferous hybrids where introgression is occurring in locations where the two species grow together in natural stands. There are, no doubt, instances where other species produce viable hybrids:

> jack pine × lodgepole pine
> red spruce × black spruce
> loblolly pine × shortleaf pine
> loblolly pine × longleaf pine
> Engelmann spruce × white spruce

Species evolution occurs as a result of continuing isolation of racial groups. Complete isolation can occur when races are separated geographically as a result of some form of geological intrusion. Evidence of a past relationship between tree species can be gained from observing whether hybrids can be produced between isolated groups of trees having different morphology.

Critchfield (1965) showed the apparent taxonomic relationship between the western big cone pines. This report pointed to the fact that

Jeffrey pine and Coulter pine can produce hybrids although these species are in two taxonomic subgroups (Figure 1.7). Introgression then occurs when two species grow together naturally, when their flowering cycle is synchronized, and when there are no major genetic barriers to embryo formation and development. Much can be learned in a practical way, as well as taxonomically from recognizing the different interspecific hybrids. Interspecific hybridization provides a way by which a trait not shown in one species can be introduced by hybridization with another species which has that trait.

Several observers have noted that loblolly pine growing west of the Mississippi River exhibits a bole form more like that of shortleaf pine than like that of loblolly pine growing east of the River. Hare and Switzer (1969) compared isoenzymes patterns of seed proteins of loblolly pine from an eastern source with that of a Texas source, and with that of Texas shortleaf pine as well. They found that not only were proteins of the two loblolly pine sources different, but the banding patterns of the Texas loblolly pine showed some bands similar to those of shortleaf pine growing in Texas. This study would support the belief that loblolly pine west of the Mississippi River has evolved differently from the races east of the Mississippi River. It could be deduced that the Texas race is introgressive between shortleaf and loblolly. The trees are identified as loblolly pine because morphologically they are more similar in the leaf and flower character to that species, and they do interbreed with other loblolly races.

Plant breeders have long recognized the potential that hybrids can hold. By crossing different genotypes, breeders have developed new hybrid varieties that satisfy a number of cultural requirements. An example of the usefulness of forest tree hybrids is the cross between Coulter pine and Jeffrey pine. These two species grow together naturally and a number of natural hybrids have been identified, although there have been no hybrid swarms located as has been the case where active introgression is occurring in other species. Jeffrey pine is a good timber tree, but Coulter pine is not so highly esteemed, being rougher and limbier. Jeffrey pine, however, is susceptible to attack by the reproduction beetle (*Cylindrocopturus eatoni*), which kills young seedlings, whereas Coulter pine is resistant to attack. Resistance appears to reside in a resin constituent that is present in Coulter pine but absent in Jeffrey pine. When individuals of the two species are hybridized, the hybrid is resistant to the reproduction beetle. The hybrid is intermediate between the parents for other traits. Backcrosses of

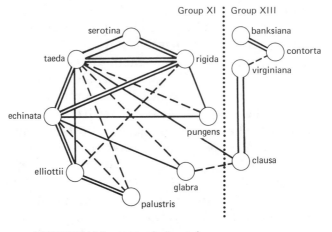

Group XI | Group XIII

serotina
taeda
rigida
echinata
pungens
elliottii
glabra
palustris

banksiana
contorta
virginiana
clausa

Little or no barrier to crossing
Cross moderately difficult to easy
Cross with difficulty (germinable hybrid seed obtained)
Failure (no germinable seed obtained)

(a)

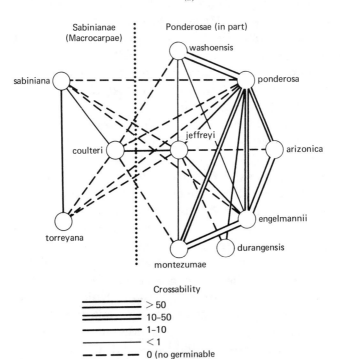

Sabinianae
(Macrocarpae)

Ponderosae (in part)

washoensis
sabiniana
ponderosa
jeffreyi
coulteri
arizonica
engelmannii
torreyana
durangensis
montezumae

Crossability

> 50
10–50
1–10
< 1
0 (no germinable
seed obtained)

(b)

the hybrid to a Jeffrey parent, (Coulter × Jeffrey) × Jeffrey, continue to exhibit resistance with an improvement in tree quality (Libby, 1958). The backcross progeny improve with each generation exhibiting more of the Jeffrey pine tree characteristics.

Usually a hybrid tree is intermediate between the two parents for most traits. Occasionally, a hybrid will show vigor—sometimes referred to as *heterosis*—by growing larger than either parent or the average of the parents. For example, Dunkheld larch, the hybrid of the cross between European larch and Japanese larch, outperforms both parents in height and diameter growth (Larsen, 1937).

Crosses between separate ecotypes or races within a species are termed *intraspecies hybrids*. For that matter, any cross between different genotypes within a species is an intraspecific hybrid because separate genotypes can be considered as individual trees, or they may be races of a species. A tree that exhibits a unique trait can be crossed with other trees not having that trait thereby introducing it into the genotype of those individuals. In this manner new forest tree varieties (cultivars) are formed. Improved timber characteristics can be developed with intraspecific hybrids; it is easier to produce new genetic combinations with intraspecific hybrids than with interspecific crosses since there are fewer reproduction barriers to the production of hybrids within a species than crosses between species.

The process of developing intraspecific hybrids for improved timber value involves locating several individuals that differ in some trait from the general population of a species. For example, a number of individual western white pines have been located that are relatively resistant to blister rust (*Cronartium ribicola*) (Bingham, Hoff, and Steinhoff, 1972). Many individuals of the species are quite susceptible to the disease. A breeding program using only resistant trees as parents will produce progeny that are relatively resistant to the rust. Individual American elms have been identified that appear to be resistant to Dutch elm disease (*Ceratocystis ulmi*). Propagating these trees could produce a resistant strain of American elm. Individual high gum

---

FIGURE 1.7. Crossability between some of the subsections in genus *Pinus*. (*a*) Crossability between subsections Austrailie (in part) and Contortae (Critchfield, 1963). (*b*) Crossability between subsections Sabinianae and Ponderosae (in part) (Critchfield, 1965).

---

yielding slash pines are used as breeding stock to produce trees with greater yields of gum (Squillace, 1965). Among the high gum yielding group, individual fast growing trees were used to produce a tandem increase in gum yield and in timber yield. Loblolly pine trees that have high density wood are crossed with one another to produce a new variety of trees with heavy wood, which produces high yields of cellulose (Zobel, 1971). Loblolly pine trees with low density wood can be crossed with trees with low density wood resulting in a wood fiber that can be used for newsprint. Trees with either high or low wood density are selected only if they exhibit straight bole form.

Individual trees showing outstanding traits can then be crossed with other trees exhibiting the same traits or different outstanding traits. In this way "new" intraspecific hybrid varieties of trees can be developed that combine a number of traits not presently found in a particular race of trees growing on natural sites.

POLYPLOIDS

Not all tree species have evolved as a result of geographic isolation and the subsequent development of sexual incompatibility. Some tree species may have had their origin as the result of the failure of miotic cell division to follow a normal pattern, thus resulting in trees having more than two sets of chromosomes or the $2n$ number. Usually in reduction division gametes (pollen or ovaries) are produced each having a haploid set of chromosomes. In rare cases, failure of the process of reduction division to follow a normal course may result in gametes with additional whole sets of chromosomes. When the union of gametes each having increased chromosome numbers occurs, the result is a polyploid individual.

The haploid number of chromosomes $(1n)$ occurs in most megaspores ($\female$) and microspores ($\male$). Upon fertilization and the formation of a zygote the chromosome number becomes $2n$, each sex having contributed one set of chromosomes. When either the megaspore or the microspore, or both, carry an additional whole set of chromosomes, a polyploid embryo is produced. Species are known to have $4n$, $6n$, and $8n$ chromosomes, indicating that in post miotic division more than one set of chromosomes was transmitted by either or both gametes.

An increased chromosome number can result in sufficient morphological difference among groups to permit a species ranking. In other instances chromosomal differences may not result in sufficient change to warrant either a species or a varietal designation.

Duplication of a single chromosome, or the deletion of a portion of a single chromosome, results in a phenotypic variation. Individuals having a replicate chromosome or a deleted chromosome portion do not usually grow as vigorously as individuals having a complete ($2n$) intact chromosome complement. Deviate forms are generally eliminated from a population through natural selection; however, some deviates have been saved as horticultural specimens. Some are preserved by placing them in special environments where they are able to survive.

Listed below are chromosome numbers for several coniferous genera.

Pinaceae	
Tsuga	$2n$ = 24
Pseudotsuga	= 26, 24
Abies	= 24
Picea	= 24
Larix	= 24
Cedrus	= 24
Pinus	= 24
Cupressacae	
Thuja	$2n$ = 22
Juniperus	= 22

Within the gymnosperms, polyploid species are not frequent and only three natural polyploids are recognized: redwood ($6n$), Pfitzer juniper ($4n$), and golden larch ($4n$) (Wright, 1976). A number of individual trees that are polyploid have been identified in Pinaceae. This characteristic does not appear to transmit any superior adaptive ability to these individuals; in fact, such individuals may be inferior in vigor to $2n$ individuals (Mergen, 1958).

Listed below are some hardwood genera in which polyploid series have been identified (see Wright, 1976 for a more complete listing).

Betulaceae	
Betula lenta	$2n$ = 28

Betula populifolia		= 28
Betula papyrifera	$2n$	= 56 or 112

Other series have been noted.

Fagaceae

Quercus	$2n$	= 24

Tetraploids noted.

Populus grandidenta	$3n$

Ulmaceae

Ulmus	$2n$	= 28

Triploids noted.

Ulmus americana	$4n$

Aceraceae

Acer rubrum	$2n$	= 78 or 104

Tetra, hexa, and octaploids noted.

Species within *Salicaceae, Juglandaceae, Magnoliaceae, Rosaceae,* and *Tiliaceae,* as well as the species *Seqoia sempervirens* and *Pseudolarix amabilis,* show differences in chromosome numbers from the $2n$ number of other species in the genus.

The occurrence of polyploids within a species exposes individuals in which it occurs to the same selection situation as occurs with a new gene mutation. If an increased chromosome number enhances the chance of an individual's survival, it can be expected that such an individual will perpetuate the genotype in a breeding population. With time, more individuals with the new chromosome number will be produced, and a new population eventually develops within the parent population. On the other hand, if there is no increase in survival potential as a result of an increase in chromosome number, then it can be expected that polyploid individuals would occur infrequently in a population. The factor that caused the formation of duplicate chromosome number would have to be removed entirely from a breeding population before polyploid gametes could be eliminated. If overtime selection is absolutely effective against individuals that produce the polyploid form, then the deviate form can be eliminated from a population.

Any genetic change, whether gene mutation or chromosomal aberration, that does not place individuals in which it occurs at a severe disadvantage can continue to persist within a breeding population. The frequency of occurrence of such types can be expected to be relatively low. If the deviation is a gene or chromosome structure that severely impairs an individual's growth or its ability to reproduce, the deviate form may continue to appear in a population, but it will occur in only a few individuals in a population. Mutant genes in a naturally breeding population are very infrequent. Wright (1976) indicated only one mutant gene may occur once in every 10,000 to 1,000,000 cells.

EVOLUTION OF SPECIES—A SUMMARY

When individuals of a species extend its range into new geographic provenances, selection pressures are manifest and as a result different geographic races may evolve. There are other evolutionary mechanisms that affect the genetic structure of a species:

Occurrence of mutations that create new genes

Introduction of new genetic material through hybridization

A severe reduction in number of individuals in the breeding population as a result of inbreeding that allows genetic drift

Taking the last of these first, the number of individuals in a breeding population must be considered as a factor of evolution, for if this number becomes very small, genotypic variation is reduced as a result of inbreeding, and although an inbred genotype may be adapted to a particular environment, should the environment change, it is possible that the inbred population would not survive and would, in time, be replaced. Inbreeding—the continued crossing among related individuals—results in the reduction of both vegetative vigor and fertility among species where outcrossing is the normal reproductive procedure. (There are plants where selfing is the normal reproductive procedure—wheat, for example.) It can be assumed that small isolated populations of trees in which inbreeding has occurred for many generations could develop traits different from other populations within the species. Provenance tests and comparison of morphological traits can be used to evaluate differences that have evolved within a species as a result of inbreeding.

Mutations regularly occur in any biological population. Most mutations are deleterious because they may disrupt the genetic balance within the individual in which they occur, and, as a result, these individuals die before reaching breeding age, or the gene is eliminated from the population later because individuals having the mutant form produce fewer progeny. The breeding value of an individual is then diminished, and after several generations the mutant gene disappears from the population. Mutations can occur at another time, but it can be expected that if breeding value is not enhanced by the mutation, it will be represented in only a few individuals in the population. On the other hand, mutations that enhance an individual's chance of survival will survive. With each generation additional individuals having the mutant form are produced, and in time the mutant can represent the predominant genotype in a particular situation.

The occurrence of interspecific hybrids in natural populations takes place with sufficient regularity at certain locations between two closely related species that new genetic combinations are available for natural selection. Again if the new form resulting from hybridization is better adapted to an environment than the parental forms, its frequency will increase in future generations.

Natural selection is a particularly effective method of sifting the many genotypes that arise as a result of sexual reproduction and the process reduces the frequency of those genotypes that are not best fitted to survive in a particular environment at a particular time. Although one genotype may not be adaptable to a particular environment at a particular time, it can continue to reoccur as a result of sexual reproduction. If the environment changes, currently nonadapted genotypes might increase and replace those genotypes that are currently adaptable. Immigration of trees into new locations exposes them to different environments, and it is necessary for the survival of a species that new genotypes be available. This can only occur through recombinations among existing genotypes.

As populations within a species emigrate and become more isolated from the base population, and as the migrants become adapted to new environments, it can be expected that barriers to sexual reproduction may occur. A good example is that of the difference in time of flower development and receptivity of pollen. When a barrier to pollination becomes sufficiently complete to prevent crossing among related groups, then a species evolution has occurred. In the process of evolving, the development of sexual isolation occurs slowly and backcrosses to the central population take place so that when a new species does finally arise it is similar to the central group in a number of anatomical, morphological, and physiological traits.

Geographic Isolation

Conifers in the north temperate region are separable taxonomically into a number of species, but some have not evolved to a stage of isolation where they cannot be hybridized. Among the conifers there are a large number of closely related species that maintain their integrity only because of geographic rather than sexual isolation.

Geographic isolation occurs when one segment of a species becomes separated from the parent population by a physical barrier—a mountain range, a section of grassland, an ocean. European tree species show the effects of past glacial intrusions that forced them to retreat up against the Alps. As a result of too severe a stress placed upon some species by the severe climate, they were unable to evolve and thus disappeared. A number of taxonomic species of forest trees had their origin in this way.* Geological intrusions have physically separated tree species into a number of separate geographic populations. For example, the larches of the world are similar morphologically, physiologically, and anatomically, and they will interbreed if brought together in a single arboretum. Yet they are recognized as several separate species because they grow in widely separated geographic areas. Other genera with a similar pattern of geographic separation are some pines, firs, spruces, yellow-poplars, beeches, and oaks. It is interesting to speculate on the origin of different forest tree species when there are so many taxonomic species found on separate continents. It can be concluded that the ancestral origins of these conifer and deciduous species were one or two locations, and that their present distribution has been the result of migration of populations away from their centers of origin.

Physical separation of species has no doubt occurred as a result of geological intrusions. The intrusions caused continents to divide and drift apart; land masses arose and receded, causing mountains and seas to be formed. Tree populations, as well as other forms of plants and animals, were forced into isolated geographic breeding groups. These trees continued to interbreed in isolation and in some instances

---

*Taxonomic is used in this instance to designate species that, although morphologically separable, are not biologically separable. Biological species are those groups that are incapable of crossing with other groups within the genus.

new sexually isolated species developed. However, in many species the evolutionary pattern has not reached the point where the geographically separated species cannot interbreed; their separate species rank is recognized only on the basis of their occupying separate geographic areas. On the other hand, sexual isolation has occurred among groups and, as a result, species integrity is assured should individuals of different species be brought into the same geographic location.

Reproductive Isolation
A tree species is a biological entity whose individuals do not freely interbreed with those of other species within a genus. Races within a species produce offspring that are morphologically, anatomically, and physiologically similar to offspring in other groups and that freely interbreed with other races within the species. Species can arise, however, as a result of reproductive isolation as well as from geographic isolation. There are several types of reproductive isolation depending upon the step at which the reproductive process is affected. The different types are (Stebbins, 1950):

1. *Sexual*—extension of pollen tube not possible because of physiological barriers
2. *Seasonal or temporal*—differences here arise as a result of different time of flowering
3. *Mechanical*—insect pollinated as compared with wind pollinated
4. *Chromosome difference*—chromosomal differences prevent gametes from producing a viable embryo (different subgroups within the hard pines would fit this situation as could polyploids)
5. *Hybrid inviability*—resulting hybrid individuals not capable of producing functional gametes
6. *Hybrid breakdown*—fertile seed produced, but resulting individuals die before reaching seed-bearing age, thus are not capable of producing second generation seed.

The western and eastern yellow pine species provide examples of trees that apparently have a common ancestry but have segregated into individual breeding groups, some of which cannot cross with other groups. These groups are designated as taxonomic subsections. The eastern subsection is represented as Austrailie, the well-known southern pines; the other section is Contortae, which contains both lodgepole pine and jack pine as well as sand pine and Virginia pine (Figure 1.7). Here are represented both geographic and sexual isolation. For the western yellow pine species where the two subsections are Sabinianae, in which Coulter pine is placed, and Ponderosae, which contains the ponderosa pine complex, there is again geographic and sexual isolation represented.

GENETICS OF INDIVIDUAL TRAITS*

Individual trees within a species exhibit different anatomical, morphological and physiological traits that, from a taxonomic standpoint, do not provide sufficient variation to place them into different races. Also the differences are such that they may occur among trees in the same stand and even among trees within the same family. The traits concerned may or may not contribute to the adaptability of the trees to the environment. It can be assumed that not all the genetic variation observed within a tree population is the result of variability due to environmental adaptability. Traits that contribute to environmental adaptability will determine which genotypes survive; however, there may be some genes contained in the genome of a population which contribute nothing, or only slightly, to environmental adaptability. All of the traits that trees exhibit can be changed in some degree in the breeding process. Genes that affect the economic value of trees can be increased in their frequency by controlled breeding with the result that wood production can be increased significantly by using improved varieties.

For example, a family of fast-growing trees might survive equally as well as a family of slow-growing trees in a particular situation. Likewise, a family of trees having wood of high density might survive as well as a family of trees having wood of low specific gravity. In a location where there is a minimum of selection pressure placed on trees with broad crowns, these trees might be more effective in production of carbohydrates, and as a result could grow faster than narrow crowned trees. In terms of total wood fiber production per unit area, however, it might be better to grow trees with somewhat narrower crowns so that more trees could be grown on an area.

*See J. L. Brewbaker [*Agricultural Genetics,* in Foundation of Modern Genetic Series, Prentice-Hall, Englewood Cliffs, N.J. (1964)] for a more detailed presentation of genetic improvement of domestic animals and plants.

In individual tree selection the concern is to determine the range of genetic variation in traits that affect the economic requirements of forest trees, and, within an environmentally adapted population, to select trees that best exhibit the desired traits. Select trees are then placed in a single location and allowed to interbreed in isolation from other trees; the progeny will incorporate into their genomes the traits that made their parents outstanding individuals. On this basis the genetic variability of the species can be exploited to produce trees that have the most desirable economic growth forms.

Genetic control of a number of individual traits in any breeding population of trees is not generally absolute. Environmental differences cause variation in the expression of most traits. Traits in forest trees such as the presence of chlorophyll, the arrangement and kinds of xylem, cambium, and phloem elements, leaf pattern and form, and the stem and roots are under rigid genetic control and do not vary with change in environment as it normally occurs. Of the examples given for developing new hybrids, there is further evidence of the degree of genetic control over traits. For example, in the Jeffrey pine × Coulter pine cross, there is a distinct difference in the genetic complement of the two species with regard to the structure of their resins. This difference in structure apparently affects resistance to attack by a reproduction beetle and shows no variation (or at least none to date) within either species. One species, Jeffrey pine, is susceptible to attack and the other, Coulter pine, is resistant. Such a difference is another example of rigid genetic control within each species. On the other hand, with western white pine there is apparently a considerable degree of variation in the amount of resistance exhibited by individuals to attack by white pine blister rust spores. Whether this varies with the location of trees have not been demonstrated. However, a number of individual western white pines have been identified as resistant in some degree to attack while many trees are quite susceptible. In the case of slash pine, gum yield is definitely a combined function of genetic complement and environment. Even high gum producing slash pine trees vary in the amount of gum that they produce as a result of soil, weather, and other environmental differences.

For a number of species, characters like bole form, branching habit, root habit, and height and diameter growth are not so strongly controlled by genetic makeup, and these are modified in their expression by a change in environment; but they are traits over which there is some degree of genetic control. They are not totally influenced by environment. Geneticists have developed a measurement termed *heritability* to express the relative degree of genetic control exercised on a particular trait.

Expressing heritability as a proportion, it is possible to indicate the relative degree of genetic control a trait will exhibit. A proportion of 1.0 implies complete genetic control while a proportion of 0.0 would imply no genetic control, or complete environmental influence in variation of expression. Knowing the range of heritability for a particular trait provides a tree breeder with part of the information required to predict the results of a breeding program. The breeder must know, in addition, the range in variation in a trait in order to estimate the expected gain to be obtained from a selection program designed to modify the character.

An unbiased estimate of heritability is difficult to obtain since estimates vary from one experimental area to another, from one age of tree to another, and with differences in experimental material. It is, therefore, impossible to state that a particular trait has a heritability of precisely so much; a range is stated instead. Table 1.2 illustrates the variation in estimated heritability of some traits encountered in slash pine.

Heritability estimates are obtained by observing differences among families for a number of traits, or they can be made by observing the amount of variation of one trait among members of several families.

TABLE 1.2 Narrow-sense Heritability Value for Open-pollinated Slash Pine Progenies Expressed as a Percent (Barber, 1966)

		HERITABILITY PERCENT			
STUDY	AGE (years)	HEIGHT	DBH	NATURAL PRUNING	CROWN WIDTH
102	6	25	16	—	—
102	8	20	6	36	—
103	5	35	37	—	19
103	7	34	34	52	—

Trees without stem cankers or fusiform rust

102	6	13	2	—	—
102	8	3	22	50	—
103	5	36	34	—	16
103	7	37	27	64	—

In any case, it is necessary that measurements be made on a number of individuals in a family and on a number of families.

The effect of gene action on a particular trait may be linear with small increments of change in the trait being associated with gene substitutions at alleles on a number of chromosomes. This type of multigenic response is termed *additive* and it is from this type of genetic change that estimates of heritability are derived (Figure 1.8). A nonlinear form of response, resulting from gene substitution at one or several loci, can produce a discrete change in phenotypic expression of the trait. In this type of response, there is a *dominance* shift and heritability estimates cannot be made for this type of gene substitution (Figure 1.8). An extreme change in character expression of a trait may be the result of the substitution of genes at a number of alleles, or of a dominance substitution. It is very difficult to determine how many gene substitutions are involved in a change in phenotypic expression.

An approach used to gain insight into the breeding response of individual members of a breeding population involves a number of control pollinated families. In a controlled pollination program, if an at-

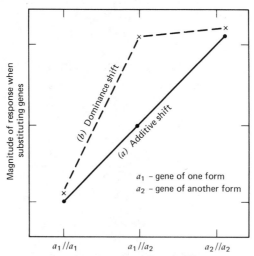

FIGURE 1.8. Effect of the substituion of genes at a single locus where there is (*a*) an additive effect; (*b*) where there is a dominant effect.

tempt is made to cross all males—using a tree as both a male and as a female—with all females (this is called a diallel crossing program), a measure of the combining ability of the different male and female crosses can be made by observing the behavior of their progeny. If most of the difference in variation for a particular trait is due to additive effects, then the gains will be linear. Progeny performance will be intermediate between that of the parents. This is termed *general combining ability* within the population, and future estimates of genetic gain can be made by using estimates of heritability made from progeny performance. If, however, difference in a trait is due, in part, to dominance effects, then the gains will result from *specific combining ability*. Crosses between specific individuals in the breeding population where dominance occurs will produce progeny that are outstanding (extreme varieties) in their expression of the trait under observation. The progeny mean will exceed the mean of the performance of the parents, although there can be instances where progeny performance can be less than that of either parent. This is called under-dominance, where in the case of superior performance the term used is over-dominance. At times the term heterosis, or hybrid vigor, has been used to indicate examples of over-dominance in specific combining ability.

There is a further need to designate whether there is interaction between alleles at one locus with alleles at another locus, for there are different numbers of genes that affect most tree traits. The alleles may act independently or they may interact with one another. If they do interact, this is called *epistases*. To unravel the epistatic effect of gene action requires very good control of breeding populations and a better understanding of chromosome structure than is currently available for forest tree species. It will probably be some time before it will be possible to identify individual genes on chromosomes of forest trees, as has been done with a number of other organisms. Interaction between non-alleles is currently included in measurements of general and specific combining ability for a particular breeding population since epistatic effects cannot be isolated at the present time.

Trees apparently respond in an additive manner to increases in diameter growth and height growth. The current belief is that the adaptability of wild plants to changes in the environment is the result of a large number of plant types that are produced as a result of additive differences. The clinal pattern of adaptation to temperature and precipitation gradients would support this belief. Further, when it is considered that most characters are affected by more than one allele, multi-genic inheritance is apparently more common than single gene effects. This does not mean that all tree characters are controlled by multiple alleles. There are a number of traits affected by only a few alleles.

Difference in gene action can be inferred from the value of the heritability estimate. Traits that are rigid or strongly affected by genetic structure will tend to show a high degree of heritability ($h^2 = 0.5$). Such tree characters as bole form and wood density generally have high heritabilty and apparently are controlled primarily by genetic structure rather than environment. Genetic control of disease and, perhaps, insect resistance, could be the result of changes in gene structure at only a few alleles. Therefore, it can be expected that one tree or one family may show resistance to disease attack where all the other trees in a locality are susceptible to attack. On the other hand, volume production and diameter growth have low heritability ($h^2 = 0.25$) and are said to be plastic, influenced more by environment than by genotype.

The number of characters that can be changed by breeding is quite long, and the list below probably does not include all those that will be modified in future breeding programs (Neinstaedt and Snyder, 1974). Numerous characteristics in the following list are of potential use in a tree breeding program. Survival after field planting heads the list, exemplifying many of the highly complex characteristics in which components are interacting to produce the measurable effect.

Survival and growth factors
  Survival percentage of planted
    seedlings
  Top-root ratio of seedlings
  Total height of trees
  Total diameter of trees
  Size of vegetative buds
  Amounts of pubescence on twigs
  Response in rate of growth to:
    Temperature
    Photoperiod
    Fertilizers

Phenological factors
  Date of flushing
    Effect of temperatures
    Effect of photoperiod
    Effect of bud-chilling
  Date of growth cessation
  Period of growth
  Date of fall coloring
  Date of leaf drop
  Timing of secondary needles on
    seedlings
  Age at first flowering

Photosynthetic efficiency
Mineral content of foliage
Pregermination requirements of seeds
Warm stratification period
Cold stratification period
Seed germination
Effect of temperature
Effect of photoperiod
Wood properties
Fiber length
Specific gravity
Summerwood percentage
Yield of extractives
Fatty acids
Resin acids
Terpenes
Phenols
Coniferous needle characters
Length
Cross-sectional area
Dry weight
Number of needles per fascicle
Number of resin ducts per needle
Hypodermal structure
Stomatal pattern
Color
Hardwood leaf characters
Surface area
Dry Weight
Shape and serration
Color of blade and petiole
Stomatal distribution
Amount of pubescence
Density of surface hairs
Amount of glandular secretion
Vein patterns

Form
Deviation from a straight vertical bole
Number and size of branches
Shape of crown
Shape of root system
Resistance of root system to be pulling
Fruits (including cones) and seeds
Frequency of good fruit crops
Abundance of fruits per tree
Structure of fruits
Seed size
Number of cotyledons in embryo
Resistance to climatic extremes
Minimum winter temperature
Needle desiccation in winter
Spring frosts
Fall frosts
Drought
Leaf-scorch
Sun-scald
Resistance to pests
Leaf diseases and rusts; e.g.
Rhabdocline
Hyperdermella
Phacidium
Cronartium fusiforme
Stem cankers
Insects
Defoliating
Gall-forming
Shoot-boring
Aphids
Bark-chewing animals; e.g.
porcupines

GENERAL REFERENCES

Dobzhansky, T. 1961. *Genetics and the Origin of Species,* 3rd ed. rev. Columbia Univ. Press, New York.

Dorman, K. W. 1976. *The Genetics and Breeding of Southern Pines*, U.S.D.A. Forest Service Agr. Hnbk. No. 471.

Falconer, D. S. 1960. *Introduction of Quantitative Genetics*, Ronald Press, New York.

Stebbins, L. 1950. *Variation and Evolution in Plants*, Columbia Univ. Press, New York.

Wright, J. W. 1976. *Introduction to Forest Genetics*, Academic, New York.

Chapter 2

SEXUAL REPRODUCTION

Sexual reproduction is the means whereby an almost infinite number of recombinations of genotypes can be produced. For every allele on a chromosome at meiosis there is a separation of the gene pairs when the chromosomes separate and are incorporated into separate cells; these cells then form separate gametes. From the time meiosis begins until the male and female gametes recombine into an embryo nucleus, there is a potential, too, for exchanges to occur between whole chromosomes and their parts. When exchanges occur, new gene combinations of a different kind result. It is not possible to illustrate here each of the ways exchange of genes can occur as a result of crossing-over; however, a brief review of the manner genes recombine as a

Reproduction of Forest Trees

result of miotic division will aid in understanding the evolutionary importance of sexual reproduction in trees.

At a single locus in meiosis the paired genes on each chromosome separate with the migration of the chromosomes to the end of the dividing cell. The genes are then included in separate gametes, each containing half the number of chromosomes of the diploid cell. As an example, the gene pair for a single allele can be identified as a_1 and a_2. These genes would represent gene pairs on homologous chromosomes in somatic tissue of a mature plant. At the onset of flowering, the chromosomes on which a_1 and a_2 are located are separated to form part of separate gametes; the gamete can be either male or female, depending on the type of organ on which the genes and chromosomes were located. The lattice below illustrates the segregation of one set of genes and the recombinations in which they can occur.

Female Genotype	Male Genotype	
$a_1//a_2$	$a_1//a_2$	
Gametes	Gametes	
	$a_1/$	$a_2/$
	Genotypes Produced	
$a_1/$	$a_1//a_1$	$a_1//a_2$
$a_2/$	$a_2//a_1$	$a_2//a_2$

One result of recombination was the production of two new genotypes of the gene pair and a new ratio of genotypes:

1 $a_1//a_1$: 2 $a_1//a_2$: 1 $a_2//a_2$. The original $a_1//a_2$ genotype is now represented in only half the offspring.

A mature tree has many hundreds of genes on the chromosomes contained in each cell, and it can be seen that a large number of recombinations will occur as a result of sexual reproduction. Although new genotypes are continually formed, it must be pointed out that not all of them will have the same or a better adaptive ability to survive as will the parental genotype. It can be assumed that the parental genotype was adapted to the environment sufficiently to reach the stage where it can reproduce; however, not all the new genotypic forms that are produced will necessarily be so well adapted, and many may die before reaching reproductive age. This may be particularly noticeable where ecotypes are maintained in populations. In these instances there is a continuing need for adapted genetic combinations to be produced to fit the single population that is best suited to a particular environment. Nonadapted forms, although they continue in the population, will be less frequent and may be eliminated over time.

It was assumed, in the above example, that the frequency of the a_1 and a_2 genes is equal in the population, $0.5a_1$ and $0.5a_2$. This proportion is often not the case and a different ratio between a_1 and a_2 results. For example, if the frequency in a population of the a_1 gene is 0.9 and the a_2 gene 0.1, [which when crossed] will produce $0.81a_1a_1$, $0.18a_1a_2$, and $0.01a_2a_2$. In this case the majority of the recombinants are of the more frequent gene, that is, a_1. Gene frequency in a natural population can vary from one location to another; this results in racial change within a tree species. Keep in mind that most traits are affected by more than one pair of genes and, as a result the binomial $(2)^n$ must be expanded to estimate the combinations that can occur for n number of genes.*

Reproduction Cycles

The cycle of reproduction can be characterized as the sequence of changes in individual trees that occurs from the time a tree reaches a state of physiological maturity when flowering begins to the time seeds germinate. The cycle can be represented as follows (Krugman, Stein, and Schmitt, 1974):

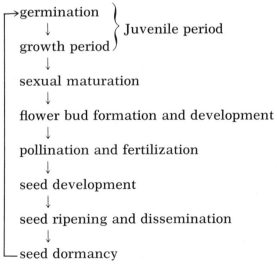

*By expanding the binomial $(2)^n$ it is possible to determine the number N of individuals required to represent all possible genotypes of a particular form, that is, if two genes segregate at one allele $N = (2)^2 = 4$; if four genes segregate $N = (2)^4 = 16$; and so on. However, the assumption is that the alleles occur on separate chromosomes, that linkage is not involved.

Juvenile Period

Species vary in the number of years required from seed germination to the point where both male and female flowers are produced—the period of time needed to complete a life cycle. Apparently, the rigorous competition most forest trees encounter necessitates that infant and early juvenile growth take the form of rapid vegetative increase followed each year with vegetative bud formation. It is only later in life that sufficient food reserves accumulate and physiological changes take place permitting reproductive growth to occur. It can be speculated that trees, and other biological organisms, must undergo some type of physiological change before they reach full reproductive potential. This change assumes the form of the production of a hormone, or hormones, or the achieving of a balance among two or more hormones that stimulates the onset of flowering as well as a capacity to produce sufficient growth substance. Environmental factors, as well, influence the time of flowering.

The length of the juvenile period is quite variable. Species such as pitch and Virginia pine, jack and lodgepole pine, and several species of cottonwood, aspen, and willows are noted for their early seeding habit. Jack pine seedlings have been observed to bear cones while still in the nursery seedbed.

Beech exhibits a definite juvenile characteristic in leaf drop. On very young trees the leaves of previous years are not shed but are carried over until the following spring; as trees begin to mature leaf drop occurs in the upper crown, but they are still retained in the lower crown area. It would appear that this trait is associated with the production of absiscic acid, which is involved in the formation of the abscission layer that results in the shedding of leaves.

Flower Bud Formation and Development

Flower buds of forest trees formed during the current growing season will not bloom until the following spring (Figure 2.1). Pollination, fertilization, and embryo development with some species occurs shortly after flowering and mature seed is disseminated within several weeks; however, with most species fertilization and seed maturity are spaced over a wider interval. For some, seed matures in the late summer or autumn following spring fertilization; in other species, seed may not mature until autumn of the following year. With the pines, fertilization does not occur until the spring following pollination (Figure 2.1).

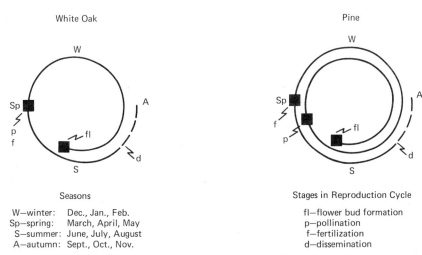

White Oak | Pine

Seasons

W—winter: Dec., Jan., Feb.
Sp—spring: March, April, May
S—summer: June, July, August
A—autumn: Sept., Oct., Nov.

Stages in Reproduction Cycle

fl—flower bud formation
p—pollination
f—fertilization
d—dissemination

FIGURE 2.1. Reproduction cycle of white oak and pine.

FLOWER DEVELOPMENT AND STRUCTURE

Flower bud primordia develop in most species in mid to late summer as winter buds begin to form. With pines, the flower primordia are contained within the terminal bud itself and develop with the onset of growth the following spring. With Douglas-fir flower buds form separately on the current growth and are visible at the end of the current season, but they too will not develop until the following spring. With oak the male flowers develop from leaf axis on the new growth.

Flowers may appear before the leaves develop in the spring, as in poplar and red and silver maple; as the leaves unfold, as in walnut; or, after leaves are fully developed, as beech and buckeye.

The reproductive organs of trees vary in two major respects. First, there are a few species that produce complete flowers such as magnolia, locust, and tulip trees (Figure 2.2b). Most species bear only incomplete flowers where part of the structure is lacking either the male or female part. These are born separately as with the conifers and many deciduous species (Figure 2.2a).

Another major difference among species is whether both male and female flowers are born on the same tree (monoecious) or on separate trees (dioecious). Pines tend to produce female parts in the upper crown area on fast growing shoots, while male parts are located on slower growing shoots in the lower crown (Figure 2.3). Oaks produce male parts separate from female parts, but these can grow near each

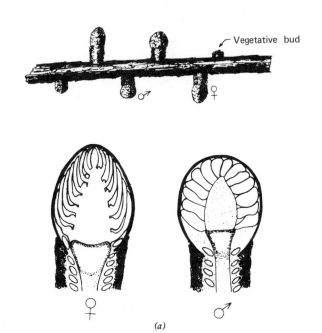

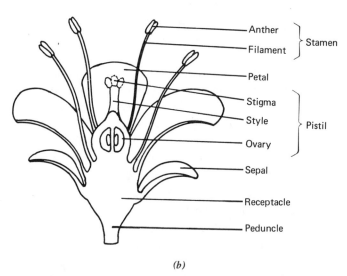

(b)

FIGURE 2.2. (*a*) Sketches showing ovulate (☿), staminate (♂), and vegetative winter buds of western larch
and the longitudinal section of ovulate and staminate
buds (Schmidt, Shearer, and Roe, 1976). (*b*) Structure
of a complete angiosperm flower.

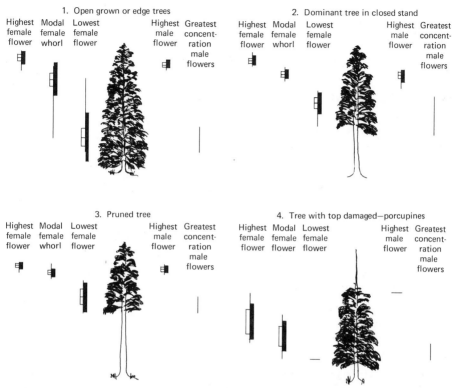

FIGURE 2.3. Position of male and female flowers in the crowns of red pine trees of four crown types. The vertical line represents the sample range; the short horizontal line is the sample mean; the black rectangle to the right of the range line is the standard deviation above and below the mean; and the hatched rectangle to the left of the range line two standard errors above and below the mean (from Fowler, 1965).

other. Genera such as ash, holly, poplar, willow, persimmon, and juniper are made up mostly of dioecious species with the flowers of the two sexes born on separate trees.

Many tree species bear male and female flowers on separate organs (imperfect) as contrasted with plants having complete (perfect) flowers (Figure 2.2). This is true of all the conifers and includes most of the commercially important deciduous species. In some genera, however, female and male flowers as well as perfect flowers can be found on the same tree (polygamomonecious); buckeye (*Aesculus sp.*) and hackberry are examples. Other species may be polygamodioecious, as buckthorn (*Rhamnus sp.*). A forester must be familiar with the reproduction habit of the species with which he will be working. The manuals *Seeds of Woody Plants in the U.S.* (USDA, 1974) and *The Silvics of North America Trees* (USDA, 1965) are good sources of this material, as well as various dendrology texts.

Pollination and Fertilization

Pollination occurs in all species shortly after the flowers reach maturity. Receptivity duration is usually several hours to two weeks (Kozlowski, 1971). Pollen grains become attached through different devices to the stigma of angiosperm flowers, or, in the case of gymnosperms, pollen grains fall between the scales of the female gametophyte and are protected by the integument that forms after fertilization (Figure 2.4).

The period between pollination and fertilization does differ among genera and species. Following pollination elm, red and silver maple, and willow undergo a relatively rapid sequence of fertilization and embryo development. Most angiosperms undergo pollination and fertilization in about 24 to 28 hours, except for some of the red oaks which require 12 to 14 months. With gymnosperms fertilization occurs within 1 to 2 months for fir and larch, and for pine and true cedars there is a delay of 13 months between pollination and fertilization.

The following is a listing of genera and species according to the length of time required between flowering and seed dissemination.

SPRING AND SUMMER SAME YEAR	FIRST AUTUMN OR WINTER		SECOND AUTUMN	THIRD AUTUMN
Aspen	Ash	Larch	Red oak	True cedar
Cottonwood	Birch	Spruce	Pine	
Elm	Fir	Sugar maple		
Red maple	Gum	Sycamore		
Silver maple	Hemlock	Walnut		
Willow (some species)	Hickory	White oak		

Seed Development

At fertilization the way begins for the development of both the embryo and endosperm. The endosperm may continue to develop as in the gymnosperms (Figure 2.5*a*) or, as in the case with some angiosperms, it is consumed by the embryo and assumes the form of cotyledon food

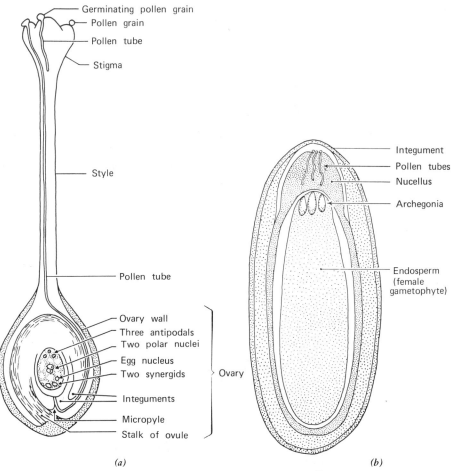

Germinating pollen grain
Pollen grain
Pollen tube
Stigma

Style

Pollen tube

Ovary wall
Three antipodals
Two polar nuclei
Egg nucleus
Two synergids } Ovary
Integuments
Micropyle
Stalk of ovule

Integument
Pollen tubes
Nucellus
Archegonia

Endosperm
(female
gametophyte)

(a)   *(b)*

FIGURE 2.4. (*a*) Longitudinal section through a typical pistil just before fertilization. (*b*) Longitudinal section through an ovule of *Pinus* during the period of pollen tube development preceding fertilization.

storage tissue (Figure 2.5*b*). Douglas-fir exhibits a form of seed where the embryo and food storage are in separate systems, while the red oak illustrates the form of seed where the storage tissue is directly associated with the embryo in the cotyledons.

Seed Dispersal

Forest tree seeds are dispersed primarily by wind and gravity, although animals and water can be instrumental in distributing seed of some species after it is detached from parent trees. The following is

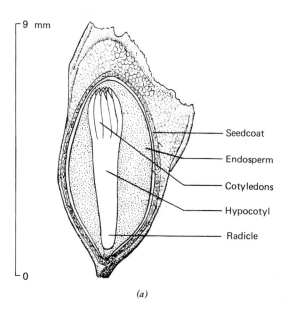

(a)

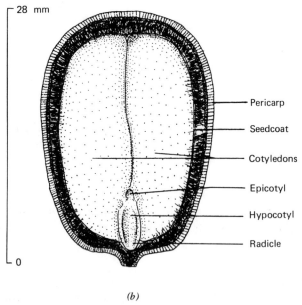

(b)

FIGURE 2.5. (a) Douglas-fir: longitudinal section through a seed, 8 ×. (b) Northern red oak: longitudinal section through an acorn, 2 ×.

a partial listing of species dispersed by several agents. Some species may be dispersed in a secondary fashion by another agent. For example, acorns once detached from the tree may be carried away by rodents who bury them at a distance from the parent tree. In the early spring the "cotton seed" from aspen can be carried some distance by running water.

			ANIMALS	
WIND		GRAVITY	BIRDS	RODENTS
Pine	Cottonwood	Oak	Juniper	Oak
Spruce	Birch	Hickory	Cherry	Hickory
Fir	Maple	Walnut		Walnut
Douglas-fir	Ash	Beech		Pine
Larch	Elm			Spruce
Hemlock	Sycamore			Douglas-fir
Willow	Poplar			and others

Wind dissemination is a common method for distribution of seeds over a distance; however, even here the distance may not be great. Longleaf pine produces relatively heavy seed that can be carried only a distance of 1 to 1½ times the tree height from the parent tree. This is common among gymnosperms. With angiosperms such as willow, aspen, and cottonwood, the wind disseminated seed can be carried miles from the parent tree. During the winter it is not uncommon to see yellow birch seed being blown along the surface of crusted snow. The distance over which seed can be distributed, even in wind disseminated species, varies (Figure 2.6). As the lightest seed of the four species western larch seed is carried farther from the seed source than any of the other three species.

Seed Dormancy
Seed production does not end the reproduction process since a seed falling from a tree is not always in condition to germinate. In fact, its life expectancy could be quite short if it were to germinate too soon. Spring and summer ripened seed can germinate and grow as soon as they ripen because the weather conditions in these seasons are favorable for growth. Autumn ripened seed, however, may need to lie dormant until environmental conditions for growth improve, or, if seeds do germinate, the weather must not be so severe that seedlings will

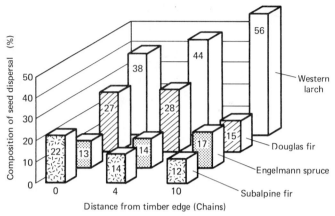

FIGURE 2.6. Composition of seed dispersed at 0, 4, and 10 chains from timber edge (Schmidt, Shearer, and Roe, 1976).

be killed by cold. As a result of this need for a dormant period, trees have acquired several adaptive mechanisms that prevent seeds from germinating. Embryo immaturity, sometimes called *embryo dormancy* is one mechanism; another is *seed coat dormancy*, which results in seeds with a seed coat impermeable to water and other elements required for growth. Seeds subject to either seed coat or embryo dormancy will not germinate at first, but must lie in the ground for a period of time until natural weathering causes a breakdown of the impermeable seed coat or until the embryo is capable of germination.

In some cases species that exhibit embryo dormancy do so because there are growth inhibitors present in the inner seed coats or in the endosperm that prevent germination. Exposure to cold stratification apparently results in a breakdown of the inhibitors permitting the seeds to begin germination with the resumption of warm weather. It should be noted that in the case of longleaf pine and species of the white oak group even though seed is disseminated in the autumn, there is no apparent need for after-ripening since germination begins soon after the seeds are disseminated. However, longleaf and some white oaks grow in regions where winters are relatively mild, and newly germinated seeds are not exposed to long periods of below-freezing temperature. Not all white oaks grow in warm regions and germinating acorns can suffer some root damage, but they are able to recover in spring.

The following is a partial listing of species showing various types of dormancy.

| SPECIES GROUPS REQUIRING NO AFTER-RIPENING | SPECIES GROUPS REQUIRING AFTER-RIPENING | | |
	SEED COAT IMPERMEABILITY	EMBRYO DORMANCY	SEED COAT AND EMBRYO DORMANCY
Willow (some species)	Locust	Fir	Black cherry
Poplar (aspen and cottonwood)	Redbud	Spruce	Basswood
Elm	Ash	Pine	
White oak		Red oak	
Longleaf pine		Hemlock	
		Sugar maple	

A few species produce seed that can endure, under special conditions, several years in the dormant state; however, seed of most species must germinate the spring after dissemination because they will be eaten or they will rot if they do not grow. From the listing above, some general conclusions regarding the adaptability of the species to cold can be drawn. For example, willows and aspen are found at or near tree line in high mountains and bordering tundra. It should seem that a short regeneration cycle would be essential in such an environment. True fir and spruce seeds in a northern environment undergo a long period of winter dormancy. Care must be exercised not to assume too much about the adaptive relationship between a species and the need for a type of seed dormancy since there are no doubt races among a number of species adapted to differing periods of dormancy. Fowler and Dwight (1964) found that the seed of eastern white pine from northern stands did not require as long an after-ripening period as did seeds from southern white pines. It might be assumed that the early onset of cold weather in northern stands which lasted until spring would keep seed from germinating until suitable germination conditions occur, whereas southern seed would be exposed to short periods of warm weather during the winter, and therefore would require a longer dormant period to prevent the seeds from germinating prematurely. On the other hand, Kriebel (1958) found sugar maple seeds from northern sources had longer periods of dormancy than southern sources. Barnett (1976) determined that 69% of the variation in germination rates of five southern pine species could be related to seed coat weight. Seed with heavier seed coats were delayed in germination.

GERMINATION

Inhibitors
Germination of seed of oak and birch is delayed or prevented by the presence of a chemical inhibitor in the cotyledons in oak and in the seed coats in yellow birch. Ash seed apparently is prevented from germinating because of inhibitors located in the embryo (Kozlowski, 1971, vol. 1). Inhibition in the case of ash could be the result of the presence of abscisic acid (ABA), a hormone also associated with leaf senesence and final leaf drop. The normal stratification of most tree seeds results in a breakdown of the substances that inhibit embryo development, or otherwise inhibit germination.

Promotors
The onset of germination in some species can be associated with the increase in giberrellic acids (GA). In fact, dormancy of some tree seed can be broken by exposing seed to GA. It appears that GA acts on the enzyme systems that transform stored food into a form usable for embryo growth (Kozlowski, 1971, vol. 1).

When germination begins, the radicle emerges through the micropyle. In gymnosperms and some angiosperms the extension of the radicle and growth of the hypocotyl causes the seed coat and cotyledons within to be raised above the soil surface (epigeal germination, Figure 2.7a). With other angiosperms the radicle and the plumule appear at nearly the same time, although the radicle does appear first; the seed coat and cotyledons are not displaced from the soil surface (hypogeal germination, Figure 2.7b).

Species that undergo epigeal germination are handicapped in that they can be easily smothered by a covering of loose litter, and, in general, the food reserves in the endosperm tissue is quickly exhausted in germination so many seeds do not become established. Hypogeal species on the other hand seem to have better success in germinating under a covering of litter.

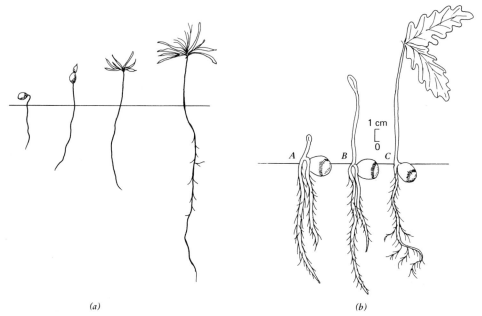

FIGURE 2.7. Douglas-fir; seedling development 2, 5, 8, and 22 days after emergence from peat moss-vermiculite potting mixture, 0.5 ×. (*b*) Bur oak: seedling development at 1, 5, and 12 days after germination.

Most epigeal species produce seed that is wind dispersed. Seed of this character can undergo some period of drying prior to germination; however, moist stratification and adequate moisture during germination are required in order to have successful germination, regardless of the type of emergence.

Seed Year Cycles
It can be expected that the amount of seed produced in a particular year will depend on the size of the seed crop the previous year. Usually there is a significant reduction in the amount of seed produced the year following a very heavy seed year (Figure 2.8). The amount of seed produced in any particular year will depend also on the weather conditions that prevail during the period of flower bud differentiation, during the period of pollination, and during the period of embryo maturation. In addition, insects and disease also affect the number of flowers reaching maturity and the amount of seed finally set. The result of all of these factors is that heavy seed years do not occur at

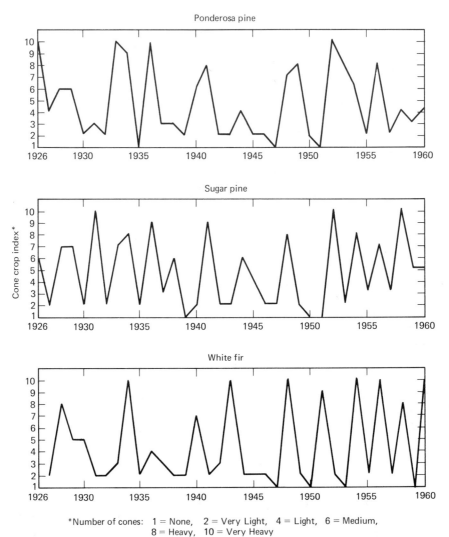

Ponderosa pine

Sugar pine

White fir

*Number of cones: 1 = None, 2 = Very Light, 4 = Light, 6 = Medium,
8 = Heavy, 10 = Very Heavy

FIGURE 2.8. Average cone crop index for dominant frees larger than
19.5″ dbh Stanislaus National Forest 1926–1960 (Fowells and Schubert,
1956).

frequent intervals, but are usually offset by succeeding years of light
to very light crops. Generally the cycle of most species is for heavy
seed crops to occur at three to four year intervals. Because of various
environmental disturbances a heavy crop occurs at less frequent intervals. In fact, with eastern white pine, because of insect damage to

developing conelets, heavy seed crops occur at only seven to eight year intervals in some locations.

Seed Production of Individual Trees

Often it is necessary to designate the seed production potential of different trees in a stand. Research shows that within a stand, trees with the largest crowns are the ones that produce the largest number of seed; further, it has been demonstrated that there is apparently a genetic relationship for fecundity (Figure 2.9). Some trees within a stand have the capacity to produce particularly large numbers of seeds; while others, even when they are of the same size and growing on the same site, are not so fruitful (Bilan, 1960; Hocker, 1962). There appear to be two major reasons why these results were obtained: (1) in order to produce large numbers of seeds a tree first must have a large number of growing tips on which flower buds can be formed; (2) with some species the female flowers are produced in the upper crown with male flowers being more confined to the lower crown. A large crown therefore enhances female flower production which results in increased fruitfulness. From research with other plants it has been demonstrated that individual plants or individual families are capable of greater seed and fruit production than are other plants and lines. Forest trees are no exception; trees of the same size, age, and vigor in a stand vary in the number of seeds they produce.

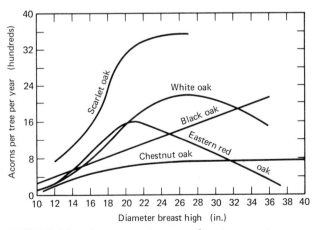

FIGURE 2.9. Average acorn production per tree per year by species and diameter based on 7-year averages of 210 trees (Downs and McQuilkin, 1944).

ASEXUAL REPRODUCTION (VEGETATIVE REPRODUCTION)

Plants, unlike most animals, have the capacity to undergo asexual reproduction. In some of the more elemental plant forms, asexual reproduction is an integral part of the reproduction cycle. In trees asexual reproduction is involved in the production of gametes; however, new individuals do not normally develop from individual pollen grains or egg cells.

Apical and lateral meristem tissue can be made to reproduce whole plants (Romberger, 1969; Brown and Sommer, 1975). Potentially every living meristematic cell has the capacity to generate a replica of the individual in which it occurs. Cuttings made of roots and shoots of some species can regenerate either a shoot or root and in time produce a replica of the original genotype. This, perhaps, is the most important fact to note: that asexually propagated individuals have the same genotype as the ortet from which they were taken.* No new genotypes can be created through asexual reproduction methods, and in an evolutionary sense asexually propagated plants are limited to the environments to which they are adapted.

Vegetative reproduction by sprouting is an important means by which many species can perpetuate themselves, especially when growing shoots are in some manner destroyed before they can bear seed. One type of sprout growth is layering—the development of roots on lower branches of trees as a result of the branches coming into contact with the soil and being covered by litter. A more common type of sprout growth is the growth of new shoots from roots and from tree stumps. The following is a list of species that reproduce well asexually.

ROOT SUCKER	LAYERING	STUMP SPROUT	BOTH ROOT AND STUMP
Sassafras	Black spruce	Oak	Locust
Aspen	Northern white	Hickory	Beech
Locust	cedar	Basswood	Aspen
Beech		Red maple	Poplar

*Ortet designates a particular tree from which cuttings (scions) are taken. The scions can be grafted onto new root stock, or they themselves may be rooted and in either case they are termed *ramets*. All ramets from a single ortet are designated a *clone*.

Redwood and pitch and pond pine are conifers that can be depended upon to develop stump sprouts. With the exception of black spruce and northern white cedar, most conifers apparently have lost the ability to layer. Lower branches of black spruce and northern white cedar may be covered by organic debris, with the result that root initials can form on the stem section that is so covered.

The ability to reproduce vegetatively is used to advantage in tissue transplanting. Growing tips of trees can be grafted to roots of other trees with relative ease, and various techniques have been developed to facilitate this operation. The technique is quite useful in preserving and multiplying a special genotype for breeding work. A number of tree species can also be propagated by rooting cuttings of vegetative material using specially developed techniques. Research indicates that it is possible to propagate whole trees of some species from small sections of meristematic tissue (Winton, 1968; Durzan and Campbell, 1974).

Tissue culture holds an attractive prospect for forest genetics because by using small segments of meristematic tissue it may be possible to replicate superior genotypes indefinitely. The techniques for tissue culture are not overly complex, but take time to perfect reliable procedures. It is first necessary to identify meristematic tissue that develops callus tissue and from the undifferented callus cells buds and root initials can then be produced (Figure 2.10). The proper combination of light, temperature, and growth substrate must then be found.

REPRODUCTION OF FOREST STANDS

The silvicultural methods used to reproduce forest stands take advantage of the reproductive behavior of the various tree species. Basically, there are two methods of reproduction that are used as a means of separating trees into groups: (1) species that are reproduced primarily from seed and (2) species that are reproduced vegetatively (although species in this group can also be reproduced from seed). Within the seed-reproducing group, species can be further designated into groups that produce light wind-disseminated seed and that can colonize open areas such as old fields, burns, and sand bars in large streams. On the other hand, there are other seed-reproducing species that require a more protected area, or that produce heavy seed that is not carried far

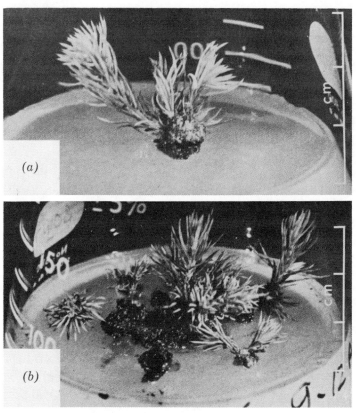

(a)

(b)

FIGURE 2.10. Multiple elongated shoots from two different hypocotyl explants. The shoots in (b) came from a single explant that was divided up after elongated needles and buds had formed. Photo after 175 days (total) in culture (Campbell and Durzan, 1975). Reproduced with the permission of the National Research Council Canada, Publishers of the Canadian Journal of Botany.

from the parent trees. Species colonizing open areas can be grown in even-aged groups made up of similar species (pure even-aged stands); while species requiring protection are often found in stands of different aged trees and of mixed species (uneven-aged mixed stands).

On the basis of reproductive habit and on stand age structure and condition of the regeneration site, the following six regeneration systems are recognized:

Seed Reproduced

Even-aged Systems   *Uneven-aged Systems*

Open   1. Clear cutting
situation   2. Seed tree

Closed   3. Shelterwood   4. Selection
situation

Vegetative Reproduced

5. Simple coppice   6. Coppice with standards

Details of the use of the different regeneration methods is included in silviculture courses. The sections that follow will expand on the silvical requirements to be considered when planning silvicultural methods to meet the needs of various tree species.

GENERAL REFERENCES

U.S.D.A. 1974. *Seeds of Woody Plants in the United States.* Agriculture Handbook No. 450. U.S. Government Printing Office, Washington, D.C.

U.S.D.A. 1965. *Silvics of Forest Trees of the U.S.* U.S.D.A. Forest Service Agriculture Handbook No. 271. U.S.D.A., Washington, D.C.

Chapter 3

Tree form includes the shape and size of the crown, the bole, and the roots. Each segment of the tree is affected in its development by a number of genetic and environmental factors.

CROWN FORM AND VOLUME

A crown of a tree is that portion of the stem above ground that consists of the main stem, or stems, and branches with leaves.

The general conformation of the crown of a particular tree depends upon its species, variety, and age, and the composition, density, and site quality of the stand in which it is growing. In addition, the condition of the crown when damaged by wind, snow, or other agents

Form and Growth of Forest Trees

needs to be considered. The characteristic form of the crown of any species may be observed on trees growing in open situations where the crown has not been modified by competition or damaged. Within a stand where a tree is competing with other trees, crown form is modified to a considerable extent. The total extent of the modification will depend upon the density of the stand, how many other trees are growing in the stand, and the quality of the site. Bella (1967) showed that there is a relationship between crown diameter and bole diameter at breast height, and that the relationship is not the same for all sites or regions (Table 3.1). Such studies help provide guidelines for determining the density under which various species should be grown. Substituting bole diameter into the equation where crown diameter = a + b (dbh) provides a means of estimating crown width. An estimate of tree area can then be obtained and from this the number of trees that can be grown on an acre of land can be approximated.

If competition for crown space occurs between trees of the same species, crowns of the individual trees will develop differently than if there are other species in the stand. If trees in a stand are of different ages (uneven-aged), crown form will be different from that of trees in stands where the trees are all of the same age (even-aged). The tendency is for larger trees in uneven-aged stands to have broader more spreading crowns than trees of similar crown position in even-aged stands.

TABLE 3.1. Crown Diameter—Diameter Breast Height Regression Statistics for Open-grown Jack Pine in Southeastern Manitoba and in Quebec (Bella, 1967)

SITE TYPE	LINEAR COEFFICIENTS		CORRELATION COEFFICIENT	dbh RANGE (IN.)	SAMPLE SIZE
	a	b			
Dry	2.13	1.80	0.90	3–12	44
Fresh, nutrient medium	1.83	1.84	0.88	3–12	46
Fresh, nutrient poor	4.20	1.59	0.90	3–15	38
Dry sub type	3.13	1.68	0.90	4–13	20
Manitoba	2.77	1.73	0.90	3–15	148
Quebec	1.76	2.04	0.96	1–11	84

In general, most conifer and certain hardwood species develop a single undivided stem producing a crown that in youth at least is spirelike. The term used to express this type of crown form is *excurrent* (Figure 3.1). Other species, mostly hardwoods, they may exhibit excurrent form when young; however, in later life their crown breaks up into a series of lateral branches no one of which dominates. The term used to express this type of crown form is decurrent or *deliquescent* (Figure 3.1). Deliquescent species when grown in dense stands have a restricted crown, which, as a consequence, may resemble that of an excurrent species. Another form, columnar or extreme fastigate, occurs infrequently. Trees of this form are usually mutant forms of deliquescent species. Excurrent species are also said to be geotrophic in that they tend to maintain an exact form even after the terminal is removed. Deliquescent species, on the other hand, are said to be phototrophic growing toward the direction where there is more light on the crown. This habit results in unequal development of the crown of a multiple stemmed tree.

Zimmerman and Brown (1971) note that lateral buds on current growth of excurrent species may develop more than lateral buds on decurrent species. Growth in subsequent years results in more vigorous growth of lateral buds and branches of decurrent species. They suggest the use of the term *apical control* to designate the difference in growth habits between decurrent and excurrent species. Excurrent species initially exhibit weak apical dominance since the lateral buds may not be completely suppressed, but lateral bud growth never overtakes the growth of terminal bud. Conversely, decurrent species exhibit strong apical dominance whereby current lateral bud development is suppressed by growth of the terminal bud; but in subsequent years lateral buds may surpass growth of terminal buds. The authors

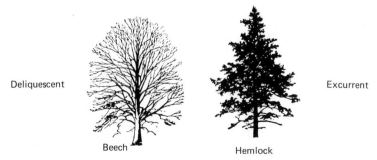

Deliquescent — Beech   Hemlock — Excurrent

FIGURE 3.1. Illustration of excurrent and deliquescent crown forms.

suggest that the concept of weak apical control be coupled with excurrent crown form, and strong apical control with decurrent crown form.

There are several notable differences between leaves of deciduous species and those of conifer species. Deciduous species, as the name implies, shed their leaves each year. The time of leaf drop varies from tropical to temperate species. Temperate zone deciduous species shed their leaves in early to late autumn and produce new ones the following spring. Tropical deciduous species may shed their leaves during a dry season and produce new ones with the onset of a rainy season. Some tropical species shed their leaves in a seasonal manner similar to temperature zone species. Keep in mind that some broadleaf species are considered evergreen. Coniferous species, except larch and bald cypress, hold their leaves for two or more years; at the end of the second or third season old leaves are shed in the same manner as are leaves of deciduous species, while the newer leaves are retained until succeeding years. Evergreen character would appear to offer an advantage to species having the trait in that they can take advantage of suitable growing conditions at anytime during the year when proper conditions for photosynthesis occur. However, endogenus rhythms restrict growth of conifers in late autumn and early winter. Deciduous species are limited in photosynthesis to the period when leaves are present each year, with the exception of aspen, which has chlorophyl in the bark and is thus able to carry on limited photosynthesis.

The interior structure of deciduous and coniferous leaves are somewhat different (Figure 3.2). Deciduous leaves have two epidermal surfaces, whereas coniferous leaves can be either flattened or lanceolate with two or more surfaces (Figure 3.2a). Deciduous leaves have a specific arrangement of palisade and mesophyll cells, while conifer leaves may contain chlorenchyma cells and resin ducts (Figure 3.2b). The thickness of epidermal layers and the condition of cuticular material determine the degree of drought that trees can endure. The number and size of stomates determines the amount of water loss and, to some extent, the sites where different species can grow. The amount of moisture stress that a tree can endure depends on its ability to reduce water loss through the leaves. Species and ecotypes with waxy, thick epidermal leaf surfaces and few stomates are capable of enduring extreme drought and can occupy drier sites.

More important to understanding production differences that occur between groups of trees is a recognition of differences in leaf volume

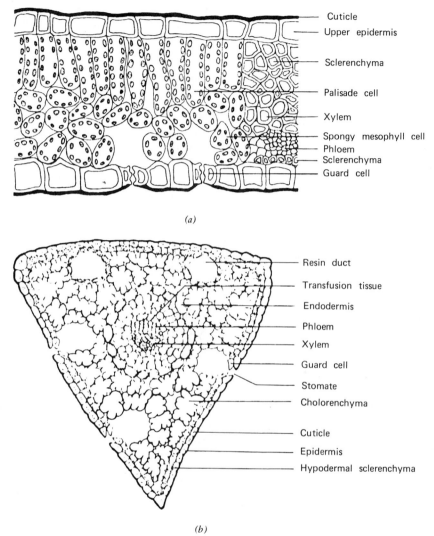

Cuticle
Upper epidermis

Sclerenchyma

Palisade cell

Xylem

Spongy mesophyll cell
Phloem
Sclerenchyma
Guard cell

(a)

Resin duct

Transfusion tissue

Endodermis

Phloem

Xylem

Guard cell

Stomate

Cholorenchyma

Cuticle

Epidermis

Hypodermal sclerenchyma

(b)

FIGURE 3.2. (*a*) Transection of a portion of a leaf blade from an angiosperm tree. (*b*) Transection of secondary needle of eastern white pine.

which occurs among species groups. Although foliage weight may not vary between stands of different density for the same species, evidence indicates that stands of different species attain different leaf mass. Bray and Gorham (1964) found, after considering a number of studies involving leaf production of different species, that deciduous species produced an average 2.8 metric tons per hectare per year; for conifers production was 2.9 tons. There were differences noted among deciduous genera, with oak sites producing 3.7 tons per year, beech 3.0, willow 2.9, elm 2.6, ash 2.5, poplar 2.5, and birch 2.4. Scotch pine sites averaged 2.8 tons per year and larch 3.0 tons per year. These studies were the result of measurement of the weight of annual leaf litter produced each year.

Kittredge (1944) indicated that leaf mass of individual trees could be estimated by the function: $\log w = a + b \log D$ where w = leaf weight and D = stem diameter at breast height (Figure 3.3). In studies

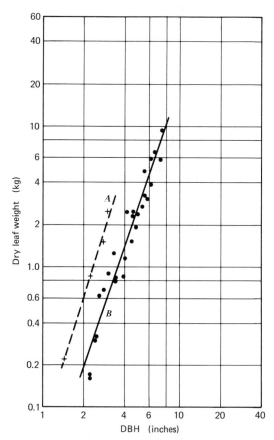

FIGURE 3.3. Relation between dry leaf weight and diameter of red pine and Douglas-fir: Douglas-fir, $\log W = 1.96 \log D -$ 0.91, line A; Red pine, $\log W = 2.56 \log D -$ 0.94, line B (Kittredge, 1944).

of energy transfer and biomass production of forest stands, it is necessary to measure foliage volume. Rogerson (1964) showed that foliage weight of loblolly pine was related to the tree diameter and projected crown area (the basal area in square feet of a horizontal crown profile when projected onto the soil surface). These two measures accounted for 84% of the variation in the data when the relationship was expressed as: kilograms of foliage per tree = 0.01273 (projected crown area) + 9.92227 (dbh) − 0.84208.

From an economic standpoint, it seems that within a species tree families that have long slender crowns should be favored over families with broad crowns. Ideally, a long-crowned tree will have sufficient leaf area for maximum photosynthate production, but would require less growing space than a broad-crowned tree. There is considerable amount of variation within a number of species in the width and length of crown which different families will exhibit.

BOLE FORM

The bole of a tree is that portion of the stem that has no live branches. The bole supports the crown aloft and is the lower trunk of the tree. Bole form refers to the number of dead branch stubs that are evident and the amount of taper from the ground to the base of the crown, or even higher; also, the nearness to a circular cross-sectional shape that the bole attains needs to be considered in judging form, as well as the amount of sweep and the number of crooks.

The ideal bole form from an economic point of view is a straight stem that is nearly cylindrical, has minimum taper from base to crown, and has no protruding branch stubs or other defects. This form is economical to log and manufacture, either into boards, chips, or veneer. Economically desired tree species are those that have the capacity to self-prune at an early age so that dead branches are shed and the branch stubs are covered with wood as a result of diameter growth. On the other hand, less desirable are those species with crooked stems that retain their branches, resulting in a large number of knots in the wood.

By altering stand density (by increasing or decreasing the number of trees in a stand) it is possible to increase or decrease the rate at which trees prune themselves. Trees on the better sites tend to compete for bole and crown space at an earlier age than trees on poor sites, so pruning rate should be faster, other factors being equivalent.

It has been shown that some genotypes within a number of species produce more desirable bole form than other genotypes.

Bole taper—the rate at which the bole decreases in diameter up the stem—depends upon the volume of the crown. Trees with long spreading crowns show more taper than those with short narrow crowns. For this reason, trees in dense stands, although smaller in diameter, have boles with less taper than trees of the same age but larger diameter growing in open stands. The rate at which a tree sheds its branches and reduces crown size will determine the relative rate of taper. Therefore, the degree of taper primarily depends upon the factors of species and stand density. As a tree gets older, it will tend to fill out its bole throughout the entire length.

Trees with short crowns have less taper because of differences in the relative rate of diameter growth along the stem. The widest points of the annual sheath of xylem growth along the bole are the section just below the crown where carbohydrates accumulate as a result of stress created by wind action, and the section at ground level where growth is accentuated also as a result of wind stress but also because of root spread. The points of maximum stress along the bole are at the base of the crown and at the stump, and as a response xylem growth, hence diameter growth, increases in response to the stress in these two areas.

Differences occur between coniferous and deciduous species in degree of roundness of the bole; part of this is due to arrangement of branches around the upper bole within the crown. Conifers having a relatively uniform arrangement of branches around the bole produce boles that tend to be circular because the distribution of carbohydrates is relatively uniform, and the branches being of nearly the same size do not cause uneven stress. The uniform arrangement of branches can be traced to the terminal meristem. Each autumn, conifer species produce a single terminal meristem bud that develops into the central stem when growth resumes in the spring. At the base of the terminal bud are arranged a whorl of lateral buds that are generally uniformly spaced around the bole. These buds will develop to form the lateral branches. Hardwoods tend to have a few large branches and as a result produce boles of less than circular form because there is an uneven stress and uneven distribution of carbohydrates along the bole. This is a result of an absence of a uniform arrangement of buds along the terminal axis at the branch terminal, and also results from a lack of a terminal bud, or a lateral bud cluster in many deciduous species. Because of their phototrophic nature and difference in bud development a number of deciduous species are characterized as having unequal branch and bole development.

Trees growing in locations where unequal stress—both tension and compression—occurs around the bole have eccentric growth rings; as a result boles of trees in such locations are not round but may be oval or fluted. Trees growing on hillsides show response to stress in the development of elliptically shaped boles. Wood on the downhill side of the bole is under compression stress and the wood on the uphill side is under tension stress. As a result, the cells produced react to the stress action and growth rings may be wider on the uphill or downhill side than the portion of the ring that is at right angles to the slope direction. In general, this eccentric growth is toward the upper sides of leaning trees and branches of deciduous species, but on the under sides of conifers. Because of these differences in position of occurrence, reaction wood in deciduous species is termed tension wood; in conifers it is termed compression wood (Panshin and DeZeeuw, 1964). Compression and tension wood, referred to as reaction wood when manufactured into chips, boards, or some other product, produces wood and fiber that does not react the same to pulping, drying, and machining as does wood produced where there was minimum stress. Reaction wood causes boards to split and warp; pulp yield from this wood is less than for normal wood. So, it is important when managing for timber production to produce trees that are straight and that have a minimum of stress placed on the bole.

Individual stems that form a part of a vegetative sprout clump originating from a single stump will tend to develop eccentrically shaped boles because of uneven crown and branch development. For that matter, it has been shown (Zobel and Aldin, 1962) that loblolly pines that "lean" or have "crooks," developed either from genetic or environmental influence, will show compression wood formation. Wood of either conifers or deciduous species that forms around the base of limbs develops under stress and, as a result, shows compression or tension eccentricity.

Physical damage to the crown, bole, and roots can occur as the result of breakage from weather factors and from attack by disease, insects, and animals. Damage as a result of weather and biotic agents is quite common in the forest and can be expected in most stands.

VOLUME INCREMENT OF THE BOLE

Volume increment of the bole depends upon the rate of height and diameter increment and the amount of taper. Variation in growth rate at these places causes species to vary in the volume of wood each produces. Diameter increment is measured as basal area growth; hence, tree volume is proportional to D^2. The basic volume equation is:

$$V = (hb)f$$

where V = cubic foot volume

h = height, in feet (may be total height or some measure of merchantable height)

b = basal area of bole in square feet, at breast height (4.5 feet)

$$= \frac{\pi D^2}{4(144)} = 0.0054D^2$$

f = rate of taper in proportion to that of a cylinder of equal base diameter and height

Growth in diameter is traceable to increase in the cell mass around the lateral meristen (cambium), which each year produces xylem cells to the inside of the stem and phloem cells to the outside. Layers of new xylem are added on the outside of the previous year's sheath, whereas layers of new phloem cells are added to the inside (Figure 3.4).

Height growth takes place as the apical buds at the end of the terminal meristem expand and grow each spring; root growth occurs when the cells at the base of the root cap divide (Figure 3.10). The many apical and lateral meristems contained within the crown expand and grow adding to the overall crown mass of the tree. As long as a central apical meristem is produced each year, a tree maintains an ability to form a single undivided bole; however, if a single dominant stem does not form as a result of damage or from loss of terminal dominance, then a tree's crown may assume a deliquescent shape. Some species are so dominant in their growth habit that a lost apically leader is replaced with an equally dominant leader from one of the lateral branches.

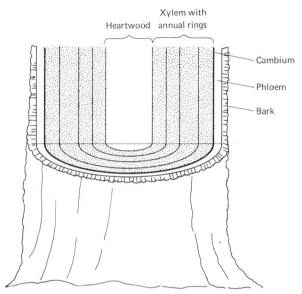

Heartwood

Xylem with
annual rings

Cambium

Phloem

Bark

FIGURE 3.4. Stem vascular system showing relationship of the major elements.

TRANSLOCATION OF WATER AND SOLUTES THROUGH TREES*

Water and minerals enter the conducting system of a tree through the root cells as a result of difference in potential of cell water and soil water, and in the case of mineral ions, a corresponding difference between cell ionic potential and ionic potential of the soil solution. (Absorption of water and minerals is discussed in more detail in Chapter 8.) Active movement of water in the tree takes place in the xylem elements where a negative water pressure may occur as a result of evaporation of water from the leaves. The result is a net movement of water, and some mineral ions, from the roots up the xylem and into the mesophyll cells in the leaves to replace water loss through evaporation. The movement is the result of negative potentials that develop in the xylem system. When a positive pressure occurs within the xylem, there can be a net movement of water out of the tree roots and into the soil, and through the leaves.

Movement of water upward in the xylem can be quite rapid with rate of flow during periods of active transpiration reaching 20–30 meters per hour. Minerals can travel passively along with the water.

*See Kramer (1969) and Kramer and Kozlowski (1960).

When the transpiration stream is diminished, minerals move as a result of differential potential between cells, as does water. Movement is rapid within the xylem because their elements (tracheids in conifers and vessels in deciduous species) are open tubelike cells that have no protoplasm and, as a result, are able to conduct solutes and water with little resistance. The cell walls of tracheids contain numerous pits that facilitate movement of water and solutes (Figure 3.17a). Vessels in deciduous species have perforated end walls facilitating movement of water and solutes (Figure 3.17b). Tracheids that occur in some deciduous species do not function as effectively in water and solute movement as do the vessel elements, because tracheids are smaller and their end walls are not as permeable.

Carbohydrates and other complex organic compounds produced in the leaves move downward in the ploem cells, although there is evidence that upward movement also occurs. Nevertheless, for the roots and the lateral meristematic tissue of the bole to receive the materials needed for cell division and growth requires translocation of these from the leaf tissues. This can occur primarily through the phloem system since the xylem system in a metabolically active plant is involved primarily in upward translocation. Movement in the phloem occurs primarily as a result of activated diffusion, resulting primarily from cell osmotic differences, that is slower than movement in the xylem during periods of active transpiration.

Solutes produced in the leaves generally move toward those areas where metabolic activity is greatest, developing reproductive buds, leaf buds, and lateral meristematic cells. Translocation takes place within the most recently formed phloem cells since older cells become inactive when the sieve tubes collapse. Older vessels and tracheids become filled with insoluble materials or, in the case of vessels, are closed off by tyloses.

Lateral movement of solutes and water can occur within the ray cells and the ray parenchyma, and since they are alive they are important in food storage as well.

ROOT FORM

As in the case of crown and bole form, the shape and extent the root mass of a tree assumes are the result of genetic structure and environment. A tap rooted species—one in which a main root extends into the soil directly below the bole—will exemplify varying degrees of tap rootedness as a result of the soil environment (Figure 3.5). A high water table or impermeable layer close to the soil surface will inhibit development of a tap root; whereas a deep soil of open structure permits the full development of a tap root. The lateral extent of a root system depends upon the amount of carbohydrate supplied by the leafy crown; the size of the leafy crown depends upon the amount of water and minerals supplied by the roots.

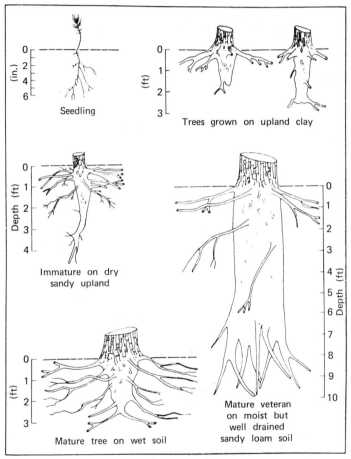

FIGURE 3.5. The inherent character of root systems, particularly the depth of penetration, is materially influenced by both age and site. The deep-seated feeding roots on the veteran loblolly pine extended to the water table. The seedling is two years old (Ashe, 1915).

Root form, therefore, does not assume as definite a pattern as does the crown or the bole. The fact that roots must contend with a variety of physical and biological factors makes it more difficult to characterize than crown or bole form. Lyr and Hoffman (1967) utilized nomenclature of others to develop a method to identify root types (Figure 3.6). Primarily there are horizontal and vertical roots, each having variations in location as well as form. In addition to placement and growth orientation, they suggested that further division be made as to root diameter: finest roots <0.5 mm; fine roots, 0.5–2 mm; weak roots, 2–5 mm; firm roots 5–10; rough roots 10–20 mm; strong roots >20 mm.

Root area is a function of crown size, site quality, and stand density. Smith (1964) showed that it is possible to estimate the root space requirements of individual trees (Table 3.2). Such estimates are help-

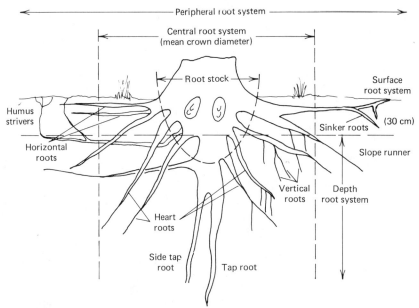

FIGURE 3.6. Schematic representation of a root system with most used nomenclature (according to Melzer). For diameter classes the classification of Grosskopf and Kreutzer is proposed (approximate equivalent English terms are used): finest roots, < 0.5 mm; fine roots, 0.5–2 mm; weak roots, 2–5 mm; firm roots, 5–10 mm; rough roots, 10–20 mm; strong roots > 20 mm. According to Grosskopf, "fine roots capacity" means root weight or length of roots from 0.5 to 2 mm per liter of soil; "finest root capacity" means roots < 0.5 mm in diameter per liter of soil (taken from Lyr and Hoffman, 1967).

TABLE 3.2. Regression Equations for Estimation of Root Spread from Diameter (dbh), Height (ht), and Crown Width (cw) of Forest-grown Trees (Smith, 1964)

SPECIES	NO. TREES	a	dbh (in)	ht (ft)	cw (ft)	R^2 OR r^2
			"b" COEFFICIENT FOR			
Douglas-fir	89	10.24	1.078	−0.120	0.249	0.411
	89	12.70	1.394	−0.134	—	0.394
	89	6.62	0.942	—	—	0.351
Western hemlock	81	4.94	1.014	−0.062	0.297	0.591
	81	1.47	0.816	—	0.303	0.579
	81	4.38	1.088	—	—	0.550
Western red cedar	61	7.68	0.700	−0.096	0.328	0.427
	61	10.78	0.972	−0.100	—	0.403
	61	7.24	0.652	—	—	0.361

ful in determining the number of stems to be grown on an acre of forest land. Using data from Stout (1956) for oak, the author developed the following relationship between tree age and crown and root area:

root area (square feet) =
$$7.16 + 4.43 \text{ (age)} + 2.69 \text{ (crown area, square feet)}$$

This is a rather efficient predicting equation having a coefficient of determination (R^2) of 0.7815.

In general, the form of the crown, bole, and roots of a forest tree depends upon the interactions between the genetic complement of the tree and the environment in which it is growing. Genotype within a species, age, stand composition, density, and injury all influence crown, bole, and root form.

WOOD PROPERTIES

Foresters have been aware of the importance that difference in wood properties has upon the products that can be produced from trees; however, until recently, little of this knowledge seemed to be of particular value except as it related to obvious differences between species. It was recognized that there were some deciduous species that produce wood of high density, and other deciduous species that produce wood of low density, as there were light wood and heavy wood conifers. As more information becomes available, it is apparent that wood density of a species is not the same from one location to another (Table 3.3). Also, wood density from two trees of the same species growing side by side in the same stand can differ appreciably. To continue to confound the issue, wood density in the same tree varies from the center to the exterior of the bole (cross section) and from the base to the top of the tree (Figures 3.7 and 3.8). Before it will be possible for foresters to produce trees with desired wood properties,

TABLE 3.3 Values for Wood Density of Douglas-fir, White Fir, Western Hemlock (From Forest Service USDA 1965 Western Wood Density Survey Report. No. 1 U.S. Forest Service Res. Paper FPL 27).

FOREST SURVEY UNIT	ESTIMATE OF AVERAGE WOOD DENSITY PER TREE		
	DOUGLAS-FIR	WHITE FIR	WESTERN HEMLOCK
Arizona	0.435	0.366	—
California			
East Side Sierra	0.435	0.366	—
West Side Sierra	0.454	0.364	—
Coast Range Pine	0.451	.0.374	—
Redwood - D. Fir	0.467	0.386	0.452
Southern California	0.494	0.363	—
Colorado	0.429	0.366	—
Idaho			
North	0.452	—	0.431
South	0.434	0.363	—
Montana			
West	0.452	—	0.444
East	0.432	—	—
New Mexico	0.430	0.356	—
Oregon			
West	0.452	0.372	0.422
East	0.446	—	—
Utah	0.436	0.381	—
Washington			
West	0.437	—	0.420
East	0.451	0.380	0.433
Wyoming	0.434	—	—

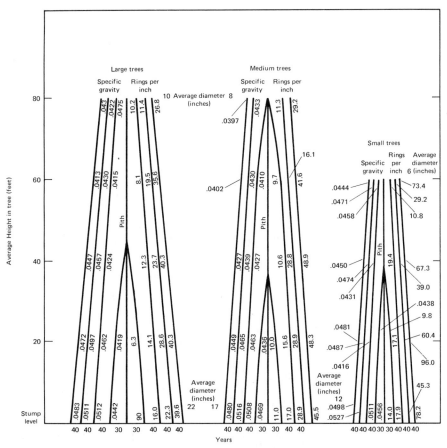

FIGURE 3.7. Variation of rings per inch and specific gravity for Douglas-fir trees of large, medium, and small diameter on a site of quality II. Position in the tree is indicated for the different growth periods, outward from the pith, and with height in the tree (Paul, 1963).

they will have to understand all of the factors that influence wood production. Wood density is of paramount importance to the use of wood since it affects not only the strength of wood, but also the yield of cellulose and the machining qualities.

The increase in wood density from pith toward the outer circumference of the bole is a general characteristic of conifers (Figure 3.8). For deciduous species there appears to be a reverse in this trend with higher density wood occurring in the center of the stem (Panshin,

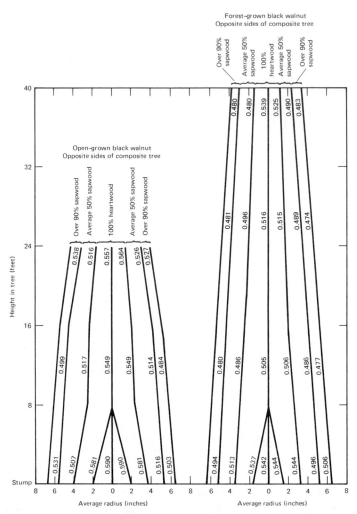

FIGURE 3.8. Composite diagrams of open- and forest-grown black walnut trees with average values for (1) moisture content (percent), (2) specific gravity, (3) shrinkage in volume (percent), and (4) width of sapwood (inches) at the respective heights and positions in the cross section (Paul, 1963).

1964) (Figure 3.9). Also note the difference in density between Douglas-fir and black walnut.

Gammon (1969) found that wood density of eastern white pine could be related most significantly to age (in the form $-1/A^2$) when

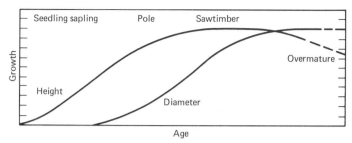

FIGURE 3.9. Generalized height and diameter growth patterns in relation to age and stage of development.

diameter was also used in conjunction with other variables, as dbh, merchantable height, Girard form class, and several other forms of each of these variables. He concluded that the patterns of density were inconsistent and not related very strongly to any of the variables used.

Wood varies not only in density but in other properties as well. Fiber length is a property that can have a profound effect on wood use. It has been shown that fiber length, like wood density, is influenced by genetic structure as well as by environmental factors (Zobel, 1961). In addition, wood varies in the amount and kinds of extractive substances as resins, terpenes, and lignins, which can be removed from the wood substance and which also affect wood use. These substances represent by-products that have commercial value, and they enhance the value of wood as a raw material. The sap of sugar maple can be used to produce a valuable commercial product. Individual sugar maple trees vary in their ability to produce sap of high sugar content and also in the volume of sap they produce.

Kotok (1965) lists a number of wood properties affecting wood use and suggests them as important subjects of research (Table 3.4).

VEGETATIVE GROWTH

When a seed has germinated and a tree is established with sufficient certainty that it will survive the season, a seedling tree is born. For the purpose of classification, the succeeding developmental stages of growth are sapling, pole, and sawtimber (Figure 3.9).

TABLE 3.4 Wood Properties Affecting Wood Use that Are Areas for Research (Taken from Kotok, 1965)

PHYSICAL	MECHANICAL	CHEMICAL
Color	Static bending	Percent alpha cellulose
Grain direction (spirality)	Inner-ring (annual) bond	Percent hemicellulose
Fibril angle	Intraring bond	Percent lignin
Fiber diameter		Kinds and amounts of extractives
Spring wood/latewood ratio		(a) water soluble
Sapwood/heartwood ratio		(b) benzene soluble
Pit structure		(c) alcohol soluble
Fluorescence		(d) acetone soluble
Resin content		(e) ether soluble
Light absorbence		Ash reduction-carbon ratio in pyrolysis
Sound absorbence		Ash content
Thickness of fiber wall		1 percent caustic soda solubility
Diameter of cell cavity		Methoxyl groups in wood

Stage of Development

Seedling	0–4.5′ in height (0″ dbh)*
Sapling	0.6″–4.0″ dbh
Pole	4.0″–10.0 dbh
Sawtimber	
young	10″–24″ dbh
mature	24″ dbh
overmature	When upper crown shows dead limbs

The limits of each stage may appear rather arbitrary. Some species are not capable of attaining a diameter of 24″ before the tree matures. Other species may grow quite vigorously to 40 or more inches in diameter. The size limits tend to mark the beginning and end of different stages of development, however, as the following diagram attempts to illustrate (Figure 3.9). In the early stages of development, tree growth primarily takes the form of increase in height; this predominates until the late pole stage when an increase in diameter be-

*dbh is the abbreviation for diameter at breast height. Breast height is measured at a point up the tree stem four and a half feet (1.3 m) from average ground level.

comes more apparent. The rate of height and diameter growth for most species will be for height to dominate the early stages of development with relatively more rapid diameter increases following at later stages after trees have had the opportunity to develop a large crown mass extending above the level of smaller plants in the forest. When trees become overmature the upper crown may show dead branches and become "stag headed."

The rate of development—the age at which a tree enters the various stages of development—is determined by the species involved, the density of the stand, and the quality of the site (Figure 3.10). Delayed crown closure in this instance resulted in a tree that was taller and had a larger diameter than a tree that had experienced early crown closure, but was released.

Annual Growth

Growth initiation of trees begins with the production of hormones within the apical meristems, the buds at the tip of the leader and the various branch tips (Larsen, 1973). Cell division of the central mother cells of the shoot meristem results in the production of xylem and phloem elements and cambium as well as bud and leaf primordia (Figure 3.11a). In the root promeristem, a growth region at the junction of the root cap and the central cylinder produces root cap cells, vascular system cells within the central cylinder, and cortex cells. (Figure 3.11b). At the same time that hormone synthesis increases, starch accumulates within the various parenchyma, xylem, and phloem elements along the stem. As the buds begin to expand, there is an awakening of cell activity along the lateral meristem in the cambium. Phloem cells adjacent to the cambium, which were not matured in the previous growing season, resume growth completing elongation and cell wall thickening. Cambium initials along the bole meristem lay down xylem cells to the inside to be followed later by the production of phloem initials to the outside (Figures 3.12 and 3.17). Development of new leaves and the initial production of new xylem and phloem occurs at the expense of carbohydrates produced the year before. When the leaves of the current year have reached full size, they are able to produce sufficient carbohydrates so that the cambium, xylem, and phloem continue to develop. As this transition continues during the growing season, it results in the production of summer wood cells in the stem and roots as contrasted with the spring wood growth that occurred in the early spring (Figures 3.12 and 3.13).

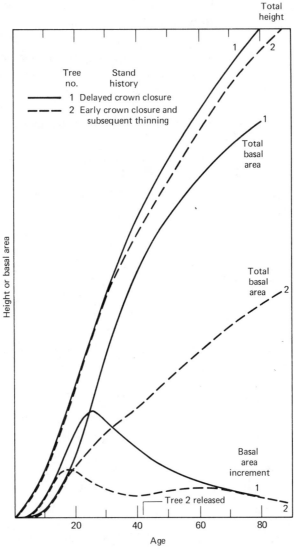

FIGURE 3.10. Effects of stand history on the height and basal area development of individual trees (Mitchell, 1975).

Shoot elongation consists of the elongation of a system that supports leaves at the terminal end. As the leaf, or needle, which contains the primitive part of the vascular system and is an integral part of the conducting tissue, becomes active, the nodal and internodal devel-

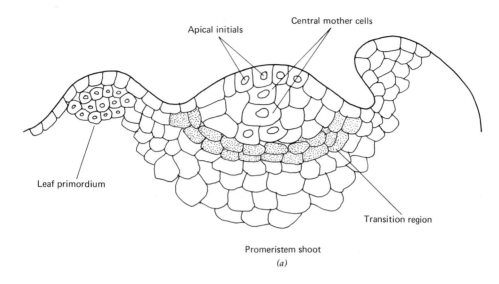

Promeristem shoot
(a)

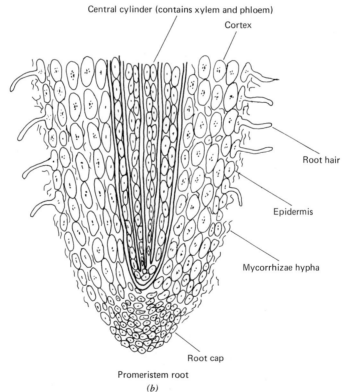

Promeristem root
(b)

FIGURE 3.11. Apical meristems of shoot and root. (*a*) Promeristem shoot. (*b*) Promeristem root.

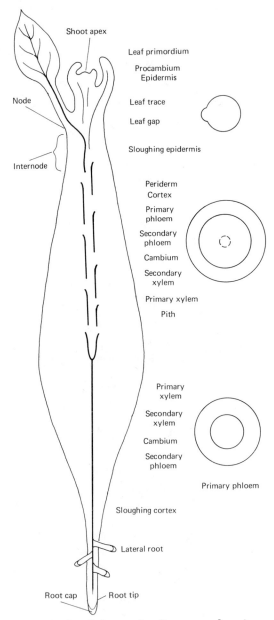

Shoot apex

Leaf primordium

Procambium
Epidermis

Node

Leaf trace

Leaf gap

Sloughing epidermis

Internode

Periderm
Cortex

Primary
phloem

Secondary
phloem

Cambium

Secondary
xylem

Primary xylem

Pith

Primary
xylem

Secondary
xylem

Cambium

Secondary
phloem

Primary phloem

Sloughing cortex

Lateral root

Root cap   Root tip

FIGURE 3.12. Schematic diagram showing arrangement of tissues in seedlings after initiation of secondary growth. 1, longitudinal section; 2, 3, and 4, cross sections at various heights (adapted from Esau, 1960).

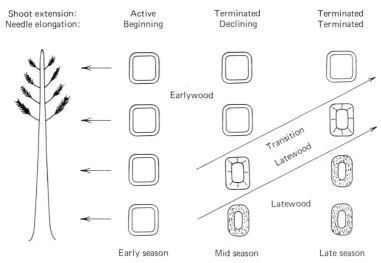

| Shoot extension: | Active | Terminated | Terminated |
| Needle elongation: | Beginning | Declining | Terminated |

Earlywood

Transition

Latewood

Latewood

| Early season | Mid season | Late season |

FIGURE 3.13. Seasonal variation in formation of earlywood, transition latewood, and latewood at different stem heights in red pine (from Larson, 1969).

opment of the new stem begins to unfold. Leaves, or needles, are nodal points with internodal spaces occurring between them (Larson, 1973) (Figure 3.12). Essentially, shoot elongation consists of the extension of a vascular system that has its origins in leaves or needles developing as nodal points along an elongating stem. Apical buds producing leaves and needles contain within their structure the primary elements of phloem, xylem, and cambium, which are essential for extension of the conductive system of a tree (Figures 3.2*a* and 3.2*b*). The branches, bole, and roots of a tree are simply an arrangement of the primary vascular system whereby the many leaves or needles and their vascular systems are arranged in a single structure.

Kozlowski (1973) classified shoots into several categories. *Determinate* genera, such as oak, hickory, five-needled pines, spruce, and fir, produce shoots that develop from a single main terminal bud and accomplish a single spurt of growth each year. *Indeterminate* species are those that produce new shoots from buds which do not occupy a terminal position; rather, new growth occurs from a strong side bud or a bud that has formed to the side of an aborted terminal, and which exhibits several growth spurts during each year. Species in these last two groups—those without terminal buds or whose new buds form to the side of the tip—are common in birch, catalpa, locust, sycamore, willow, basswood, and elm.

Not all bud primordia that occur along the apical meristem develop following formation. Some buds, particularly on deciduous species, remain dormant for many years; when, for some reason, dominant control of the stem is lost and production of IAA is reduced, some dormant buds (also called suppressed or epicormic) may develop. This is common in species that sprout from stumps. At other times, as a result of injury, bud initials may form and sprouts form along the bole or from roots. These sprouts are termed adventitious sprouts.

The pattern of height growth of different species varies during a single growing season for a given location (Figure 3.14). Species such as white oak accomplish the greater portion of annual height growth usually by mid-June; only small increments occur after that time. This characteristic seems to hold true for species having one annual flush in height growth. Other species such as shortleaf pine accomplish several periods of height growth during the growing season resulting in a multinodal terminal growth. Each period of growth is followed by the setting of buds. Temporary (summer) buds break dormancy when moisture conditions are favorable. In some situations as many as four growth periods can occur in one growing season. With longleaf pine six growth flushes in height were noted in a single year at one location (Allen and Scarborough, 1969). Winter buds, however, remain dormant until photoperiod and temperature are again favorable for growth, which occurs with the onset of warming temperature and of

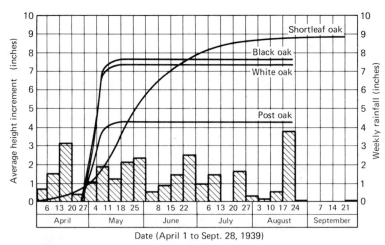

FIGURE 3.14. Seasonal height growth of shortleaf pine, and black, and white, and post oak (Johnston, 1941).

longer photoperiods the following spring. The leaves develop from winter buds set during the previous growing season.

During a single growing season some deciduous species produce two types of leaves; primarily species of indeterminate growth habit. The first type of leaf, produced from the buds laid down during the latter part of the previous growing season (winter buds), tend to be smaller than the second type of leaf, which develops from buds formed during the early spring and early summer of the current season (summer buds). The second type of leaf forms on new growth that takes place upon the elongation of terminal buds. This difference in leaf size and structure is termed *leaf dimorphism* and is quite striking in birch and aspen, but is not so noticeable on oaks. Larch, a conifer, shows some evidence of leaf dimorphism. Also, species that have the ability to undergo more than one spurt of height growth in a growing season are able to compete better in the seedling and sapling stages of growth than can uninodal species.

Kramer (1943) reported the length of growing season and time of suspension of terminal stem growth for ten tree species grown in North Carolina (Figure 3.15). This study showed that onset and termination of height growth of red and white pine have similar patterns for trees growing in North Carolina and in New Hampshire; however, the total season of growth is somewhat longer in North Carolina. Red and white pine are determinate species, having only one major flush of height growth each year.

Annual diameter growth begins in early spring and does not terminate until late summer or early fall, although the rate of growth during the growth period can vary (Figure 3.16). The sheath of wood laid down each year by temperate tree species is easily distinguished in that the spring wood is of a different character from the summer wood, and as a result it is not too difficult to distinguish between annual growth rings. Early wood cells may be larger and thinner walled than late wood cells (Figure 3.13). As a result of this difference in character a growth ring is distinguished as cell changes progress from spring growth initiation to autumn growth cessation. There are further differences between deciduous and coniferous species in the type and arrangement of the xylem cells (Figures 3.17a and 3.17b). Essentially, coniferous species are nonporous because the xylem elements are made up of nonporous tracheid cells (Figure 3.17a). Deciduous species can be either porous or nonporous (Figure 3.17b). Species with porous wood structure are either of the ring porous group,

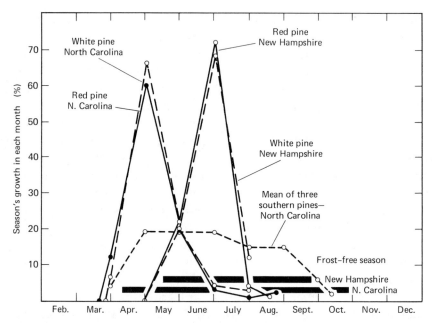

FIGURE 3.15. Variations in seasonal height growth patterns of red pine and white pine in North Carolina and in New Hampshire, and of three southern pines (lobolly pine, shortleaf pine, slash pine) in North Carolina. The northern pines have preformed shoots and usually have one annual growth flush whereas the southern pines grow in recurrent flushes (from Kramer, 1943).

the spring wood cells being made up mainly of large vessel elements that are joined longitudinally, or diffuse porous where the larger vessel elements are spaced throughout the annual growth ring and do not vary appreciably in size. There are other anatomical differences that occur between and within species.

The amount of wood laid down in any particular year by an individual tree depends on the weather conditions of the current growing season; in some instances the weather of the previous season affects current increment, as well as the species and quality of the location. The total yearly wood increment of a tree is measured by the volume of the sheath of cells laid down around the core of root, bole, and branch wood produced in previous years, and by the volume of elongating branch and root apical meristems.

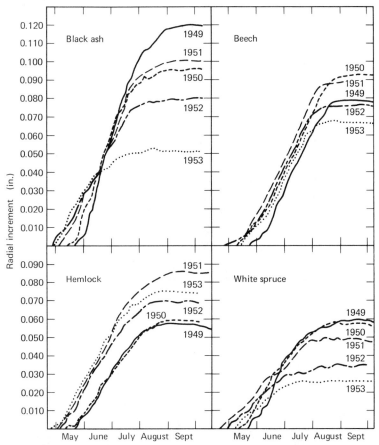

FIGURE 3.16. Variations in time of cambial growth initiation, rate of growth, and duration of growth of four species of forest trees during five successive growing seasons in Ontario, Canada (from Fraser, 1956).

Daily Growth

It is much easier to detect differences between daily measurements in diameter growth than it is to detect differences in height growth. A striking feature of a graph showing daily diameter growth is the fact that trees undergo a rather marked change in diameter (Figure 3.18). Although the general trend of growth shows an increase in diameter, a daily or hourly comparison may show rather pronounced changes, and a daily decrease in diameter can be evidenced for a

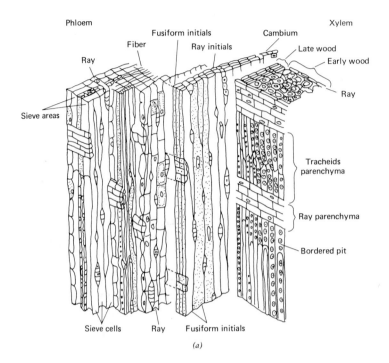

(a)

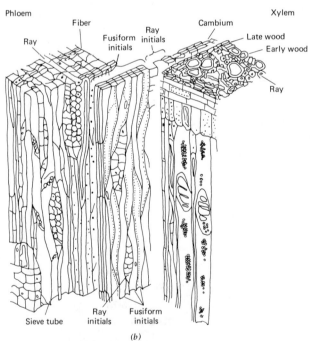

(b)

FIGURE 3.17. Anatomical arrangement of xylem, cambium, and phloem cells for (*a*). Conifer species (*b*) diffuse porous deciduous species (adapted from Esau, 1960, and later).

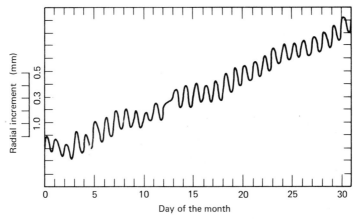

FIGURE 3.18. Diurnal variations in radius superimposed on radial increment of a Canary Island pine (*Pinus canariensis*) tree during March, 1965, in Australia. Note the effect of rain on March 13 (from Holmes and Shim, 1968).

number of days. "Negative growth" is associated with changes in transpiration demands in the leaves. During the day, as the transpiration rate increases and large volumes of water are lost from the leaves, tension in the leaf and stem cells increases. The force created is sufficient to cause the diameter of the bole to shrink. When transpiration stops during the night and water, when it is available, is replaced in the bole, branch, and leaf cells, the tree regains its original diameter plus any increment that occurred as a result of cell division during the time between measurement. During protracted periods of drought, a tree may remain "shrunken" until the soil moisture supply has been replenished by rain.

VEGETATIVE DORMANCY

Tree species that grow in temperate regions undergo annual periods of vegetative dormancy that occur during the winter season when temperatures are low and the photoperiod short. It is possible by exposing some species to long photoperiods and to favorable temperatures to cause trees to continue growth during the dormant period. However, studies show that a period of cold is apparently necessary to break dormancy of a number of species and that there can be variation within a single species as to the degree of dormancy that takes

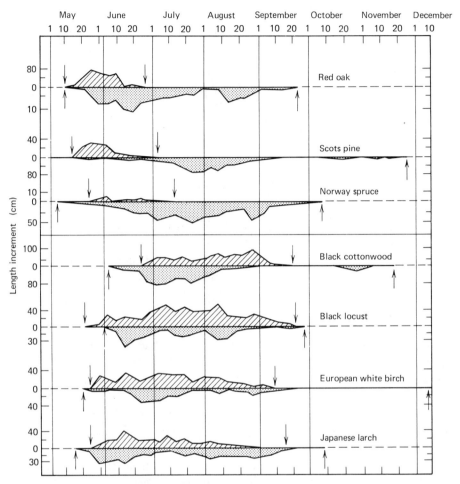

FIGURE 3.19. Variations in seasonal shoot and root growth characteristics of eight species of forest trees. Shading indicates shoot growth and solid black represents root growth. Seasonal initiation and termination of growth are indicated by arrows (from Lyr and Hoffmann, 1967).

place and the duration of chilling requires to break dormancy (Perry and Baldwin, 1966). Vegetative dormancy is an adaptive mechanism that temperate region trees have developed to permit them to survive the cold winter season. Temperate species whose geographic range extends near or into a subtropical climate do not appear to undergo as deep a condition of dormancy as do other members of the same species growing in colder climate (racial variation).

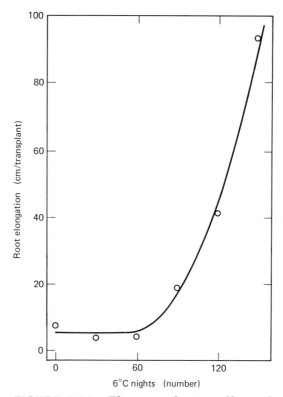

FIGURE 3.20. The cumulative effect of seedling exposure to cold nights on the root-growth capacity (from Stone, 1970).

A type of transient dormancy may occur during the growing season when soil moisture supply becomes too low for continued growth of root, stem, and crown meristems. Bilan and Wu Jan (1968) found that root elongation of loblolly pine has stopped when needle moisture content fell to about 150 percent. Range in needle moisture was found to be 200 percent when the soil was at field capacity, but dropped to 85 percent when the soil moisture reached permanent wilting.

Seasonal root growth differs between species in the time when root growth begins in the spring and with the times of cessation of growth in autumn (Figure 3.19). Stone (1970) demonstrated that to obtain maximum amount of root growth of ponderosa pine required a period of "after ripening" involving exposure of seedlings to night temperatures at or below 6°C for at least a 60-day period (Figure 3.20).

GENERAL REFERENCES

Kozlowski, T.T. 1971. *Growth and Development of Trees*, vols. I and II. Academic, New York.

Kramer, P.J. and T.T. Kozlowski. 1960. *Physiology of Trees*. McGraw-Hill, New York.

Zimmerman, M.H. and C.L. Brown. 1971. *Trees. Structure and Function*. Springer-Verlag, New York.

Part Two
Forest Stands

Chapter 4

The *stand* is a basic unit of management and is important to planning silvicultural operations, since it is by tending and reproducing individual stands that a forest is made to produce the required amount of wood. Also, by manipulating a stand, it is possible to alter wildlife populations, change the quantity and quality of water in the streams that traverse the landscape, and to alter the aesthetic appearance of an area.

In the terminology of plant ecology, a forest stand would be designated as a community, or an ecosystem. Such a designation includes the entire complex of environment, stand structure, and the plant and animals populations that inhabit the stand. Foresters use the term stand to designate the tree portion of the ecosystem—that part with

The Stand

which they are primarily concerned. By manipulating stand structure it is recognized that other populations of plants and animals that are part of the larger community will be altered. Thus, stand structure should not be altered without full knowledge of the effects this will cause to the total ecosystem in its general dimension. As a result of the need to produce the goods and services from the forest for the benefit of humans, at different times forest stand structures must be altered, but with careful planning so that at no time are other parts of the ecosystem irreparably harmed. Wildlife, soil, and plant systems must be taken into account when planning silvicultural operations.

A stand can be defined as an aggregate of trees, or other growth, occupying a specific site and sufficiently uniform in age arrangement, species composition, and density as to be distinguishable within the forest and from other growth on adjoining areas. Species composition of the stand refers to the proportion of various tree species represented in a stand. Age arrangement concerns whether the trees in a stand are of nearly the same age or whether a diversity of ages prevails. Density of the growth of a stand is important not only to growth of the stand itself, but also the growth of individual trees in the stand. Any changes in density must be tied to a specific site (location) since the elements of climate, soil, and biota determine the environment that significantly affects composition and density. The quality of the site needs to be considered in arriving at a complete description of a forest stand.

Some terms used to describe stands are:

even-aged stand	seedling stand	mixed stand
uneven-aged stand	sapling stand	pure stand
two-storied stand	pole stand	
	sawtimber stand	
understocked stand	overmature stand	
fully stocked stand		
overstocked stand		

To partially describe a stand, several terms may be used; for example, fully stocked even-aged pure stand, or uneven-aged mixed understocked stand on a poor site. To fully describe a stand *all* of the components need to be specified; for example, an even-aged, 50-year-old Douglas-fir stand, basal area 160 ft²/ac., on site quality II describes a stand very well and provides the information needed for management and silviculture. One thing missing is the volume of wood contained in the stand; however, this may be obtained from the information given in the description since this information is available for Douglas-fir. Stand volume tables are not available for all species, however.

CLASSIFICATION OF STANDS ON THE BASIS OF AGE

The spread in age from the oldest to the youngest trees in an even-aged stand should not exceed 15–20 years in forests of the eastern United States. For older even-aged growth such as occurs in the forests of the western United States, perhaps 10–15 percent of age at rotation should be used. Rotation age is the number of years required to grow a stand from seed to maturity. At maturity the stand would be harvested. The exact age difference between trees within an even-aged stand is not the most important consideration. What is important is the fact that within an even-aged stand individual trees compete at the same stages of development at the same time. In uneven-aged stands, on the other hand, trees of several ages are present, and individual trees within the stand are competing with other trees that are at different stages of development.

To state an exact condition when a stand is no longer even-aged but uneven-aged is difficult. Is it when there are two separate age classes present or are three age classes needed before a stand can be considered uneven-aged? Perhaps it is best to require the presence of three separate age classes, because there are instances in even-aged management when the shelterwood regeneration method is practiced, and, in this case, there will be two age classes present; yet even-aged stand structure is to be maintained following the final removal of the older trees.

Trees in any stand will enter different stages of development at different times because of differences in growth rates among species. However, where the selection system is used, the stand can be considered uneven-aged for the intent is to create a stand where a range of ages exists within the stand. The reasons for maintaining stands with variations of age class distribution will be considered in courses in silviculture and management.

The developmental stages of even-aged stands are similar to those of a tree. The stand progresses through the different stages in somewhat of an orderly manner—seedling→sapling→ pole→sawtimber→ overmature. Within an uneven-aged stand most or several stages of development will be present in a stand at one time (Figure 4.1). Rather than having trees of nearly the same size, uneven-aged stands should have trees of different sizes and ages, hence, different stages of development at all times.

Certain advantages are attributed to even-aged stand structure as there are to uneven-aged structure. These advantages are:

1. A larger number of crop trees per acre at a particular time than in an uneven-aged stand
2. Bole form tends to be better and boles are longer and cleaner than in an uneven-aged stand;
3. Crowns tend to be smaller
4. Lends itself to uniform treatment; that is, silvicultural operations will be applied to the entire stand at the same time
5. Favors the development of trees that are intolerant of shade and competition
6. Product quality may be better than in an uneven-aged stand

Many of the advantages associated with even-aged stands are related to economic considerations. And it is for this reason that even-aged management will probably be used more frequently as a management system than uneven-aged management. The relative merits of one system of management—even-aged versus uneven-aged—will be argued by foresters for many years to come. Each system has its particular merits, and it should be on the basis of biological *and* economic merit that final adoption of a particular system be made. Advantages of uneven-aged natural stands are:

1. The site is protected at all times because the entire stand is never totally removed
2. Trees of many sizes are present, which insures that physical damage to the stand may not result in a complete loss
3. Favors the development of trees that are tolerant of shade and competition
4. Room for continuous reproduction so that the stand does not depend upon a seed crop maturing at a particular time

If seed from a number of trees was used to establish a stand, all of the resulting trees could not attain the same size in an equal amount of time. The size of a particular tree in either an even-aged or uneven-aged stand depends upon a number of conditions, assuming there is no difference in the time when the stand, or part of the stand, was established from seed or when the sprouts began to grow. The conditions that cause trees to vary in size can be categorized under three headings:

1. Inherent capacity to grow
2. Amount of growing room
3. Favorableness of a particular location

For a particular situation not all trees will begin life with the same genetic structure; this will result in differences in tree size. *Inherent capacity* of a tree to attain a certain size in a specific time depends on several factors. Tree size difference can be the result of the difference in seed size brought about as a result of a seed's location with respect to its siblings. Many forest tree species produce fruit in which there are a number of seeds contained in a single flower; as a result there is competition for food from the female parent; therefore, all embryos within a flower cluster, or cone, do not receive the same amount of growth substance. Embryos within a seed cluster develop at different rates because of differences in time of fertilization. In some species self-fertilization produces trees of low vigor. In species having multiple seeded fruits, the pollen parent can be different for individual seeds developing within the same female organ, so genotype differences occur within the same fruit cluster. Further, it appears that weather conditions which prevail during seed development affect seedling vigor as well as the amount of seed produced in different years. Insects and disease, although they may not destroy the embryo, may damage the endosperm, thus reducing the vigor of the developing seedling. So if a stand of trees was to start from seed, all disseminated at the same time, size differences between trees in the stand cannot help but occur. These reasons account, in part, for the range of diameters that exists even within an even-aged stand. The age difference between trees in an uneven-aged stand makes the difference in size all the more evident.

When a tree has become established, its chances of survival will depend in part on the amount of *growing room* it has available. If the stand is crowded and there are more trees in the stand than there is room for each of them to grow unhindered by other trees, the weaker trees will die because they cannot keep up with the more vigorous trees. Not all species are endowed with the same capacity for growth, nor will all trees within a species have the same growth capacity. And, as will be pointed out in the discussion of stand density, a forest site has a limit to the number of trees it is able to produce.

The favorableness of the location with respect to trees in the same stand refers to the micro-site upon which a seed falls, and where embryo growth and seedling establishment must take place. It is possible to generalize on the site capacity of one acre of land, but it is quite difficult to evaluate the growth supporting capacity of the growing site of a small tree seedling since there are many variations in micro-site contained within a stand, and a single tree, in effect, occupies a single micro-site.

In the natural forest, seeds that are not eaten by birds and mammals, and that are not killed by disease and insects [loss of seed to animals, insects, and disease may represent more than 99% of the annual seed crop (Hocker, 1961)] must find a location that is favorable for growth. Even if a seed has the capacity to germinate, it must soon find water and minerals to support its life processes or it will lose its place in the sun; without sunlight it will surely die. Micro-site differences are common within a forest stand, so that two trees which were established from seeds having the identical genetic potential and having the same amount of growing room can develop at different rates because one seedling received more water than the other, or one received more of mineral nutrients than the other.

It is possible to illustrate the structural difference between even-aged stands and uneven-aged stands by plotting numbers of trees over diameter (Figure 4.1). Such a neat arrangement of diameters as shown in the figure does not occur in a single stand or on a single plot within a stand. It is usually necessary to measure a number of plots to obtain such patterns.

CLASSIFICATION OF STANDS ON THE BASIS OF COMPOSITION

Differences in species composition are probably the most readily recognized of all stand characteristics. Two general classes of stands are defined: pure and mixed. As with age, it is necessary to adopt a somewhat arbitrary expedient to differentiate between pure and mixed classifications. In general, if 80 percent of a stand is composed of a single species, it can be considered a *pure stand.* If the percent composition of a second species exceeds 20 percent, it will probably be necessary to recognize this second species in the stand description,

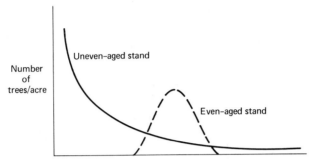

FIGURE 4.1. Diameter distribution in an uneven-aged stand and an even-aged stand.

and the stand would be designated a *mixed stand*. Then a mixed stand would be one in which two or more species occur and one species accounts for less than 80 percent of the total stems in the stand. If some other measure of stocking was used, such as basal area, at least 20 percent of the basal area would be represented by another species. When three or more species are growing in a stand there is generally no question that it is a mixed stand.

It is necessary to specify which size trees or plants are included in the measure of composition. On the basis of total numbers of plants growing in a forest stand, trees, although the largest and most evident, are by far in a minority. Classification of species composition for forest stands then is based on the composition of the dominant stand of tree species. However, a type of site classification will be presented later in which lesser vegetation is used in stand classification. When defining stand composition, if it is not obvious what the dominant tree stand might be, as can occur with uneven-aged stands, then a minimum diameter may be set below which a tree is not considered in a composition classification.

The advantages of pure stands, in contrast to mixed stands, can be designated on both an economic and a biological basis. In each, the advantage to one may be cited as a disadvantage in the other. Advantages of pure stands are:

1. Simplicity of treatment and reproduction
2. Economy of harvesting
3. Uniform development of the individual tree

Advantages of mixed stands are:

1. Maximum protection from insects, disease, and, in some instances, fire
2. Protection for "tender" species
3. Natural regeneration may be easier because if enough species are present, some are bound to survive and grow
4. Species do not make the same demands on soil mineral reserves, hence they are not likely to deplete soil minerals or damage soil structure

The frequency and abundance of trees of a single species varies throughout a locality. Two terms are used to indicate the occurrence of trees in different stands. *Frequency* refers to the number of times a species is represented in a stand; *abundance* refers to the number of stems of a species occurring in a stand. A further differentiation can be made: *dominance,* which is the percent basal area represented by a species within a stand.

Toumey and Korstian (1947) used the terms gregarious, consorting, and occasional with reference to the frequency and abundance of different tree species as they might be found in a forest stand. Although it tends to oversimplify the complex way in which trees are found in stands, this classification does not violate any evolutionary information that is available for the different species. Some species produce stands that are predominantly of a single species; species producing such stands are termed *gregarious* and are capable of producing pure stands. Other species occur frequently but are not so abundant in the stand; these species are termed *consorting* and will usually make up a large segment of mixed stands. Then there are those species that are found less frequently and that make up only a small portion of the stand; these species are termed *occasional* and occur in either mixed or pure stands. However, occasional species represent such a small proportion of the total stand in some cases that they do not affect the number of stems sufficiently to change the classification of a pure stand to a mixed stand.

Many factors determine whether a species is classed as gregarious, consorting, or occasional; some of these are considered in the discussion of succession. The following species represent various types of composition groupings.

It should not be concluded that consorting species may not form a pure stand, nor that a gregarious species are always found in pure

GREGARIOUS	CONSORTING	OCCASIONAL
Pines	Oaks	Ash (on some sites)
Firs	Beech	Black walnut
Douglas-fir	Maples	Buckeye
Aspens	Hemlock	Basswood
Cottonwood	Elm	Cucumber tree
Spruces	Tuliptree	
Redwood	Hickories	
Atlantic and		
Northern white cedar		

stands; rather, that there is a tendency for all species to be found in pure or mixed stands at some time. However, in the case of species within the occasional group, pure stands will be infrequent and not large in area.

CLASSIFICATION OF STANDS ON THE BASIS OF DENSITY

Density refers to the physical space that the trees take up in the stand and can be defined as the space occupied by either the boles or, less frequently, the roots.

The degree of crown closure was difficult to measure until the advent of aerial photography. Crown area can now be rather easily measured from aerial photographs by comparing existing stand conditions to sets of measured densities on crown density scales.

One way to express bole density is to count the number of trees on a unit area, usually an acre. Density can then be expressed, for example, as 1000 stems per acre. This is an easy way to designate the density of seedling and sapling stands, for most of the trees will be of nearly equal size. When stands reach pole size, or where stands are uneven-aged, there is less uniformity in diameter so an expression that incorporates tree size is needed. Such an expression is basal area, which is given as the sum of the square feet of bole area of the individual trees on an acre; as 120 square feet per acre or some other basal area.*

Root area is the most difficult to measure since it involves excavating the root systems of entire trees. This is generally unfeasible to

*For a stand of trees: Basal Area = $(\sum \text{dbh}^2)0.005454$.

do except in connection with an experimental investigation. However, root area of individual trees is correlated with crown area and bole diameter and thus estimates of root volume can be obtained, for some species at least, from measurement of bole or crown area. Roots of some species have commercial value, however, and it is not inconceivable that root density measurement, or root volume estimates, might be of value. Longleaf pine roots have been sold for their tupentine and resin content; other species have roots that are of medicinal value and some of these are still harvested in this day of synthetic substitutes. Technological advance made it possible to use wood from entire trees, and so it is necessary in some instances to measure root volume as well as bole and crown volume.

In order to be able to measure the volume production of a stand, a precise measure of bole density is required. Stocking, as the term implies, represents the density of a stand with respect to a standard. A number of approaches have been developed for estimating stocking. Reineke (1933) developed a measure of stocking that equates average tree diameter and number of trees per acre to density (Figure 4.2). Wilson (1951) proposed a method for measuring stocking similar to that of Reineke but used tree height in place of diameter. Schumacher and Coile (1960) utilized both height and basal area as well as age in their measurements of stocking. Their formula for estimating stocking of loblolly pine is:

$$S = B \left[0.8409 - 0.1707 \frac{(H)}{100} + 0.1062 \left(\frac{1000}{A} \right) - 0.1408 \left(\frac{H}{A} \right) \right]$$

where B is basal area per acre, H is height of dominant trees, A is total stand age, and S is stocking percent.

Each of these measures state density in relative terms such as 1.0 (or 100, or 1000) which represented the average stocking of the stands used to derive the density expression. In estimates of stocking, stands having a density greater or less than 1.0 were said to be over or understocked, and the difference in relative stocking can be related to the volume of wood they supported.

Gingrich (1967) developed a method for measuring stocking in even-aged oak stands. Utilizing the tree area ratio reported by Chisman and Schumacher (1940) and stand data from Schnur (1937), Gin-

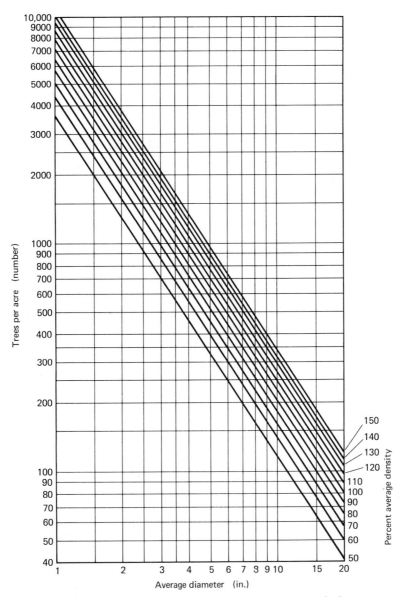

FIGURE 4.2. Relationship between average stand diameter, number of trees per acre, and stocking (Schnur, 1937).

grich was able to develop a tree area equivalent: mil acres equivalent*
$= 0.0507(N) + 0.1698 \sum D + 0.317 \sum D^2$, where N = number of trees
per acre and D = sum of the diameters of the individual trees. Gingrich
considered a stocking of 1000 stocked mil acres as full stocking.

It is not always necessary to develop separate stocking equations
for different situations since diameter is a function of site. For example, Gingrich found that a 10-in. dbh, 40-year-old tree required the
same area on a good site as a 65-year-old tree of the same size but
growing on a poor site.

Leak, Solomon, and Filip (1969) and Philbrook (1971) developed
stocking levels for even-aged northern hardwood (beech–birch–maple)
and eastern white pine stands growing in the northeastern United
States; these guides were developed using methods similar to that of
Gingrich (Figure 4.3).

From time to time the suggestion is made that bole surface area be
used as a method for measuring stand density. Bole surface area of a
stand is equal to $\sum (h \times \pi d \times f)$, where h = total height, d = dbh, and
f = form quotient. Because measuring tree bole area is such a tedious
task and other satisfactory methods for measuring stocking are available, advocates of bole surface area have few supporters. In any case,
when the trees in a stand are relatively uniform in height and vary
little in bole form, stand density is proportional to $\sum (d)$, which is the
basic form of the Chisman–Schumacher approach illustrated in Gingrich's method.

CLASSIFICATION OF STANDS ON THE BASIS OF SITE QUALITY

At times it is difficult to express the growth potential of a particular
locality because growth rate changes with stand age, stand composition, and stand density. In addition, disturbances by insect or disease
or damage from wind or ice can alter either the qualitative or quantitative yield. Ultimately, there is a need to express productive capacity of a forest stand in terms of its potential for either timber, water,
wildlife, or recreation. Timber production potential will be the primary
concern of this text. Quantitative measurements of water, wildlife,
and recreation are not readily available.

Timber productivity of a single stand can be measured by making

---

*mil acre = 1/1000 acres. Stocking is measured as the number of stocked mil acres that
the trees in a stand occupy.

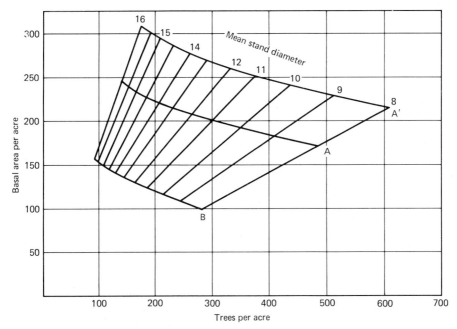

FIGURE 4.3. Stocking guides for even-aged eastern white pine, based on number of trees in the main canopy, average diameter, and basal area per acre. Stands above the A[1] line are overstocked. Stands between and A and B lines are adequately stocked. Stands below the B line are understocked (Philbrook, Barrett, and Leak, 1973).

periodic remeasurements of growth in the stand. It is often necessary, however, to estimate growth rate at different ages and for different stands, so a knowledge of past growth of a single stand has limitations.

The most commonly used measure of site quality, or site productivity, is *site index,* which is based on the height attained by the dominant tree in a stand at some base age; usually 50 or 100 years (Figure 4.4). Other base ages are used, such as 15, 25, or 35 years, for shorter rotations. Index age is chosen as the time nearest to that when the stand is to be harvested; however, there is no biological necessity to use a specific index age. For management purposes different index ages may be employed. For example, site index curves for index ages of 25 to 50 are available for some of the southern pines. The tendency in short rotation southern pine plantation management is to use 25 years for comparing sites, and for long rotations to use 50 years for the index age.

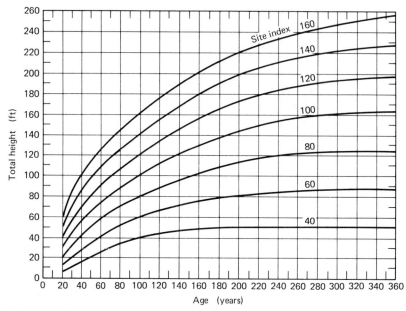

FIGURE 4.4. Site index curves for ponderosa pine (Walter H. Meyer, 1938).

If a set of base curves of height and age such as those shown in Figure 4.4 (Table 4.1) are available for a particular species, then to determine the site index of a stand one needs to measure total height and age of 5–8 dominant and co-dominant trees and to compare the average total stand height and age to the base curves. Such a comparison will indicate whether a stand is high, low, or average in productive capacity since other measurements of yield follow height growth. The example of ponderosa pine site index illustrated here assumes that the curves for height on all sites are harmonized; that they have a proportional difference in height for all of the different site and age conditions. It must be recognized that height curves for a number of species are polymorphic in character; that each site and age class has a separate curve whose form differs from those of other sites and ages. The need to recognize polymorphic site differences will be discussed in the section on site quality evaluation.

Total height of dominant and co-dominant trees in a stand is the best measure of site potential because trees on poor sites do not grow as tall as trees on the best sites. Also, height growth is not too greatly affected by stand density, as are stand diameter and stand volume, except that height growth may be affected where the stands are near

TABLE 4.1 Height of Dominant and Co-dominant Ponderosa Pine Trees of Average Breast-height Diameter (Myer, 1938)

AGE (YEARS)	HEIGHT, BY SITE INDEX (ft)												
	40	50	60	70	80	90	100	110	120	130	140	150	160
20	6	9	12	16	20	25	30	35	40	45	50	55	60
30	11	15	20	26	32	38	44	51	57	64	70	77	84
40	16	22	28	35	42	49	55	63	70	77	85	93	100
50	21	28	35	43	51	58	65	73	80	89	97	105	113
60	26	34	42	50	58	66	73	81	90	99	107	115	124
70	30	39	47	56	64	73	80	89	98	108	116	125	134
80	34	43	52	61	70	79	88	97	106	116	124	133	143
90	37	47	57	66	75	85	94	104	113	123	132	142	152
100	40	50	60	70	80	90	100	110	120	130	140	150	160
110	42	53	63	74	84	95	106	116	127	137	147	158	168
120	44	55	66	77	88	100	111	122	133	144	154	165	175
130	45	57	69	80	92	104	116	128	139	151	161	172	182
140	46	59	71	83	96	108	121	133	145	157	167	179	189
150	47	60	73	86	99	112	125	138	151	163	173	185	195
160	48	61	75	89	102	116	129	143	156	169	179	191	201
170	48	62	77	91	105	119	133	147	161	174	184	196	206
180	49	63	78	93	108	122	136	151	165	179	189	201	211
190	49	63	79	95	110	125	139	154	169	183	194	205	216
200	50	64	80	97	112	128	143	157	172	187	198	209	220

the extremes of stocking. Measurement of height of trees in overstocked and understocked stands should not be used in constructing site index curves for they will not reflect an accurate estimate of site index; the trees will be shorter in both the overstocked or understocked stands than will trees in well-stocked stands (Lynch, 1958; Curtis and Reukema, 1970). When species grow in mixed stands and occur on similar sites, it is sometimes possible to develop comparison site tables using the growth of one species as a basis for estimating site of second species (Figure 4.5). For example, Doolittle (1957) found it possible to relate site index of scarlet, black, Northern red, or chestnut oak to that of white oak as follows:

White oak site index = −1.088 + 0.929 (site index of scarlet, black, or Northern red, or chestnut oak site index)

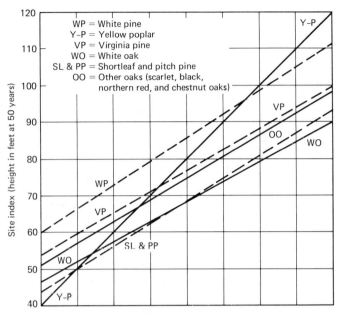

FIGURE 4.5. A comparison of site indices for 10 species on the same land in the Southern Appalachians. For example, on land that is site index 90 (1) for yellow-poplar, red down (2) and across (3) to find that this same land averages about site 82 for Virginia pine (Doolittle, 1958).

in the same manner:

scarlet, black, northern red, or
chestnut oak site index = 27.642 + 0.586 (yellow poplar site index)
white pine = 34.968 + 0.630 (yellow poplar)
virginia pine = 12.746 + 0.932 (shortleaf and pitch pine)

MEASURES OF STOCKING AND GROWTH—EVEN-AGED STANDS

When mapping forest land, foresters usually estimate the site index of each stand and the current stocking level. If tables are available, current stocking and site can be compared with the data contained in published yield tables to obtain estimates of timber yield. Yield table data that include a number of measurements of standing crop for different ages, densities, and site qualities are available for a number of species; however, most yield tables are based on pure even-aged stands.

In the tabulation of stands presented in Table 4.2, note particularly the change in number of trees per acre for red spruce as compared with the other species. Trees in the spruce stands were not measured unless they had attained a minimum size of 4 in. dbh, and even though present in the stand they were not counted. For this reason, the number of trees per acre for spruce except for site index 70 increases with stand age whereas for Douglas-fir and loblolly pine the number of trees per acre decrease with stand age. When using yield table data it is important to determine what standards of diameter were used to tabulate the data. Note that for all three species in Table 4.2 the number of trees decreases with an increase in site quality for a given age. Trees on the best sites grow more rapidly, and the slower growing trees are eliminated at an earlier age. Note also the lack of a relationship between basal area per acre and volume per acre; equal basal area per acre does not indicate proportional volume per acre. The information presented in these tables was derived from measurements taken from a number of stands and then summarized in graphic form (Forbes, 1955).

For a number of years the graphic approach was used in the construction of standard yield tables. Another approach to the measurement of stand growth and the production of stand and yield tables is the use of multivariate analysis. As contrasted to the graphic approach, multivariate analysis provides a more flexible means for determining potential production for varying conditions. To illustrate, Nelson, Lotti, Brandes, and Trousdell (1961) proposed that a form equation for estimating five-year periodic growth of loblolly pine would require the following measurements:

$$\text{Growth} = b_1\,(1/A) + b_2\,(S) + b_3\,(D) + b_4\,(D^2) + b_5\,(S/A) + b_6\,(D/A)$$
$$+ b_7\,(D^2/A) + b_8\,(SD) + b_9\,(SD^2)$$

where A = stand age in years at the beginning of the 5-year period
S = site index of stand; base age 50 years
D = density of a stand percent theoretical stocking after thinning at the beginning of 5-year period
b = coefficients to be derived from the data.

Following analysis it was found that some variables were not sig-

TABLE 4.2. Stand and Yield Tables for Douglas-fir, Loblolly Pine, and Red Spruce for Selected Ages and Site and Site Classes (Forbes, 1955)

| | NUMBER OF TREES PER ACRE SPECIES | | | | | |
| | DOUGLAS-FIR SITE CLASS[a] | | LOBLOLLY PINE | | RED SPRUCE SITE INDEX | |
(YEARS)	V	I	60'	120'	30'	70'
20	5500	490	1600	560	175	835
40	1275	203	585	205	175	835
60	670	116	360	125	660	760
80	455	81	275	95	815	482
100	352	64	—	—	870	439
120	292	53	—	—	—	—

[a]Base age 100 years of Douglas-fir and 50 years of loblolly and red spruce. Site class V corresponds to site index 80–95 feet and site class I to site index 190–210 feet.

| | BASAL AREA IN SQUARE FEET PER ACRE SPECIES | | | | | |
| | DOUGLAS-FIR SITE CLASS | | LOBLOLLY PINE | | RED SPRUCE SITE INDEX | |
(YEARS)	V	I	60'	120'	30'	70'
20	70	102	121	152	—	—
40	132	196	147	187	17	127
60	169	250	156	196	77	244
80	194	287	160	209	113	269
100	212	314	—	—	125	276
120	226	335	—	—	—	—

| | VOLUME IN CUBIC FEET PER ACRE SPECIES | | | | | |
| | DOUGLAS-FIR SITE CLASS | | LOBLOLLY PINE | | RED SPRUCE SITE INDEX | |
(YEARS)	V	I	60'	120'	30'	70'
20	0	200	1900	4400	—	—
40	0	6450	3750	9350	138	2160
60	680	12750	4750	12000	940	6850
80	2110	16900	5150	13050	1670	8880
100	3690	19820	—	—	1900	9570
120	5150	21870	—	—	—	—

nificantly related to growth, and the prediction equation for unthinned loblolly pine stands was:

$$\text{Net cubic-foot growth} = 103.99189 + 0.31055 \frac{10000}{\text{age}}$$
$$- 0.02553 \, (D)^2 + 0.05158 \, (SD)$$

Using several values of age, density, and site, and substituting these into the equation it is possible to obtain an estimate of five-year periodic growth. Using the above equation for two site conditions and several conditions of stocking, the cubic-foot yields in Table 4.3 were derived.

The study of loblolly pine illustrates a need for recognizing age, density, and site quality when determining stand growth, and in defining stand response to treatment. Composition of the stand in this case was fixed by selecting only pure loblolly pine stands, and only even-aged stands were considered. Usually, pure even-aged stands are chosen for measurement; however, there are a number of cases where mixed stands have been measured, and some cases where un-even-aged stands were measured.

TABLE 4.3 Periodic Annual Cubic-Foot Growth in Unthinned Loblolly Pine Stands (Abstracted from Nelson, Lotti, Brandes, Trousdell, 1961, Table 6)

			SITE INDEX 60		
			DENSITY PERCENT		
AGE	20	40	60	80	100
		CUBIC FEET			
40	25	57	67	58	28
60	0	31	42	32	2[a]
			SITE INDEX 100		
40	67	139	191	223	234
60	41	113	165	197	208

[a]Volume was reduced due to high mortality of trees on these relatively poor sites; on Site Index 100 areas this land will support more trees to a greater age.

TABLE 4.4 Weight in Kilograms of Immature Balsam Fir Trees from 1 to 10 inches in Diameter by Component and the Percentage of Total Weight in Each Component[a]

COMPONENT	DIAMETER AT BREAST HEIGHT IN INCHES									
	1	2	3	4	5	6	7	8	9	10
Foliage	0.05	0.23	0.86	2.13	4.40	7.85	12.88	19.87	27.03	40.60
Branches	0.05	0.22	0.82	2.00	4.03	7.39	12.11	18.64	25.45	38.24
Cones	—	—	—	—	0.18	0.32	0.54	0.82	1.13	1.72
Stem wood	0.54	2.54	6.40	12.34	20.55	29.71	44.18	59.92	74.71	99.70
Stem bark	0.04	0.32	0.81	1.67	2.95	4.58	6.72	9.34	11.88	16.24
Total aboveground	0.68	3.31	8.89	18.14	32.16	49.85	76.43	108.59	140.20	196.50
Roots (wood + bark)	0.18	1.04	2.77	5.63	9.71	15.15	22.13	30.71	38.88	53.06
Total tree	0.86	4.35	11.66	23.77	41.87	65.00	98.56	139.30	179.08	249.56
(Percent of Total Tree Dry Weight)										
Foliage	3.3	5.4	7.3	9.0	10.5	12.1	13.1	14.2	15.2	16.3
Branches (wood + bark)	2.8	4.9	6.8	8.4	9.8	11.3	12.3	13.4	14.2	15.3
Cones	—	—	—	—	0.4	0.5	0.6	0.6	0.6	0.7
Stem wood	63.9	58.8	55.0	51.9	49.1	45.7	44.8	43.0	41.7	39.9
Stem bark	6.7	7.0	7.1	7.1	7.0	7.1	6.8	6.7	6.6	6.5
Total aboveground	76.7	76.1	76.2	76.4	76.8	76.7	77.6	77.9	78.3	78.7
Roots	23.3	23.9	23.8	23.6	23.2	23.3	22.4	22.1	21.7	21.3
Total tree	100	100	100	100	100	100	100	100	100	100
Number of trees sampled:										
Above ground	3	6	10	11	22	17	12	11	6	3
Below ground	4	10	15	14	19	11	9	3	3	1

[a]From Baskerville, 1965.

DRY MATTER PRODUCTION (BIOMASS)

Stand productive capacity has been approached from different directions. In several studies dry matter (also termed biomass) production was used to measure site quality differences (Moller, 1946; Ovington, 1957, 1961; Baskerville, 1965). Dry matter increase as measured in the studies included dry weight increase of stem, branches, twigs, leaves, and fruit of the trees (Table 4.4). In some cases weight of subordinate vegetation and the animals inhabiting a stand may also be estimated. Tree dry weight increase is a more precise measure of growth than is volume growth, hence weight change can be more sensitive to site quality differences, particuarly where differences in stand density and composition need to be considered. The measurements required to make accurate predictions of dry matter increase are quite detailed and difficult to obtain, and it does not appear that they will soon replace measurement of bole volume and height growth of the standing crop in assessing growth and site potential of forest stands. Analysis of biomass increment does appear to have considerable value when combined with the analysis of mineral cycling and the comparison of growth of different species.

Swank and Schreuder (1974) discuss procedures used to determine surface area and biomass of a young eastern white pine stand. In addition to the conventional measurements of bole diameter and height, they measured foliage, branch, and bole area and weight. Fascicle area was determined by orienting the individual needles in a fascicle so that the five needles were closed into a cylindrically formed shape; they were wrapped with thread to hold them in place and then diameter and length measurements made. The form of the fascicles approached that of a truncated prolate spheroid, and not a cylinder, as was first supposed. Use of the cylinder form overestimated the volume of the fascicle, but by only 2%, so the area of a fascicle was determined as $S_1 = \pi dh$, where d is average fascicle diameter and h is total length. Total surface area of all needles was determined as $S_2 = 5hd$. Fascicle area was related to fascicle weight; therefore, fascicle area (cm^2) = 3.61709 + 0.10588 (fascicle weight). This form is different than that proposed by Madgwick (1964), who suggested a curvilinear relationship. A comparison to a curvilinear form showed the linear form to provide a better fit to the data for Swank and Schrender. With broadleaved species, leaf area is somewhat easier to determine by shadow area or by planimetering the leaf margins, and the area-weight relationship can be determined in the same manner as for pine.

Surface area of branches was determined by Swank and Schrender from measured length and diameter and by use of the formula S branch $= \pi dh$, where d is the average diameter and h is branch length. From this measurement it was possible to develop a relationship where Y (branch area in cm^2) $= 228.55 + 7.8343(x_2)$, where x_2 is branch weight. Stem surface area was estimated as $S = \pi r \sqrt{r^2 + h^2}$. Stem weight was determined by applying a density measurement to stem volume (m^3). Branch weight in turn was estimated by the equation log branch weight (gm) $= -3.2913 + 2.8681$ (log$_e$ branch basal diameter in mm). Prediction equation for estimating foliage, branch, and stem areas and weight for individual trees:

DEPENDENT VARIABLE	INDEPENDENT VARIABLE	COEFFICIENTS	
		a	b
Foliage area (m^2)	Stem Basal area (cm^2)	3.716	0.70351
Branch area (m^2)	Stem Basal area (cm^2)	0.04944	0.11122
Stem area (dm^2)	Stem Basal area (cm^2)	31.29	1.09327
Foliage weight (g)	Stem Basal area (cm^2)	221.1	35.067
Branch weight (g)	Stem Basal area (cm^2)	−1332.0	133.79
Stem weight (kg)	Stem Basal area (cm^2)	−0.3292	0.11570

Using a stratified sampling system Swank and Shrunder estimate the total area and weight of the pine stand in question as:

SURFACE AREA (ha/ha)			BIOMASS (tons/ha)		
foliage	branch	stem	foliage	branch	stem
5.3	0.76	0.126	2.71	6.83	7.01

TREE CLASSIFICATION

Trees differ because they do not experience the same environment nor have the same genetic complement. In forestry practice it is necessary for management and silvicultural purposes to be able to classify these differences, not so much as to cause of difference, but to classify trees

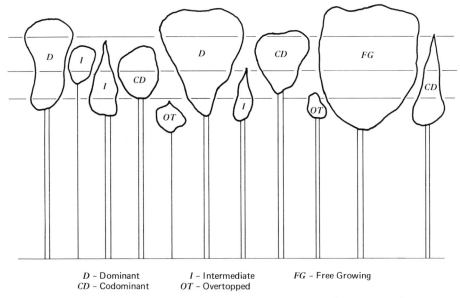

D – Dominant   I – Intermediate   FG – Free Growing
CD – Codominant   OT – Overtopped

FIGURE 4.6. Crown classes within an evenaged pure stand.

on the basis of vigor and end-product use. There are two general types of classifications used to indicate differences between trees in a stand: tree vigor classes and tree quality classes. A third class might be listed that combines vigor classes with quality classes although in principle this does not add anything different.

Vigor of trees in even-aged stands or even-aged groups within an uneven-aged stand can be determined by order of crown position. The following is a five-class system used frequently in thinning practice (SAF, 1950) (Figure 4.6).

> *Dominant.* Trees with crowns extending above the general level of the canopy and receiving full light from above and partly from the side; larger than the average trees in the stand, and with crowns well developed but possibly somewhat crowded on the side.
>
> *Co-dominant.* Trees with crowns forming the general level of the canopy and receiving full light from above, but comparatively little from the sides, usually with medium-sized crowns more or less crowded on the sides.
>
> *Intermediate.* Trees shorter than those in the two preceding classes, but with crowns either below or extending into the canopy formed by co-dominant and dominant trees, receiving little direct light from above, but none

from the sides; usually with small crowns considerably crowded on all sides.

Overtopped. Trees with crowns entirely below the general level of the canopy, receiving no direct light from either above or from both sides. Syn. suppressed, but not to be confused with younger trees in the understory of a stand which are not competing with the overstory.

Free Growing. In unmanaged stands, this class may sometimes be required. It includes trees with large spreading crowns brought about as a result of open growing conditions throughout much of the tree's life.

Chappell (1962) showed that in predicting growth of individual longleaf and slash pine trees, it was necessary to designate crown class and live crown ratio as well as site index and diameter. Live crown ratio is the relative length of live crown to total tree height:

$$\text{LCR} = \frac{\text{crown length}}{\text{total height}}$$

The prediction equation derived by Chappell for individual longleaf pine trees states:

5-year merchant volume growth (cu. ft. to 3-in. top i.b. dia.) = 0.868449 + 0.005744 (dbh) (CR) + 0.000247$(SI)(CR)$ + 0.300461 $(D + C) - (I + S)$

where dbh = diameter breast height, at the beginning of the period;
    CR = live crown ratio, as percent of total height, at beginning of the period;
    SI = site index feet—50-year base age;
    D = dominant, I = intermediate,
    C = co-dominant, S = suppressed.

An extension of the crown classification approach has been used to classify trees resistant to insect attack. Such systems as Keen's (Figure 4.7) and Dunning's (1928) for ponderosa pine and Taylor's (1939) for lodgepole utilize vigor; other characteristics used are age and crown size.

Tree quality grades have been used since large trees were cut for sawlogs. Distinctions among these trees were made if one wishes to consider that the reservation of eastern white pine mast trees for the

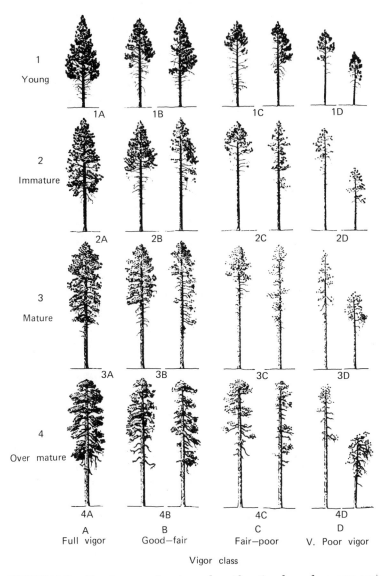

| | A
Full vigor | B
Good–fair | C
Fair–poor | D
V. Poor vigor |

Vigor class

FIGURE 4.7. Ponderosa pine tree classification based on age and vigor (from Keen, 1943).

Crown in the early years of colonial America was a recognition of tree quality. For a number of species, the value of a particular tree is directly proportional to the value of the butt, or first log. The value of this log depends upon its grade for veneer, lumber, poles, piling, or some other product. The higher the proportion of quality veneer or lumber that can be recovered from the first log, the greater its value, and thus the greater the value of the entire tree. Logs can be graded into five classes: veneer, #1 sawlog, #2 sawlog, #3 sawlog, and cull depending on the grade recovery of veneer or #1 common lumber. Other products are pulpwood and fuel wood, and poles and piling logs.

CHANGE IN STAND STRUCTURE WITH TIME

It has been shown that different species produce stands of different density and have different growth potential when grown on the same site. It can be assumed generally that as a stand becomes older it will have fewer but larger trees. There are other changes that occur and must be recognized as stands change in age. Seedling stands, even when of mixed composition, are fairly simply in structure since the trees and other vegetation can be arranged as a single stratum insofar as they occupy the above-ground space. Even as the young stand grows in height into the sapling stage, there is not a very noticeable change in the single-layered effect. As the stand enters the pole stage there can be a noticeable development in the stand when a sub-stratum appears in the form of shrubs and tree reproduction. As the stand continues to age, the subordinate stratum becomes well identified and, in many instances, requires measurement. The generally accepted practice is to recognize at least four strata (Figure 4.8). In a more complex circumstance it may be necessary to recognize an additional layer. For example, in some very old mixed stands a conifer stratum may consist of a layer of emergent trees extending above the general crown stratum. The density of stands in each stratum will differ from one site to another, the tendency being for best sites to maintain higher densities in all strata than will occur on medium sites; and poor sites will tend to support relatively open stands with low densities in all strata. Where site differences occur there will be species changes as well as density differences (Figure 4.9a and 4.9b; from Byrnes, Goss, and Losche, 1965).

In general, deciduous species in eastern forests tend to be more site demanding than coniferous species with the result that on moist sites in late successional stages deciduous species will dominate the stand (Figure 4.9a). Conifers tend to dominate on dryer sites (Figure 4.9b). Logging, fire, disease, and insects may alter this situation in a partic-

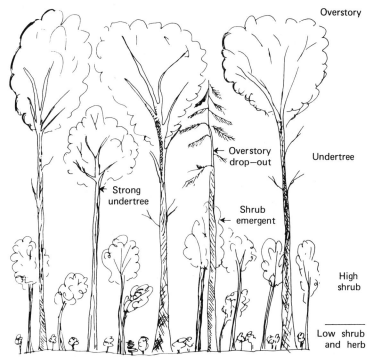

Overstory

← Overstory
✖ drop—out

Undertree

← Strong
   undertree

Shrub
← emergent

High
shrub

Low shrub
and herb

FIGURE 4.8. A typical stand structure showing vegetative layers sampled (from Byrnes, Goss, and Losche, 1965).

ular location; and, as a result, they can obscure the situation as it would develop if a stand had been left undisturbed.

With management it is possible to control stand development so that a variety of conditions results. Alexander (1972) indicates that lodgepole pine, a light demanding species, can be made to reproduce and to develop under several stand conditions (Figure 4.10). Some species require shade of an overstory stand to reproduce; the overstory protects seed and seedlings from excess light and heat, keeping the microsite around the developing trees from becoming too dry and hot. Other species, such as lodgepole pine, are able to grow in exposed sites where seed and seedlings must be able to contend with hotter and dryer conditions. These adaptations to site conditions represent a part of the definition of niche, and will be discussed in more detail in later sections. Nelson, Roach, Frank, and Ward (1975) summarized some requirements for several eastern deciduous species with respect

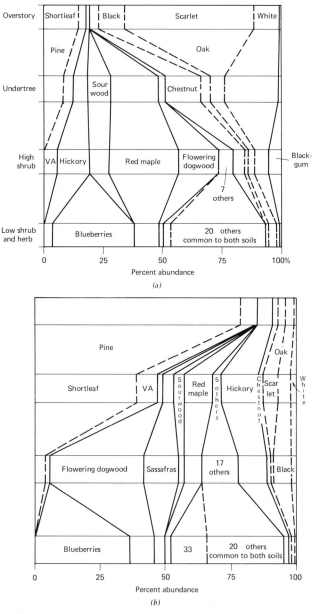

FIGURE 4.9. (*a*) Diagram illustrating percent abundance of the major species in each vegetative layer on Wellston silt loam. (*b*) Diagram illustrating percent abundance of the major species in each vegetative layer on DeKalb sandy loam.

FIGURE 4.10. An example of the crown class and size class distribution for a single species stand, showing a transition from even-aged single-story to multiple-aged multi-storied arrangement (Alexander, 1972).

to requirements for regeneration (Table 4.5). It can be concluded that hemlock, sugar maple and beech develop best in layered uneven-aged stands; that birch, black cherry, Virginia pine, yellow poplar and aspen develop best in single strata stands; ash, hickory, oak, and red maple develop best in a two strata stands.

TABLE 4.5 Characteristics Affecting Natural Regeneration of Some Important Pennsylvania Trees[a]

	HEMLOCK	SUGAR MAPLE	BEECH	ASH	HICKORY	OAKS	RED MAPLE	BIRCH	BLACK CHERRY	VIRGINIA PINE	YELLOW POPLAR	ASPEN
Time seed remains viable in soil or humus (years)	1	1	1	3–5	1	1	2–3	1–2	3–5	1	3–7	1
Mineral soil required for germination	helps	no	no	helps	no	no	helps	yes	no	yes	helps	yes
Germination and seedling survival in:												
deep shade	X[b]	X	X									
partial shade	X	X	X	X	X	X	X	X	X			
full sunlight				X	X	X		X	X	X	X	X
Seedling growth in:												
partial shade	X	X	X	X	X	X	X					
full sun		X		X	X	X		X	X	X	X	X
Establishment of advance reproduction in partial shade	X	X	X	X	X	X	X	X				
Response of established reproduction to release	X	X			X	X						
Reproduction ability from:												
root suckers			X						X			X
small stumps	X	X	X	X	X	X	X				X	X

[a]From Nelson, Roach, Frank, and Ward (1975).
[b]X indicates good to excellent response.

TABLE 4.6. Diameter Distributions for Pure Natural Yellow-Poplar Stands by Age and Stand Density per Acre on Site Index 100 (From McGee and Della-Bianca, 1967)

TOTAL TREES (NUMBER)	BASAL AREA (SQUARE FEET)	NUMBER OF TREES PER DIAMETER CLASS (IN.)															
		5	6	7	8	9	10	11	12	13	14	15	16	17	18	19	20
								Age 20									
100	35	13	20	19	16	12	9	6	4	1							
150	41	43	36	26	18	12	8	5	2	—							
200	46	84	48	29	18	11	6	3	1	—							
250	52	129	56	30	17	10	5	2	1	—							
300	58	173	63	31	17	9	5	2	—	—							
350	65	214	71	34	17	9	4	1	—	—							
								Age 40									
50	48			1	2	3	4	5	6	6	6	6	6	4	1		
100	74	1	4	6	9	10	11	11	11	9	8	6	3	—			
150	91	4	11	16	18	18	17	16	15	13	10	8	4	—	—		
200	104	11	22	26	27	25	22	20	16	13	10	6	2	—	—		
250	113	21	35	38	35	31	27	22	17	13	8	3	—	—	—		
300	122	32	49	49	43	37	30	24	18	12	6	—	—	—	—		
350	131	44	63	60	52	43	34	25	17	10	2	—	—	—	—		

Continued on next page

TABLE 4.6. Diameter Distributions for Pure Natural Yellow-Poplar Stands by Age and Stand Density per Acre on Site Index 100 (From McGee and Della-Bianca, 1967)

TOTAL TREES (NUMBER)	BASAL AREA (SQUARE FEET)	NUMBER OF TREES PER DIAMETER CLASS (IN.)															
		5	6	7	8	9	10	11	12	13	14	15	16	17	18	19	20
ON SITE INDEX 120.																	
Age 20																	
100	39	10	17	16	13	10	8	5	3	1							
150	46	33	33	26	20	15	10	7	4	2	—						
200	54	64	47	32	22	15	10	6	3	1	—						
250	62	97	59	37	23	15	9	6	3	1	—						
300	70	128	70	42	26	16	10	6	2	—	—						
350	80	154	82	48	29	18	11	6	2	—	—						
Age 40																	
50	59	—	—	1	1	2	3	3	4	5	5	5	6	5	5	4	1
100	93	1	2	6	8	9	9	10	10	9	9	8	7	5	3	—	—
150	114	2	8	14	15	15	15	14	13	12	11	9	7	4	—	—	—
200	133	6	15	21	22	21	20	18	16	14	12	9	6	1	—	—	—
250	148	11	23	29	29	27	24	22	19	15	12	8	3	—	—	—	—
300	163	16	32	37	38	36	32	29	25	21	16	12	6	—	—	—	—
350	180	20	40	46	46	43	39	34	29	23	17	11	2	—	—	—	—

Individual Trees in Stands

Stand growth is influenced by stand density, age, and site assuming a fixed composition. Density affects both diameter and height of individual trees, and, as it increases, individual trees are suppressed in diameter growth (Table 4.6). High density stands support a large number of small diameter trees. As density decreases in an age-site situation, individual tree diameters increase with a corresponding increase in range of diameters. All trees are not of the same diameter because of the development of crown class. Differences become more noticeable with age. Gilbert (1965) found that among sugar maples, yellow birch, and paper birch there were significant differences in range of diameters and that these differences increased with age. Sugar maple had the largest range in size; yellow birch had the least. The tendency is for some species to form stands where there is little differentiation into diameter crown classes while other species show a relatively wider range. Trimble (1969), in a study of diameter growth of individual trees, concluded that within a species growth could be predicted from measurements of site quality and individual crown class. The study included such other individual tree measurements as vigor class, crown length, crown width, live crown ratio, and basal area surrounding individual trees; these were not, however, as important as site quality and crown class in predicting individual tree growth.

GENERAL REFERENCES

Toumey, J.W. and C.F. Korstian. 1947. *Foundation of Silviculture on an Ecological Basis*. Wiley, New York.

Baker, F.S. 1950. *Principles of Silviculture*. McGraw-Hill, New York.

Chapter 5

PRIMARY AND SECONDARY SUCCESSION

Plant succession is the orderly replacement of one plant community by another plant community—one forest stand being replaced by another. Generally, a temporary plant community is replaced by a relatively more stable community until a dynamic equilibrium between plants and environment is attained. However, it is quite possible for a relatively stable community to be replaced by a temporary community. The present concept is that stability of the various stages is relative because disturbances are natural.

In situations where in the past natural disturbances occurred, plant communities changed at infrequent intervals; but change did occur. Man's present influence on the forest is more complex than it was

Forest Succession

before his dominance; and man's influence can be more long lasting than the influence of other biological organisms. There is an inclination to identify man-made disturbances to the forest as artificial; however, the activities of man as they relate to the forest system will be included in this text as part of the natural forest ecosystem. Disturbances of small and large magnitude no matter by what factor are not uncommon, and it can be anticipated that changes in the structure of even a seemingly stable forest community will occur. This concept of succession differs somewhat from one that inferred that succession is unidirectional, and progress is always from a temporary to a stable system.

Although a unidirectional concept recognizes the possibility of disturbance, it does not make clear the necessity to consider succession as a transition from a stable community to a temporary community, which is a type of succession. Rather than assume that forest succession is unidirectional, it seems more straightforward to consider succession as being bi-directional, that disturbance is a continuing phenomenon. One need only specify whether succession is toward a relatively more stable or less stable community.

Two general forms of succession are recognized: primary and secondary. *Primary succession* takes place on new land not previously occupied by plants. Classically, locations such as rock outcroppings, shallow ponds, and glacial moraines are considered as being open to primary succession. The supposition is that the action of the plants (action on the site) modifies the environment so that the site (reaction of the site) becomes more favorable for plant growth and in time is subject to habitation by relatively more demanding plant species than are present at the beginning of succession. The process is illustrated in Figure 5.1.

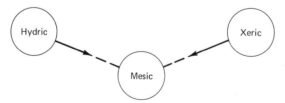

FIGURE 5.1. Action of vegetation upon xeric and hydric sites modifies the site so that in either case each tends toward a more moderate mesic site condition.

In Figure 5.1 the dotted lines show the direction toward which each series will tend. It should not be assumed that in all cases the site will be modified so completely that all wet or all dry sites become mesic. There is no doubt that over time extremely dry (xeric) and extremely wet (hydric) sites are modified by plants. There is a question, however, as to how greatly a site can be modified by plant growth, and it is doubtful whether the resulting changes in environment would result in anything more than slight modifications of very dry or very wet sites. Conceptually, a mesic site is one on which there is neither an excess of water nor a deficit of long duration. A mesic site can be considered as being ideal for tree growth. A shallow pond can probably progress farther toward a mesic site condition than a rock outcropping will progress toward a mesic site. However, if geological change is taken into account, then both xeric and hydric sites may become mesic.

Wright (1974) indicates that a short term equilibrium—assuming weathering, erosion, and vegetational stability under crustal and climatic stability—may occur; but in the long term it must be assumed that hills will be reduced and the externally controlling conditions will also change. Succession of vegetation may occur in the short term on a stable site, but then in the long term external factors change causing instability of the site. Also, it must be kept in mind that there are long (10,000 years or more) term changes in climate that have had profound effects on vegetation. The recurrence of warm and cool periods is a good example of long term change.

New land formed as a result of flooding, landslides, volcanic disturbance, and glacial retreat produce site conditions upon which forest succession can proceed. Under these circumstances soil material is transported from one location to another, or it can replace soil at a location. On these newly formed sites primary succession begins on a situation where mesic conditions could prevail at the time of initial plant growth. Along the channels of major rivers "new" land is formed as a result of shifts in the river channel. Whole islands appear almost overnight. Sites such as these are soon occupied by trees, and succession begins at a more advanced stage of site development than in the case of a rock outcrop or in pond succession. Tree growth rapidly follows glacial retreat, and if landslide locations are not too active trees soon invade slide areas. It is on sites such as these that many present mesic forest stands occur. Conceptually, it does not seem possible for all mesic sites to have been brought about solely as a result of plant action.

Secondary succession differs from primary succession in several ways. A major difference is the fact that at no time are higher plants unable to colonize a site that is experiencing secondary succession. Sites on which secondary succession occurs are never sufficiently disturbed for the occupancy by grasses, trees, or shrubs to be interrupted for any long period. Usually, in secondary succession if a disturbance causes damage to vegetation on a site, new vegetation will take the place of the previous vegetation within a year or so after the disturbance. There are, however, instances where vegetation may be slower in colonizing a disturbed site, and in these instances site degradation may have occurred.

It is important to note that there is a much greater area of land undergoing change as a result of secondary succession than from primary succession. It is, therefore, important that we be able to identify the factors that initiate it and the patterns occurring in secondary succession. In every country of the world where land has been cleared for agriculture and later abandoned, it is open to secondary succession. As much as three-quarters of the land area in sections of the eastern United States was cleared for crops at one time since human settlement and much of this land now supports trees.

During the early settlement period the clearing of land for agriculture in the U.S. was relatively slow; however, as the economy changed in the early 1800s land clearing was accelerated (Figure 5.2). Following the Civil War when the middle western states became available for agriculture, there was a general reduction of land under cultivation, a trend that, except for a brief period during World War II, continued to the present. The pattern illustrated in Figure 5.2 represents a general trend of land clearing and abandonment common to many eastern states, although the time periods may differ somewhat in individual states.

In addition to the amount of land involved, it is necessary to consider the value of the stands that occur in the initial and later stages of secondary succession. Many of the most valued timber species become established shortly after land is removed from field crop production, or following fire, or following other major disturbance. White pine in New England, the southern pines, jack pine, and Douglas-fir all occur early in the secondary succession. Extensive stands of these species resulted from disturbances that destroyed previous stands.

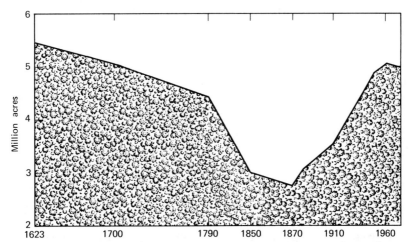

FIGURE 5.2. The trend of forest land cover in New Hampshire since 1623 (Kingsley, 1976).

RELATIVE TOLERANCE

Tolerance to environmental amplitudes is a characteristic of all plants and animals. Plants, because of genetic endowment, are able to survive and reproduce under a range of climatic and soil conditions. Stated another way, through natural selection plants and animal populations have evolved into diverse genetic populations that occupy various situations (niches). The following discussion does not concern single sets of conditions such as the capacity of trees to survive conditions of high or low temperature or conditions of dry or saturated soil. These conditions may be termed *absolute tolerance*, which is the ability of trees to survive extremes of a particular site factor, such as heat, cold, dryness or wetness. The present discussion concerns a special case of absolute tolerance in which the interaction of energy, water, and minerals affect survival and growth of trees within their natural range where they must compete with other trees and plant growth.

Relative tolerance can be defined as the ability of a tree to grow and reproduce in the shade of and in competition with other trees. This definition implies both physiological, hence, genetic amplitudes and the fact that all trees do not have the same capacity to react under similar conditions. One way to approach an understanding of the problem is to visualize a system where a number of factors such as water, light, and minerals affect the growth of a plant. Light is used in this discussion to express the total effects of solar input. It is assumed since most studies of relative tolerance have dealt with direct solar radiation that the main energy effects are the solar inputs that determine photosynthetic effectiveness. Consider two systems—A and B (Figure 5.3). In system A light is the factor that most limits growth since it is the factor in least supply; in system B water is most limiting. In either system a growth response might be obtained by increasing the amplitude of one of the other factors (water or mineral in A, or light or mineral in B), but the greatest response is gained by increasing the amplitude of the factor in least supply, light in A or water in B. This is Liebig's Law as modified by Mitscherlich (1909). Trees are frequently confronted with threshold levels at a minimal level, and these are the levels that are involved in assessing relative tolerance. Also, it should be kept in mind that physiological reactions can also have maximum limits. This is particularly true for light and temperature reactions where high levels of light can cause excessively high temperatures, which are lethal to plants. Consideration must therefore be given to factors as they might exceed optimum levels—super optimum. In these instances, the effect could be as limiting as when the factor was supplied at less than optimum.

The relative tolerance of a tree is related to its ability to carry on photosynthesis at a level sufficient to sustain growth when light, in particular, and water, to some extent, are at minimal or near minimal levels. In the past, arguments were carried on as to the greater or lesser importance of light or water. It is now generally recognized that light and water cannot be separated; roots must continue to grow to be effective in absorbing water, and carbohydrates produced in the leaves are needed for root growth; thus light and water must be considered together. However, the ability of a tree to carry out photosynthesis under reduced light conditions does provide a measure of its relative tolerance.

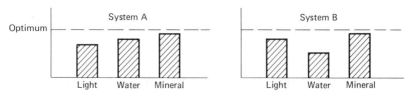

FIGURE 5.3. Two ecosystems where in A light is the most limiting factor; in B water is the factor that limits growth more than light or mineral.

The relative tolerance of a species can be determined either directly or indirectly: directly by physiological experimentation using techniques such as infra-red gas analysis of CO_2 exchange in the leaves, or by growing trees under reduced light. It probably would be enlightening if experiments were carried out with trees under several levels of moisture stress to determine whether there is a change in a species' ability to continue to function at the same relative position when water as well as light is restricted in supply. Field observations indicate that species rated as most tolerant to low light seem to require relatively high moisture levels, and species intolerant to low light seem capable of growing on dry sites.

No definitive study has been carried out that permits an absolute ranking of all species with regard to their tolerance to light and moisture. As a result, only empirical measures of tolerance have been made for a large number of species. Therefore, any ranking of trees as to shade tolerance must be relative where one specie is compared with each species occupying a position relative to other species. In obtaining a ranking of all species found in a particular region, it is necessary to array them in an order even when they do not grow in direct association.

Indirect Measures of Relative Tolerance
Indirect measures of the relative tolerance of a species can be obtained from observing the growth habit of that species as it occurs in closed stands and of the seedlings of the species in the understory. Species that can maintain long live crowns and large numbers of branch orders in closed stands are rated tolerant. Species with short live crowns and few branch orders are rated intolerant. The number of branch orders a tree produces is determined by counting the number of lateral branchlets from each of the past several years that are producing leaves or, in the case of conifers, needles. If seedlings of a species are able to grow vigorously under a closed stand, this is further evidence that the species is tolerant. Absence of reproduction is evidence that a species is intolerant, a poor seeder, or the seed was destroyed by birds or mammals; so absence of seedlings in itself does not mean much as far as judging a species' tolerance. Where seedlings are present, their evidence of vigor or lack of vigor can be used to judge tolerance.

Empirically, it is possible to construct a tolerance scale for a number of species. Graham (1954) proposed a scoring system using a numerical scale to rate the tolerance of species native to Michigan. The scale does not eliminate subjective judgments; however, it does support a consistency of approach to rating tolerance. The ratings are as follows.

		Score
I.	Crown density	
	Density comparable with hemlock or sugar maple	2
	Density comparable with elm or red maple	1
	Density comparable with aspen or paper birch	0
II.	Ratio of leaf-bearing length to total length of branches	
	50 percent or more	2
	35–49 percent	1
	1–34 percent	0
III.	Ratio of crown length to total height of trees growing in an unbroken forest	
	50 percent or more	2
	35–49 percent	1
	1–34 percent	0
IV.	If a species is found growing in the underbrush of a stand of:	
	sugar maple or hemlock	2
	elm or yellow birch	1
	aspen or paper birch	0
V.	Degree of tolerance observable from other characters	
	tolerant	2
	mid-tolerant	1
	intolerant	0

Relative tolerance is a continuous variable and does not lend itself well to separation into discrete classes. Graham's method permits quantifying on a continuous scale; however, one should not think of the weight given to each of the five characters used in the scale as equal, nor of the 2, 1, 0 scoring weights as representative of a natural weighting. Further study is required before a natural weighting scheme can be devised.

When using the Graham system a separate rating for each of the five categories is made for a species and the five numerical ratings are added. Supposedly, eastern hemlock, which is the most tolerant species receives a rating of 10 while a very intolerant species like aspen could receive a rating of 0. Graham, using the system, arrived at a listing of species native to northern Michigan (Table 5.1).

TABLE 5.1 Tolerance of Species Found in Northern Michigan (From Graham, 1954)

TOLERANT	SCORE	LOW-MID-TOLERANT	SCORE
Hemlock	10.0	Black cherry	2.4
Balsam fir	9.8	Black ash	2.4
Sugar maple	9.7	Red pine	2.4
Basswood	8.2		
HIGH-MID TOLERANT	SCORE	INTOLERANT	SCORE
White spruce	6.8	Jack pine	1.8
Black spruce	6.4	Paper birch	1.0
Yellow birch	6.3	Tamarack	0.8
Red maple	5.9	Aspen	0.7
White cedar	5.0		
White pine	4.4		

Although the number of classes of relative tolerance can be infinite because of the lack of ability to define discrete tolerance boundaries, it is possible to recognize three broad classes of tolerance with ease, and, perhaps, four or five with some degree of precision. If three classes are recognized, they are tolerant, intermediate, and intolerant; if five classes are used, they are very tolerant, tolerant, intermediate, intolerant, and very intolerant (Table 5.2).

Direct Measures of Tolerance to Light.
Physiological studies of seedling growth that determine the compensation point—the point where photosynthetic output equals respiration of the plant—provide a way to quantitatively rank tolerance. In these studies the number of foot-candles of light (or some other measurement of energy input) required to meet the respiration needs of the species is determined. To evaluate the tolerance of all commercial species in this manner is too laborious and might be of doubtful worth since satisfactory judgments can be made using subjective scales.

Bates and Roeser (1928) carried on a limited study in which they determined the compensation point for a few species (Table 5.3). This study supports the order of tolerance that resulted from observation of trees as they grow in the forest.

DIVERSITY AND STABILITY OF STANDS
The number of species present on a site undergoing succession changes in time (Figure 5.4). Each stratum experiences an initial

TABLE 5.2. Relative Tolerance to Shade of Some North American Tree Species. (Arranged in Order of Tolerance Among Groups but Not Within Groups. The Arrangement is a Combination of that of Baker, 1950, and Toumey and Korstian, 1947.)

EASTERN CONIFERS	EASTERN DECIDUOUS	WESTERN CONIFERS	WESTERN DECIDUOUS
VERY TOLERANT			
Balsam fir	American-beech	Western redcedar	Vine maple
Eastern hemlock	American hornbeam	Alpine fir	
	Flowering dogwood	Mountain hemlock	
	American holly	Western hemlock	
	Eastern hophornbeam	California torreya	
	Sugar maple	Pacific yew	
TOLERANT			
Northern white-cedar	Basswood	Alaska yellow-cedar	California laurel
Red spruce	Rock elm	Incense cedar	Madrone
White spruce	Blackgum	Port-Orford cedar	Big-leaf maple
	Sourwood	Grand fir	Canyon live oak
	Red maple	Noble fir	Tanoak
		California red fir	
		Silver fir	
		White fir	
		Redwood	
		Englemann spruce	
		Sitka spruce	
INTERMEDIATE			
Eastern white pine	Ash spp.	Douglas-fir	Red alder
Black spruce	Sweet birch	Monterey pine	Oregon ash
	Yellow birch	Sugar pine	Golden chinkapin
	Buckeye	Western white pine	California white oak
	American elm	Blue spruce	Oregon white oak
	Sweetgum	Giant sequoia	
	Hackberry		
	Hickory spp.		
	Cucumber magnolia		
	Silver maple		
	Black oak		
	Northern red oak		
	Southern red oak		
	White oak		

Table 5.2   (cont.)

EASTERN CONIFERS	EASTERN DECIDUOUS	WESTERN CONIFERS	WESTERN DECIDUOUS

INTOLERANT

Bald cypress	Paper birch	Bigcone Douglas-fir	
Loblolly pine	Butternut	Juniper spp.	
Pitch pine	Catalpa spp.	Bishop pine	
Pond pine	Black cherry	Coulter pine	
Red pine	Chokecherry	Jeffrey pine	
Shortleaf pine	Kentucky	Knobcone pine	
Slash pine	coffeetree	Limber pine	
Virginia pine	Honeylocust	Lodgepole pine	
	Pin oak	Pinyon pine	
	Scarlet oak	Ponderosa pine	
	Pecan		
	Persimmon		
	Yellow-poplar		
	Sycamore		

VERY INTOLERANT

Jack pine	Aspen spp.	Subalpine larch	Aspen spp.
Longleaf pine	Gray birch	Western larch	Cottonwood spp.
Sand pine	River birch	Bristlecone pine	Willow spp.
Eastern redcedar	Black locust	Digger pine	
Tamarack	Post oak	Foxtail pine	
	Turkey oak	Whitebark pine	
	Blackjack oak		
	Willow spp.		

TABLE 5.3.   Light Required To Attain Compensation Point Where Photosynthesis Equaled Respiration[a]

SPECIES	PERCENT OF NOON WINTER SUN	PERCENT ARTIFICIAL LIGHT
Ponderosa pine	30.6	17.0
Douglas-fir	13.6	7.6
Eastern red oak	13.3	7.4
Eastern white pine	10.0	
Eastern hemlock	8.0	4.7
Beech	7.5	4.1
Sugar maple	3.4	2.0

[a]In this study compensation point was judged to have occurred when there was no net increase or decrease in seedling dry weight.

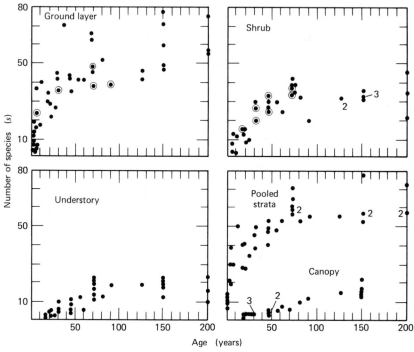

FIGURE 5.4. Number of species (S) in the various strata plotted against succession. Circled points indicate samples consisting of $\geq$ 50% exotic species by stems (Nicholson and Monk, 1974).

rapid increase followed by a decrease in new species. Finally, a point is reached as the stand approaches climax when stability of species number is attained. Initially in succession, species occupying a site are limited in vertical space to the ground stratum. As succession proceeds, taller plants occupy the site and a stratified stand is formed. Species may be intermixed in the lower strata of a pioneer stand; however, there will be a variety of species in all strata in the subclimax and climax stages. Subclimax species will reach the canopy layer as a stand advances, and the number of pioneer species will diminish. Eventually, climax trees will dominate the upper stratum. At the climax stage the stand attains its greatest diversity in terms of the number of species represented in each stratum.

The number of species on a site will vary not only as a result of successional change but also with site quality. Good sites have the capacity to support more species than poor ones with the result that there is usually greater species diversity on the better sites. Some poor sites produce stands that are composed of only a few species. Diversity does not necessarily represent maximum production of the site; maximum net annual production usually occurs toward the end of the dominance of pioneer or subclimax species (Figure 5.5). After that point a stand will decline in annual production; however, as indicated total biomass production will stabilize at a later time. It is assumed that a stand will maintain biomass production until or unless it is disturbed by some climatic or biotic agent. As has been pointed out, the level of biomass production is also a function of site quality.

It is not certain whether stability of a stand—in the sense that a stand can regain its species mix in a minimum time following disturbance—is a characteristic of diversity. Horn (1975) argues that complex stands—those with the greatest species diversity—may be the least stable. He believes that stands with the least number of species can attain their former state more quickly following disturbance than can stands that show maximum diversity. Complex sys-

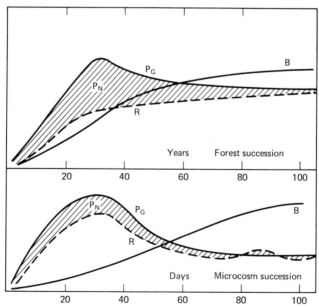

FIGURE 5.5. Comparison of the energetics of succession in a forest and a laboratory microcosm, B—net biomass produced; P_G—gross photosynthesis; P_N—net photosynthesis; R—respiration (Odum, 1969).

tems represent a delicate balance among the species present and a disturbance that upsets this balance will bring long lasting effects requiring many years for a stand to regain its previous diversity. A forest stand that has many strata and many species represents a situation where niche structure has reached its maximum development.

During succession, species occupy various niches within the stand. At first there are few niches because the site is dominated by the extremes of the environment and there is little chance of niche differentiation. As plants begin to influence the environment of the site, niche differentiation occurs and new species can occupy the site as succession progresses. In time, then, after a long period of uninterrupted development, the site will support the maximum number of species that are capable of finding their required niche structure. The relative tolerance of tree species will determine to some degree the stage in succession a particular species will occur. Forest stands can be viewed as a collection of species on a site where the individual and collective needs are met by the environmental conditions that prevail as a result of energy, moisture, mineral, and biotic interactions.

CLIMATIC AND BIOTIC FACTORS THAT INITIATE OR MODIFY SECONDARY SUCCESSION*

Disturbance of any existing plant community is to be expected. Of the environment factors affecting a site, climatic and biotic agents are responsible for major and minor disturbances that modify community structure. Some climatic factors that can initiate secondary succession are wind, lightning, drought, frost, and snow.

The so-called hurricane belt along the Atlantic and Gulf Coasts is a good example of a region that is subjected to periodic high winds capable of destroying whole forests. The hurricanes of 1938 and 1954 are classic examples of the destructive force of high winds. Hurricane Camille in 1969 struck in the vicinity of Gulfport, Mississippi, completely destroying many forest stands as well as causing extensive damage to others.

There were 107,160 lightning fires on National Forest Land in the United States between 1945 and 1966 (Komarek, 1967). This represented 64% of all wildfires on these lands for the period. There was a high of 9344 lightning fires in 1961 and a low of 2445 in 1948. The

---

*These factors are discussed in more detail in Chapters 7 and 11.

22-year average was 4871. The southwestern United States had the most fires, followed by the northwest with fewest lightning fires occurring in the east.

It is not difficult to imagine the size of a fire that can occur following lightning strikes when one has seen the extensive Douglas-fir stands west of the Cascade range in Oregon and Washington. Accounts of the Great Idaho Fire in 1910, which devastated a large portion of Idaho and western Montana, indicate how fire was instrumental in shaping the forest landscape in that region (Bradley, 1974). In 48 hours this holocaust destroyed 3 million acres and left a path of burned timber 160 miles long by 50 miles wide. The conditions that produced this wildfire were a combination of a long dry spring and summer, frequent dry lightning storms during July and August, which started a number of small fires, and a Chinook wind on August 20 that caused many fires to join into one large catastrophic conflagration.

Many of the lodgepole pine stands that originated since this fire are now at a stage where they are overmature, subject to bark beetle attack, and contain a large number of dead trees that have produced a condition where fire could again be an important factor in succession.

New forests occupying hundreds of thousands of acres have been brought into existence by wide-spread fires caused by either lightning or humans. Contemporary methods of forest fire suppression limit the size of fires, with the result that frequency of large forest fires should continue to decrease. Fire can be used, however, as a silvicultural tool to enhance the site conditions for the regeneration of valuable timber species that are adapted to fire. Judicious use of fire can assist in the continued production of valuable timber crops on the same site for many generations.

Drought is a more subtle element in its effect upon stand structure than wind or fire since its effects are not often evident for several years. The advance and retreat of the grassland-forest fringe can be directly traced to the succession of dry winters and dry summers. It is difficult to imagine, however, how drought conditions would destroy whole forests. There are instances where prolonged dry spells can cause the death of groups of trees on soils where the effect of low precipitation is most severe. However, it does not seem possible that short duration drought (three months) will result in the death of entire stands, although seasonal drought can cause the death of individual trees or groups of grees, particularly new seedlings and possible decadent and overmature trees. The thought to be conveyed here is that unlike fire and wind, succession following drought would be confined to relatively small areas—a part of a stand—or to sites where soil moisture is chronically in short supply.

Winter kill, a condition where conifers are killed as a result of winter drying, can be considered as a form of drought damage. In this instance, trees lose water during warm, clear days and cannot replenish the loss because soil water is frozen.

Perhaps the most striking effects of frost are demonstrated in seedling stands that occur in "frost pockets." These pockets are topographic depressions where radiation cooling is greatest; the cold air is trapped in the "pocket." The resulting low temperatures can cause young trees considerable damage—particularly in the spring when new growth has just begun to show, and can delay succession for many years by repeatedly killing the tops, thus preventing a new stand from becoming established.

Wet snow and ice can literally break trees and if the soil is saturated and unfrozen, can uproot them. Most damage from ice and show occurs from breaking off of trees and of limbs.

Biotic factors represent a second group of agents that are responsible for creating disturbances resulting in secondary succession. These agents are insects, disease, animals, and humans. It is not difficult to see the change in an oak stand which results from a gypsy moth attack; in a red spruce-balsam fir stand infested by spruce budworm; in a ponderosa stand following a severe outbreak of bark beetles. Whole forest areas can be affected and stand composition and density can be drastically changed as a result of insect attack. When a severe insect attack that kills large numbers of trees is followed by fire, the resulting destruction is greater than that from either insects or fire alone.

The littleleaf disease of shortleaf pine is an interesting example of a soil-borne pathogen that becomes extremely virulent on a particular set of sites. The passing of the chestnut blight through the eastern oak-hickory (formerly the oak-chestnut) forest is an example of how a virulent pathogen can change an entire forest region. Blister rust is responsible for preventing the development of both eastern and western white pine in "hot spots," where pines and *Ribes* spp grow together in relative abundance.

Animals, other than man, are active in changing stand structure. Impoundment basins behind beaver ponds result in the death of rather extensive stands. If and when the beavers leave and the dam breaks or the impoundment fills with silt, the land is available for the return of trees with perhaps an additional couple of inches of silt which increases surface soil depth. Overpopulation of deer and rabbit causes considerable change in forest tree composition.

Humans in their various agricultural and other land-based pursuits have changed the face of vegetation perhaps more than any other animal. Fire, disease, insects, land clearing, logging, and grazing of domestic and game animals are all connected with human activity. Humans are more capable of altering their environment to suit their needs than any other organism and for this reason they can contribute to their own destruction.

CONTINUING CAUSES OF SECONDARY SUCCESSION

When community structure on a particular site is modified so that openings occur in a stand, plant species that are successful in occupying the openings are those best adapted to the environment created. The ability of a particular species to occupy a particular site depends upon its *proximity to the site* and *amount of seed* that is disseminated on the site, and upon the species ability to become *established on the site*.

The *proximity* of seed trees to a site to be invaded is a first prerequisite to invasion. Assuming a species to be growing in the vicinity of an opening in a stand, the rate of appearance of seedlings of that species in the opening will depend, in part, upon the amount of viable seed that gets into the opening, or upon the sprouting ability of the species to reproduce vegetatively. Root sprouts can occupy an opening perhaps faster than any other type of growth.

The *amount* of seed produced by a species will influence invasion success. Species producing large amounts of seed are more successful than species that produce infrequent crops, or annual light crops. However, light seed crops can be offset partially by a larger number of seed trees of a species, each tree producing a small amount of seed. Invasion success then depends partly upon the quantity of seed. Generally, seed of most tree species will not germinate the second year following dissemination because viable seed that does not germinate the first year will rot or be otherwise destroyed. There are exceptions to this: Atlantic white cedar is a species that produces large seed crops, but the seed, because of a wet sterile surface soil on which it is deposited, can be held for several years without its losing viability or being destroyed. Chokecherry, yellow-poplar, and basswood seeds can lie dormant for several years in the litter. However, these species are exceptions, and for most species to be successful in invading an area requires that their seed germinate the growing season following dissemination.

Seed weight is important in determining the distance mature seed travels from the mother tree and determines whether seed from some species can be carried to a site that has been disturbed. Light seed that is wind disseminated increases the effective area of a seed tree while heavy nutlike seeds do not get far from the mother tree, except as they may be carried by animals or other means, for example, water. Birds carry cherry pits and juniper seeds in their digestive tracts and then excrete them later at some distance from the parent trees. Squirrels and other ground rodents may bury acorns and nuts in the ground, which they do not later find. These seeds germinate and produce a major segment of the understory beneath pioneer stands.

The following list of tree species groups them on the basis of seed weight, abundance of seed, and sprouting ability:

LIGHT SEED (FREQUENCY OF SEEDING)		HEAVY SEED	SPROUTS
Frequent	Infrequent		
Birch	Pine	Oak	Aspen
Aspen	Hemlock	Hickory	Locust
Poplar	Spruce	Walnut	Poplar
Ash	Fir	Beech	Cottonwood
		Cherry	Beech

The *establishment* of a seedling depends upon having the seed survive until conditions are favorable for it to germinate. The odds are greater that a seed will not survive to germinate. First a seed must have the capacity to grow—the embryo must be normal. Some seeds are produced that do not have a normal embryo, and some normal appearing seeds have no embryo or endosperm. These are called blind seed, or "pops," because the seed is hollow. In some seedlots only about 60% of the apparently full seed will have embryos (Hocker, 1969; N.C. State, 1977). Assuming each of these seeds has the capacity to grow, they must not be eaten by birds or animals, nor must they become infected with disease or infested by insects. With eastern white pine, seed to seedling ratios of 700 to 1 under natural conditions are not uncommon; this means that 0.14% of the seed produced seedlings, or that 99.86 percent of the seed was destroyed or was not viable (Hocker, 1961). Yocom and Lawson (1977) report tree percents between 1.41 and 0.46% for naturally regenerated shortleaf pine. Burning as a site preparation treatment increased tree percent from 0.42 to 0.98% in one year. Seeds that do germinate are still subject to the rigors of the site. If seedling roots find a suitable medium for growth, chances of survival are enhanced; but the litter of a forest floor is very unsuitable to the light seeded species. If a seed is successful in germinating, the maternal food reserve contained in the endosperm cannot become exhausted before seedling roots reach a suitable medium for supporting the tree's vital functions. In the spring, the surface litter of soil drys out very rapidly and seedlings whose roots have not penetrated to mineral soil soon die from lack of water. On the other hand, on exposed mineral soil, a seedling's succulent stem can be girdled by excessively high soil temperatures. Damping-off fungi can attack the stems of young seedlings in any environment. However, for a number of commercially important light-seeded tree species an exposed mineral soil seedbed has been shown to provide the best moisture and light conditions for germination and initial establishment.

It is important that nutlike seeds be covered by litter so that they will not dry out before germination. If an acorn loses too much water, the embryo may be damaged. Many of the winged seeds can undergo extended periods of drying before the embryo is destroyed. Sugar maple is a notable exception because it is very tolerant of shade and produces a new radicle that emerges from the seed with the ability to penetrate layers of litter as it grows toward the mineral soil. Most epigeal species lack the ability of their roots to penetrate through a litter layer. A nut has a much larger food reserve than is contained in small seeds and, as a result, seedling roots of species with nut-type fruit can penetrate quite far into the soil before the upper soil and litter reach critically low moisture levels. The condition of the embryo in nut fruits is also important. Many acorns are destroyed by nut weevils and as a result the embryo cannot germinate (Korstian, 1927).

A germinating seedling is subjected to damage from a number of insects, diseases, and browsing animals. Even when a new seedling has survived all the rigors of germination, it must still compete with other plants that already occupy the site or are attempting to do so.

Competition for survival begins when two organisms occupying the same site make demands on a single factor in excess of the supply of that factor. Species that survive on a particular site are those capable of maintaining a growth superiority so that their crowns dominate the site, or that can grow sufficiently well in the shade so that they can eventually push their crowns into the canopy. In other words, intolerant species must be able to maintain a height growth superiority in order to survive within a closed stand while tolerant species can survive because of their ability to grow even under conditions of low light.

Once trees and other plants successfully occupy a site they begin to affect the environment of that site. Initially, the effects on the physical site are confined primarily to changes in moisture, temperature, and light. Temperature changes, however, will also affect the moisture conditions. Action and reaction modify the site so that species not capable of occupying an open site because of too severe temperatures find conditions more suitable for germination and establishment in a closed stand. Generally, species that invade open sites do not find conditions in the closed stand suitable for their continued survival.

Soils of sites that once supported forests and that have been cultivated for agricultural crops, or that have been burned over as a result of grass or forest fire, undergo a considerable change when trees again occupy them. In addition to changes in temperature and moisture, organic buildup is one of the most striking characteristics of forest soils. The organic debris that begins to accumulate on the soil surface as soon as the trees dominate the site prevents water from running off the surface and permits it to infiltrate into the soil. A change in the micro and macro organismic populations also takes place as a result of organic matter accumulation, and the increase in moisture and decrease in temperature extremes that occur. As a result, physical soil structure is also altered. The nature of these changes, as well as chemical changes, will be discussed in later chapters; however, it must be noted that the changes in the physical and biological environment that occur as a result of succession have a profound effect upon the trees and other organisms that later will occupy the site.

CLIMAX CONCEPT

If a site that is undergoing succession, either primary or secondary, is not disturbed for a period of several decades, a succession of plant communities will occupy it. Transient communities may be of relatively short or long duration. Usually communities that occupy the initial stages of succession, those which follow a major disturbance, are short-lived intolerant species, while those communities occupying the later stages tend to be long-lived and more tolerant. The difference in successional stages may be termed as *temporary* (pioneer) if the species are short-lived and do not occupy the site for a long period, or if long-lived or if the species continue to occupy the site because of continuing disturbance then the stage may be termed *subclimax*. In time the vegetation will accomplish an equilibrium with the site and a relatively stable long-lived *climax community* occurs. Continuous interruptions, such as repeated burning during the successional process, assures that a subclimax community will occupy the site for an indefinite period.

In general, the dominant vegetation that occurs as climax on a particular site will be composed of the most tolerant species capable of occupying the site, and the climax vegetation is capable of being replaced by a vegetation of similar composition if disturbed, provided the productive capacity of the site has not been altered.

The successional pattern on an old field in the spruce-fir region was shown by Whitford (1901) to have the following pattern:

Moss-meadow grasses → aster-firewood → sedge meadow → willow-birch-aspen → spruce-fir

If such a field is burned at intervals, mowed for hay, or pastured, the communities can be altered and the duration of any one may be prolonged. However, if disturbance is not excessive and a spruce-fir stand should develop, it will maintain itself for an extended period as a stable community, hence, a climax stand. Leak (1976) reported that where red spruce and balsam fir occur, the soils on the sites are rocky, outwashed, compacted, or poorly drained. Hardwoods, primarily beech and sugar maple, are most common on open glacial till soils. The presence of these species was related to climate and to the resistance of the soil substratum to weathering.

Grigal and Ohman (1975) identified 13 upland plant community types in the Boundary Water Canoe Area of Minnesota. These occurred on logged and virgin undisturbed areas. Identified as successional were lichen; jack pine-oak; red pine; jack pine-black spruce; jack pine-balsam fir; black spruce-feather moss; maple-oak; aspen-birch; aspen-birch-white pine; maple-aspen-birch; maple-aspen-birch-balsam fir; and balsam fir-birch. Northern white-cedar was considered climax; however, white-tailed deer were considered to have an impact on restricting the occurrence and reproduction of the white-cedar type.

Following the sedge stage, the stands that occur on a site in the hemlock-hardwood region were shown by Lutz (1928) to progress toward the climax as eastern redcedar-grey birch → hardwood → hardwood-hemlock. Species present at the different stages were as follows:

	REDCEDAR GRAY BIRCH	HARDWOOD	HARDWOOD HEMLOCK
Gray birch	X	—	—
Redcedar	X	—	—
Silver maple	X	X	X
Black oak	—	X	X
White oak	X	X	X
Scarlet oak	X	—	X
Red oak	X	X	X
Beech	—	X	X
Red maple	—	X	X
Dogwood	X	X	X
Basswood	—	—	X
Chestnut oak	—	—	X
Yellow birch	—	—	X
Hemlock	—	—	X

In the western white pine type of northern Idaho the pattern of dominant species can be, according to Larson (1929):

western larch-western white pine - Douglas-fir → Douglas-fir-w. white pine →
→ Douglas-fir-w. white pine-w. hemlock-white fir → w. hemlock-white fir-w. white pine.

Terms such as climax, subclimax, temporary, and pre- and post-climax are found in eoclogical literature. For forestry purposes climax refers to a stand occupying a site where both the trees and site have achieved a dynamic equilibrium and, barring disturbance, where no major changes in composition will take place. If successional change is not anticipated on a particular site, the climax stand ends succession. A stand that occupies a site at a particular time during the period of succession will survive for varying lengths of time. Forestry terminology currently uses both the term subclimax type and the term temporary type when referring to such stands. Subclimax refers to stands whose composition is maintained for indefinite periods because of special circumstances of the environment that delay the attainment of a climax, or to those stands that are composed of long-lived species which give way very slowly to successional pressure. Temporary stands are relatively short-lived and in a period of 40–50 years may be replaced by more tolerant species unless some disturbance causes them to be perpetuated.

It is important for foresters to recognize the successional stage a particular stand represents because it will be necessary to either prolong the presence of the stand by modifying it or the site, or to hasten or retard the successional process that brought about present stand structure. To reproduce a particular stand requires that a forester recognize the factors that brought it into existence and be able to duplicate in some way the conditions created by those factors. Kimmins (1972) has shown that in the Douglas-fir region successional patterns will vary with different sites. For example, on a dry site (Figure 5.6) a disturbance such as logging may create a mild disturbance to a climax hemlock-cedar stand with the result that in a relatively short time Douglas-fir succeeds hemlock and cedar. A major disturbance in logging that results in a loss of valuable organic matter from the soil may require a hundred years to restore the stand to its climax condition. Wetland sites are also fragile since a major disturbance may require many years to restore them to a former condition. Although fresh sites are not immune to disturbance, they can perhaps recover from a "major" disturbance more quickly than either dry or wet sites.

Many of the more valuable commercial forest types occur in either temporary or subclimax stages of stand development. It was estimated that 50 percent of the sawtimber volume of the United States was made up of Douglas-fir, loblolly, longleaf, slash, shortleaf, and other southern pines (U.S. Forest Service, 1958). All of these species are representative of subclimax species that follow disturbance and whose presence is prolonged by continued disturbance. Valuable hardwoods, such as aspen, cottonwood, yellow-poplar, red gum, and paper

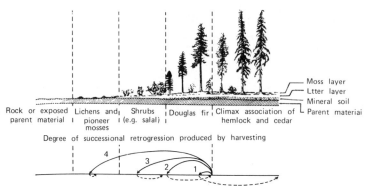

Moss layer
Ltter layer
Mineral soil
Parent materiai

Rock or exposed parent material | Lichens and pioneer mosses | Shrubs (e.g. salal) | Douglas fir | Climax association of hemlock and cedar

Degree of successional retrogression produced by harvesting

4 3 2 1

Extent of successional development occurring over (e.g.) 100 years post—harvesting

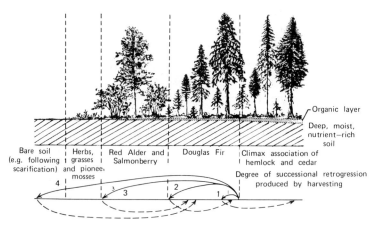

Organic layer

Deep, moist, nutrient—rich soil

Bare soil (e.g. following scarification) | Herbs, grasses and pionee, mosses | Red Alder and Salmonberry | Douglas Fir | Climax association of hemlock and cedar

Degree of successional retrogression produced by harvesting

4 3 2 1

Extent of successional development occurring over (e.g.) 100 years post—harvesting

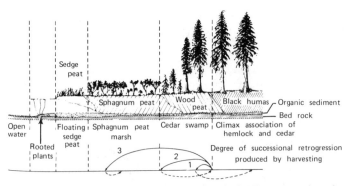

Sedge peat

Sphagnum peat

Wood peat

Black humas

Organic sediment

Bed rock

Open water | Floating sedge peat | Sphagnum peat marsh | Cedar swamp | Climax association of hemlock and cedar

Rooted plants

Degree of successional retrogression produced by harvesting

3 2 1

Extent of successional development occurring over (e.g.) 100 years post—harvesting birch occur in temporary stands. The white pines are subclimax while hemlock types and spruce-fir are climax.

Preclimax and postclimax are terms that probably have little application in day to day forestry practice. These terms refer to special conditions of climax where the type is not representative of the one that prevails for the region. *Postclimax* is represented by a climax that is more mesic than the prevailing soil and climate support. Such types as sugar maple along the rivers of Oklahoma, paper birch in the river canyons of Nebraska, and oak in the prairies of Texas are examples of postclimax (Weaver and Clements, 1938). *Preclimax* types are more xeric than the climate and soil of a region will generally support. The presence of oak-hickory within the beech-maple region is considered an example. Preclimax conditions are not as easily recognized as postclimax ones. The environmental amplitude of many forest species is quite great; trees have migrated long distances in the

FIGURE 5.6. Diagramatic representation of the successional consequences of different levels of disturbance during harvesting of the penultimate successional stage on three different types of succession in British Columbia. Arrow 1 represents either minimal disturbance permitting succession to proceed to climax or disturbance that accelerates the successional process. Arrow 2 represents the degree of disturbance necessary to maintain the penultimate successional stage in perpetuity. Arrows 3 and 4 represent increasing levels of disturbance. The variation in ecological significance of heavy disturbance between the three successions can readily be seen (from Kimmins, 1972).

In the dry-land succession, successional changes accompanying disturbance include loss of nutrients, loss of nutrient and moisture storage capacity, and a general drying of the area. Recovery from excessive disturbance is slow and cannot readily be accelerated.

In the fresh-land succession, successional changes accompanying disturbance are modest and recovery is generally rapid. Recovery can be easily accelerated when it is undesirably protracted.

In the wetland succession, the major successional changes accompanying excessive disturbance are increasing wetness and loss of soluble nutrients. Recovery from disturbance tends to be slow but can be accelerated by such techniques as drainage.

past, so it can be expected that many forest species can occupy, at the limits of their botanical range, sites that are particularly favorable to their survival. As a result it does not seem necessary to give special recognition to these stands since it is apparent that the environmental conditions which prevail on the sites are sufficient to meet the species needs, otherwise the trees would no longer be able to continue to grow there.

GRADIENT ANALYSIS

During the period of development of the science of ecology, two rather distinct schools developed with regard to the manner in which individual species are associated in communities. One school accepted the hypothesis of Clements (1916), which assumed a dependency of species one to the other as they occurred in the various assemblages within an association and as they represented various successional stages including the climax association. An opposing view was presented by Gleason (1939) who proposed that plant associations, regardless of their successional state, represent only an expression of an opportunity of individual species to find a suitable niche in which to grow and to reproduce; that there was no apparent obligate relationship among species in a particular community.

This latter view has been expanded by Curtis and McIntosh (1950, 1951) and Whittaker (1956). Support for the Gleason hypothesis was obtained from the manner in which plant distribution could be identified from climate and soil zones. Climate-soil zones represent environmental conditions to which individual species have become adapted. If one were to define a series of temperature or moisture gradients, then it would be possible to identify on the axes of the system the abundance of various species as their presence is noted in field sampling (Whittaker, 1967). The abundance of a species will have a continuing distribution conforming to the environmental needs of the species (Figure 5.7).

As more factors are combined in the analysis it is possible to refine the definition of the environment and of species occurrence; eventually, if the system is refined sufficiently, a definition of niche results. If all of the environmental factors can be precisely defined, it is theoretically possible to define a single niche structure into which single species can be located.

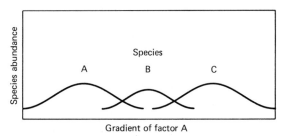

FIGURE 5.7. Location and abundance of three species along a gradient of a single factor. To illustrate a two factor set requires a two-dimensional figure.

ALLELOPATHY

Whether individual species and individual trees are a simple random mixture seems questionable because there is basis for assuming obligate relationships among species when allelopathy is considered. *Allelopathy* is the suppression of germination and growth, or the limitation of the occurrence of plants as the result of the release of chemical inhibitors by some plants. The widespread occurrence of the phenomenon has been documented by Whittaker and Ferny (1971). Foresters have been aware of apparently antagonistic relations between various types of plants. The fact that the substance juglone found in the leaves of black walnut inhibits the growth of plants in areas where the leaves fall has been recognized for many years. Horsley (1974) reports black cherry seeds were inhibited in germination by extracts from bracken fern, wild oak grass, goldenrod, and flat-topped aster. Goldenrod and aster are fairly common in old field succession in the eastern United States. Mueller (1966) showed that terpenes produced by *Artemisia californica* and *Salvia leucophylla* were apparently capable of limiting growth of other species, and in fact the presence of *Artemisia* may be related to the presence of *Salvia,* and the occurrence of *Salvia* in turn might be related to an autotoxic accumulation of terpenes. Forcier (1973) and Tubbs (1976) reported that the growth of yellow birch is restricted by the presence of sugar maple in the stands and proposed a succession pattern where sugar maple and yellow birch maintain a relationship in which yellow birch can grow in the presence of beech but not in the presence of sugar maple (Figure 5.8). The abundance of each of these species in

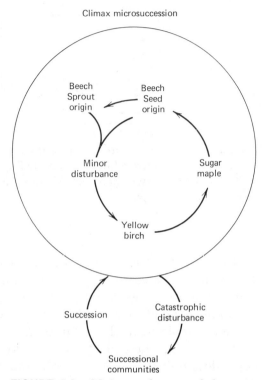

Climax microsuccession

Beech Sprout origin

Beech Seed origin

Minor disturbance

Sugar maple

Yellow birch

Succession

Catastrophic disturbance

Successional communities

FIGURE 5.8. Major pathways of the microsuccessional scheme hypothesized for the climax forest. The microsuccession also has several alternative routes and is expected to vary in rate, end point, and even stages on different sites within the forest, with the interaction of disturbance and with stochastic aspects of any species' regeneration (from Forcier, 1973).

the beech-birch-maple climax association varies depending on the abundance of the other two species. Peterson (1965) reported that a water-soluble substance from sheep larvel (*Kalmia angustifolia*) leaves inhibited primary root growth of black spruce a common associate, and DeBell (1971) found that exudations from cherry bark oak leaves limited the growth of neighboring trees. Most studies show that when allelopathy comes to play, succession is delayed rather than promoted or hastened. In the case of the beech-birch-maple complex, an obligate relationship appears to exist.

Daniel and Schmidt (1972) studied the germination of tree seed from several species that had been treated for surface pathogens. Seed from Englemann spruce, Douglas-fir, subalpine fir, and lodgepole pine was collected and stratified in litter from stands of the same species. Part of each seed lot was treated to control seed coat pathogens. All species were greatly reduced in germination rate when germinating in litter from their own trees; however, some species were not greatly affected by litter of another species. For example, lodgepole O-horizon was neutral to all other species except lodgepole. This study thus indicated another factor when considering succession of tree species.

It can be concluded that the presence of a particular species in a particular place at a particular time is related partly to chance, but also and more important it is related to the fact that the physiological needs of the species are met by the environment prevailing at the time of germination and establishment, a factor that continues to prevail during later development. That there are certain obligate relationships between plants cannot be denied, but within a natural environment there are probably sufficient microsites available so that all species in a locality have an opportunity to invade suitable areas. Species that succeed in colonizing newly exposed sites do so because their niche requirements are met to a greater degree than are the requirements of species that fail to colonize those sites. Within climax communities, it can be assumed that changes in species composition will occur as a result of continuing disturbance of either small or large magnitude.

GENERAL REFERENCES

Drury, W.H., and I.C.T. Nisbet. 1973. Succession. *J. Arnold Arboretum* 54(3):331–368.

Kershaw, K.A. 1973. *Quantitative and Dynamic Plant Ecology*, 2nd ed. American Elsevier, New York.

Kormondy, E.S. 1969. *Concepts of Ecology.* Concepts of Modern Biology Series. Prentice-Hall, Inc. Englewood Cliffs, N.J.

Rice, E.L. 1974. *Allelopathy.* Academic, New York.

Spurr, S.H. and B.V. Barnes. 1973. *Forest Ecology*, 2nd ed. The Ronald Press, New York.

Whittaker, R.H. 1970. *Communities and Ecosystems*, in Current Concepts in Biology Series. MacMillan, Toronto, Canada.

Toumey, J.W. and C.F. Korstian. 1947. *Foundations of Silviculture—on an ecological basis.* Wiley, New York.

Chapter 6

APPROACHES TO CLASSIFICATION

The fact that forest stands that are uniform in species composition occupy extensive areas within a climatic or geographic region has tempted ecologists to assign a particular importance to classify these stands on a taxonomic basis. Foresters find it important to classify forest stands for inventory and other management purposes and are interested more in the total classification of the vegetation of a particular locality than are plant ecologists. This is because the forester must determine the usefulness of each acre of land within his or her management unit and, therefore, must be able to put each stand into some type of category that fits the overall management objectives. The plant ecologist, on the other hand, may only be interested in the

Forest Classification

successional direction of a particular stand or whether a stand is representative of the climax potential of the region or for some type of single use. The ecologist can disregard stands or sites that at the moment do not suit his or her needs. The forester must first classify each stand and site and then decide whether to defer, temporarily at least, management decisions for certain stands in the classification.

Such systems of classification as proposed by Brewer (1874), and those used in statistical articles of the United States by Sargent (1884) based on geographic distribution of forest, are of general interest. The major efforts of ecologists, however, have been directed toward either a climatic or a synecological classification. Several studies are based upon various indices of temperature and precipitation resulting in a classification of climate and vegetation on a rather broad scale (Köppen, 1923; Merriam, 1898; Thornthwaite, 1931). Such classifications are useful for regional or area planning, but lack detail required in day to day planning.

The classifications of Shreve (1917) and Shantz and Zon (1924) are based on the similarity of composition of the dominant vegetation. These synecological studies, which are examinations of terrestrial ecosystems, lead to a more detailed classification into which forest stands can be placed and can be refined to suit the needs of management. Any classification system requires that in moving from a lower level to the next higher level of classification a certain amount of abstraction must occur and detail must be reduced, otherwise there would be the same number of classes at each level. For the sake of a beginning presentation it seems best to start with the more generalized types of vegetation, and work down, adding detail.

On the North American continent are seven climax vegetative formations: tundra, forest, woodland, scrub, desert, grassland, and tropical (Oosting, 1950) (Figure 6.1). A *formation* represents a vegetational type occupying a particular climatic and soil provenance, and is part of the biome that includes all of the biotic groups.

In Figure 6.1 Pitelka did not designate areas of tropical forest recognized by others to occur in North America. These are located in South Florida, the Baja Peninsula of California, and along the west coast of Mexico. The last two are part of areas designated by Pitelka as creosote-bush desert. Braun, and Weaver and Clements classify some of the ecotones differently from Pitelka; these represent different approaches in recognizing transition between formations.

The types in Figure 6.1 fit into the systems of climatic classification

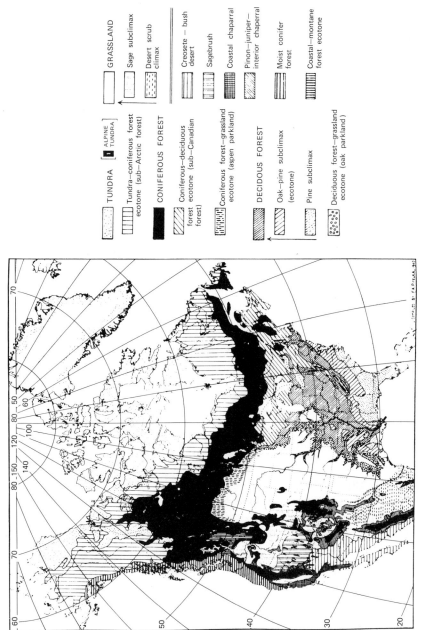

GRASSLAND

▭ Sage subclimax

▨ Desert scrub climax

▥ Creosote – bush desert

▦ Sagebrush

▧ Coastal chaparral

▨ Pinon–juniper– interior chaparral

▤ Moist conifer forest

▦ Coastal–montane forest ecotone

TUNDRA ▣ ALPINE TUNDRA

▤ Tundra–coniferous forest ecotone (sub–Arctic forest)

CONIFEROUS FOREST ■

▨ Coniferous–deciduous forest ecotone (sub–Canadian forest)

▨ Coniferous forest–grassland ecotone (aspen parkland)

DECIDOUS FOREST

▨ Oak–pine subclimax (ecotone)

▨ Pine subclimax

▨ Deciduous forest–grassland ecotone (oak parkland)

FIGURE 6.1. Major vegetative formations of North America including tundra, forest, grassland, scrub, woodland, and desert (from Pitelka, 1941).

proposed by Merriam, Köppen, and Thornthwaite, for they used the vegetation as a means of identifying differences in climate. A more detailed division of the forest formations is a classification of the climax *associations* that make up a formation. Using the classification of Braun (1950) for the Eastern Forest Formation (Figure 6.2) and that of Weaver and Clements (1938) for the Western Forest Formation (Figure 7.12), the major associations are:

Eastern Coniferous Formation
  White spruce–red spruce–balsam fir association; called boreal forest (Note the similarity to the subalpine in the western forest)
Eastern Deciduous Formation
  Mixed mesophytic association
  Beech–maple association
  Maple–basswood association
  Hemlock–hardwood–white pine association
  Oak–chestnut association
  Oak–hickory association
  Oak–pine association
  Southeastern evergreen forest association

Western Coniferous
  subalpine   Engelmann spruce—alpine fir association
  montane   Douglas-fir association
               ponderosa pine association
  woodland   piñion—juniper association
               oak—mountain mahogany association
  Pacific slope
               western hemlock—arborvitae—grand fir association

In each instance, the name of the association is based upon the predominance of a particular climax species grouping. Detail is still lacking at this level but the approach to an increasing degree of detail becomes possible. The problem is to fit into the climax array subclimax and temporary stands. Subclimax associations vary according to formation and geographic position within the formation by the similarity of the dominant species. *Subclimax* associations represent successional stages that will eventually be replaced by the representative climax association. For example, the successional stages of the hemlock-hardwood-white pine association of the Eastern Deciduous Forest Formation can be:

FIGURE 6.2. Forest regions of the deciduous forest formation (from Braun, E. L. 1950. *Deciduous Forests of Eastern North America.* Blakiston, Philadelphia, Pa.).

DECIDUOUS FOREST FORMATION

1. **MIXED MESOPHYTIC FOREST REGION**
   a. Cumberland Mountains
   b. Allegheny Mountains
   c. Cumberland and Allegheny Plateaus
2. **WESTERN MESOPHYTIC FOREST REGION**
   a. Bluegrass Section
   b. Nashville Basin
   c. Area of Illinoian Glaciation
   d. Hill Section
   e. Mississippi Plateau Section
   f. Mississippi Embayment Section
3. **OAK–HICKORY FOREST REGION SOUTHERN DIVISION**
   a. Interior Highlands
   b. Forest-Prairie Transition Area
   NORTHERN DIVISION
   c. Mississippi Valley Section
   d. Prairie Peninsula Section

4. **OAK–CHESTNUT FOREST REGION**
   a. Southern Appalachians
   b. Northern Blue Ridge
   c. Ridge and Valley Section
   d. Piedmont Section
   e. Glaciated Section
5. **OAK–PINE FOREST REGION**
   a. Atlantic Slope Section
   b. Gulf Slope Section
6. **SOUTHERN EVERGREEN FOREST REGION**
   a. Mississippi Alluvial Plain
7. **BEECH–MAPLE FOREST REGION**

8. **MAPLE–BASSWOOD FOREST REGION**
   a. Driftless Section
   b. Big Woods Section
9. **HEMLOCK–WHITE PINE–NORTHERN HARDWOODS REGION GREAT LAKES–ST. LAWRENCE DIVISION**
   a. Great Lake Section
   b. Superior Upland
   c. Minnesota Section
   d. Laurentian Section
   NORTHERN APPALACHIAN HIGHLAND DIVISION
   e. Allegheny Section
   f. Adirondack Section
   g. New England Section

B. BOREAL OR SPRUCE–FIR FOREST FORMATION
G. GRASSLAND OR PRAIRIE FORMATION

grasses → shrubs → grey birch-white pine → white pine
→ white pine-hardwood → white pine-hardwood-hemlock → hemlock-hardwood-white pine

In the association white pine stands may represent stands of long duration and are considered subclimax; however, stands on mesic sites will be replaced by mixed stands containing hardwoods and hemlock.

There is the need to recognize a diversity of climax associations that can occur within a specific locality. The vegetation of a particular locality will not be made up of a single climax association; instead, there will probably be several associations represented on one locality. It is not uncommon in northern Maine to find stands of beech-maple intermingled with spruce-fir. However, within a particular formation, there is a tendency for particular associations to be more frequent than they are represented within another association. Outlying stands of spruce-fir are found at high elevations in the southern Appalachian Mountains (Figure 6.2, section 4a) where they remain because of climatic conditions at the high elevations, although the regional classification of the surrounding country, according to Braun, is oak-chestnut. One can find beech-maple stands within the oak-hickory association (Whittaker, 1956), and there are oak-hickory stands in the beech-maple association (Braun, 1950).

Simplicity of classification will not affect complexity, which still prevails. One should keep in mind that the environment of a locality is not uniform with changes occurring abruptly or gradually; as a result, the vegetation will respond to pressure from the environment. If the genetic amplitude of a species is not exceeded by changes in environment, then the species will remain; however, its form may be modified, or its abundance or frequency may be altered. It is the gradual shift in one or several of the factors of the environment that is hard to measure but is reflected in the frequency, abundance, and dominance of particular species or species groups. Perhaps if the classifications of Merriam, Köppen, and Thornthwaite were extended beyond their present levels, then the synecological approach to classification could receive the assist it needs. The continuum concept of Curtis and McIntosh (1950, 1951) and the gradient approach of Whittaker (1956) emphasize the need to consider the gradient changes in community structure that take place within a regional association. An association presents different floral aspects in different environments thus necessitating the need to evaluate the relative importance of the characteristic species of the association throughout its regional extent.

One cannot help conjecturing about not only the striking similarity of the forms of vegetation that represent similar temperature-moisture niches, but also the similarity among the species that occur within each niche. An interesting relationship appears to exist among the spruce species that are representative of the eastern boreal forest and the western alpine forest. Research has shown that where the ranges of white spruce, black spruce, red spruce, and Engelmann spruce overlap, hybrids are produced. It has not been demonstrated that each of the spruce species will cross with all other species, but there are a sufficient number of natural crosses on record to indicate a common ancestry of the species as a group. The western white pine and eastern white pine occupy similar niches and these two species hybridize when crossed artificially. Critchfield (1965) has illustrated the nature of the taxonomy of the Genus Pinus. These studies indicate that speciation has occurred in pines as a response to the evolution of *niche*. It would appear that the differentiation of species among the conifers has not yet reached its final evolutionary limit.

On the other hand, there are a number of species, like ponderosa pine, lodgepole pine, and Douglas-fir, whose geographic ranges are so extensive that there are several recognized races within each species. These races have niche requirements that differ rather distinctly from other races in the species. The need to recognize the west coast and inland races of Douglas-fir is one example. The west coast race grows in a relatively warm subtropical rain-forest climate, moist coniferous forest (Figure 6.1), and the inland race grows in a colder, drier temperate continental climate, coastal-montane ecotone (Figure 6.1).

Waring (1970) used moisture stress and temperature to define conditions for reproduction of four western conifers. Hemlock reproduction occurred on sites that were relatively cool and where moisture stress was minimal, while ponderosa pine reproduction was found on hotter and drier sites (Figure 6.3). White fir and red fir were intermediate between hemlock and pine with white fir reproduction occurring on sites warmer and drier than red fir using the same moisture-temperature gradients.

Siccama (1974) was able to associate climate with difference between the beech-birch-maple deciduous forest at a lower elevation in the Green Mountains of Vermont and the balsam fir forest that oc-

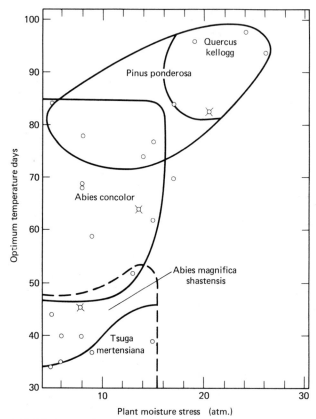

FIGURE 6.3. Distribution of natural regeneration in relation to gradients of moisture and temperature. Starred data points indicate supplementary stands to verify precision of original data (Waring, 1970).

curred at higher elevation. He noted a nonlinear decline at 2600 ft elevation in both the length of the frost-free season and the occurrence of frequent cloudy days, which in winter resulted in ice buildup on vegetation. The division between the deciduous forest and fir forest occurred along this line of demarcation of climate difference. It would appear from this study that temperature and moisture gradient, although adequate to explain vegetative difference in moderate climate, would not be entirely adequate where climate is harsh, particularly where an extremely low temperature can occur with a resulting change in form of precipitation.

FOREST COVER TYPES

Foresters working somewhat independently of other groups have developed a system for classifying individual forest stands that utilizes the uniformity of the dominant species as a major criteria for designating different groupings. The system is called forest cover type classification. It does not recognize climax potential as a criteria of classification; however, climax and subclimax species groupings are identified. Because the system begins by classifying individual stands as they occur in the forest, it satisfied the need for detail for day to day management planning.

The rules for recognizing *forest cover types* (taken from the Society of American Foresters, 1954, Forest Cover Types of North America) are as follows:

1. The cover type must actually be found occupying large areas in the aggregate. It does not require that it should cover any single large area in a solid stand, but it should be the characteristic composition found typically through a considerable range of country.
2. The cover type must be distinctive and easily separated from other types that most closely resemble it.
3. Within the foregoing limitations every important combination of cover must be recognized as a forest type.

A forest type is defined as:

A descriptive term used to group stands of similar character as regards composition and development due to given ecological factors, by which they may be differentiated from other groups of stands. The term suggests repetition of the same character (similarity of dominant species) under similar conditions. A type is temporary *if its character is due to passing influences such as logging or fire;* permanent (subclimax) *if no appreciable change is expected and the character is due to ecological factors alone;* climax *if it is the ultimate stage of a succession of temporary types. A* cover type *is a forest type now occupying the ground, no implication being coveyed as to whether it is temporary or permanent.*

Although the type name uses only the species present (in the written description of the type), there are sections entitled "Nature and Occurrence" and "Transition Forms and Variants" that contain statements about the successional position of the type and the succeeding types. Thus, the use of forest cover types has utility beyond its primary intent.

In eastern North America, there are 106 recognized major forest types; in western North America, there are an additional 50 types. The types are listed numerically 1–106 for the east and 201–250 for the west (see Tables 6.1 and 6.2).

Forest cover types should not be considered as a different system of classification from the synecological system of the plant ecologist. The forest cover type approach is an extension of the synecological system fulfilling the need for a complete cataloging of individual forest stands as they actually occur in the forest, and recognizing the need to consider successional stages of stand development in classifying stands.

SITE TYPES (PLANT-INDICATORS)

In addition to the need to classify forest stands on the basis of composition, there is the additional need to place them into productivity groupings. Site index classes can be used, but these are rather cumbersome and tedious to make and the proper index curves may not always be available. A discussion of site quality measurement, within which site index is a method of classification, will be presented later; however, an approach to quality classification using "lesser" vegetation as site indicators has been used with some degree of success to fulfill the need for site quality classification. It has been recognized that the presence of certain groupings of indicator species characterizes sites of differing productivity. Although not so definitive as site index, classification by site type does permit a way of arranging stands into productivity and vegetative classes.

The site type method is credited to Cajander (1926) who developed it for the boreal forests of Finland (see Jones, 1969; Carmean, 1975). The method has been used in eastern Canada and the northeastern United States where the spruce-fir forests are not too different from those in the Scandinavian countries; in the ponderosa pine forest (Daubenmire, 1961); in the Douglas-fir forest in British Columbia (Eis, 1962); and in the western larch forest (Roe, 1967).

The type classes are based on the frequency and abundance of key species, or species group in the understory. The species that occur with fidelity upon a site represent a particular ecological situation. Ray (1941), for example, presented a summary for the Lake Edwards

TABLE 6.1. List of the Forest Cover Types of Eastern North America Showing Arrangement by Forest Regions and Moisture Relationships (Soc. Am. For., 1954)

TYPE NUMBER	COVER TYPE	S.A.F.	CANADIAN	U.S. FOREST[a] SURVEY
			TYPE GROUPS	
		Boreal Forest Region		
Dry				
1	Jack pine		Coniferous	
2	Black spruce—white spruce		Coniferous	
3	Jack pine—paper birch		Mixedwood	
Fresh to moist				
4	White spruce—balsam fir		Coniferous	
5	Balsam fir		Coniferous	
6	Jack pine—black spruce		Coniferous	
7	Black spruce—balsam fir		Coniferous	
8	Jack pine—aspen		Mixedwood	
9	White spruce—balsam fir—aspen		Mixedwood	
10	Black spruce—aspen		Mixedwood	
11	Aspen—paper birch		Hardwood	
Wet				
12	Black spruce		Coniferous	
13	Black spruce—tamarack		Coniferous	
		Northern Forest Region		
Dry				
1	Jack pine		Coniferous	11
14	Northern pin oak			16
15	Red pine		Coniferous	11
42	Bur oak		Hardwood[b]	16

TABLE 6.1. List of the Forest Cover Types of Eastern North America Showing Arrangement by Forest Regions and Moisture Relationships (Soc. Am. For., 1954) — Continued

TYPE NUMBER	COVER TYPE	S.A.F.	TYPE GROUPS CANADIAN	U.S. FOREST SURVEY[a]
Fresh to moist				
16	Aspen	⎫	Hardwood	20
17	Pin cherry	⎬ Aspen—birch—pin cherry	Hardwood	20
18	Paper birch	⎪	Hardwood	20
19	Gray birch—red maple	⎭	Hardwood	20
20	White pine—northern red oak—white ash	⎫	Mixedwood	11
21	White pine	⎬ White pine	Coniferous	11
22	White pine—hemlock	⎭	Coniferous	11
23	Hemlock		Coniferous	12 or 19
24	Hemlock—yellow birch	⎫	Mixedwood	19
25	Sugar maple—beech—yellow birch	⎬ Northern hardwood—hemlock	Hardwood	19
26	Sugar maple—basswood	⎪	Hardwood	19
27	Sugar maple	⎭	Hardwood	19
28	Black cherry—sugar maple		Hardwood[b]	19
29	Black cherry		Hardwood[b]	19
30	Red spruce—yellow birch		Mixedwood	12
31	Red spruce—sugar maple—beech		Mixedwood	12
32	Red spruce	⎫	Coniferous	12
33	Red spruce—balsam fir	⎬ Spruce—fir	Coniferous	12
34	Red spruce—Fraser fir	⎭		12
35	Paper birch—red spruce—balsam fir		Mixedwood	12 or 20
36	White spruce—balsam fir—paper birch		Mixedwood	12
5	Balsam fir		Coniferous	12

		TYPE GROUPS		
TYPE NUMBER	COVER TYPE	S.A.F.	CANADIAN	U.S. FOREST SURVEY[a]
Wet				
12	Black spruce		Coniferous	12
37	Northern white-cedar	Northern conifer swamp	Coniferous	12
38	Tamarack		Coniferous	12
39	Black ash—American elm—red maple		Hardwood	18
	Central Forest Region			
Dry				
40	Post oak—black oak	Oak—Hickory		16
41	Scarlet oak			16
42	Bur oak		Hardwood[b]	16
43	Bear oak			16
44	Chestnut oak			16
45	Pitch pine	Pitch pine—oak		11 or 14
75	Shortleaf pine			14
76	Shortleaf pine—oak	Shortleaf pine—oak		14 or 15
77	Shortleaf pine—Virginia pine			14
78	Virginia pine—southern red oak	Virginia pine—oak		14 or 15
79	Virginia pine			14
46	Eastern redcedar	Eastern redcedar	Coniferous[b]	15 or 16
47	Eastern redcedar—pine			15 or 16
48	Eastern redcedar—hardwoods			15 or 16
49	Eastern redcedar—pine—hardwoods			15 or 16
50	Black locust			16
51	White pine—chestnut oak			11

TABLE 6.1. List of the Forest Cover Types of Eastern North America Showing Arrangement by Forest Regions and Moisture Relationships (Soc. Am. For., 1954) — Continued

TYPE NUMBER	COVER TYPE	TYPE GROUPS		
		S.A.F.	CANADIAN	U.S. FOREST[a] SURVEY
Fresh to moist				
52	White oak—red oak—hickory	⎫ Oak—	Hardwood[b]	16
53	White oak	⎬ Hickory	Hardwood[b]	16
54	Northern red oak—basswood—white ash	⎭	Hardwood[b]	16
55	Northern red oak	⎫ Northern	Hardwood[b]	16
56	Northern red oak—mockernut hickory—sweetgum	⎬ red oak	Hardwood	16 or 17
57	Yellow-poplar	⎫ Yellow-		16
58	Yellow-poplar—hemlock	⎬ poplar	Hardwood[b]	16
59	Yellow-poplar—white oak—northern red oak	⎭		16
60	Beech—sugar maple			19
61	River birch—sycamore		Hardwood	18
62	Silver maple—American elm			18
63	Cottonwood		Hardwood	18
64	Sassafras—persimmon			16
65	Pin oak—sweetgum		Hardwood[b]	16 or 17
95	Black willow		Hardwood[b]	18
Wet				
97	Atlantic white-cedar			17
	Southern Forest Region			
Dry				
66	Ashe juniper ("Mountain cedar")			n.c.

| | | TYPE GROUPS | | |
TYPE NUMBER	COVER TYPE	S.A.F.	CANADIAN	U.S. FOREST[a] SURVEY
67	Mohrs ("Shin") oak			n.c.
68	Mesquite			n.c.
69	Sand pine			14
70	Longleaf pine	Longleaf pine		13
71	Longleaf pine—scrub oak			13
72	Southern scrub oak			16
73	Southern redcedar			14
74	Sand live oak—cabbage palmetto			16
Fresh to moist				
75	Shortleaf pine	Shortleaf pine—		14
76	Shortleaf pine—oak			14
77	Shortleaf pine—Virginia pine			14
78	Virginia pine—southern red oak	Virginia pine—oak		14 or 15
79	Virginia pine			14
80	Loblolly pine—shortleaf pine	Loblolly pine		14
81	Loblolly pine			14
82	Loblolly pine—hardwood			14
83	Longleaf pine—slash pine			13
84	Slash pine			13
85	Slash pine—hardwoods			13 or 15
86	Cabbage palmetto—slash pine			13
87	Sweetgum—yellow-poplar			16 or 17
88	Laurel oak—willow oak			16
89	Live oak			16 or 17
90	Beech—southern magnolia			16 or 17
91	Swamp chestnut oak—cherrybark oak			17
63	Cottonwood			18

TABLE 6.1 List of the Forest Cover Types of Eastern North America Showing Arrangement by Forest Regions and Moisture Relationships (Soc. Am. For., 1954)—Continued

			TYPE GROUPS	
TYPE NUMBER	COVER TYPE	S.A.F.	CANADIAN	U.S. FOREST[a] SURVEY
92	Sweetgum—Nuttall oak—willow oak	⎫ Bottomland		17
93	Sugarberry—American elm—green ash	⎬ hardwoods		17
94	Sycamore—pecan—American elm			18
95	Black willow			18
96	Overcup oak—water hickory	⎭		17
Wet				
97	Atlantic white-cedar			17
98	Pond pine			14
99	Slash pine—swamp tupelo			13 or 15
100	Pondcypress	⎫ Cypress—		17
101	Baldcypress	⎬ tupelo		17
102	Baldcypress—water tupelo			17
103	Water tupelo	⎭		17
104	Sweetbay—swamp tupelo—red maple			17
	Tropical Forest Region			
Dry				
105	Mahogany			
106	Mangrove			

[a]The number shown here represents the *usual* Forest Survey type groups into which the S.A.F. types fall; "n.c." indicates noncommercial.
[b]Present extent of these types very limited in Canada.

178

TABLE 6.2. List of the Forest Cover Types of Western North America Showing Arrangement by Natural Groups.[a]

COVER TYPE	TYPE NUMBER	U.S. FOREST SURVEY NUMBER
Northern Interior		
White spruce	201	
White spruce—birch	202	
Poplar—birch	203	
Black spruce	204	
High Elevations in the Mountains		
Mountain hemlock—subalpine fir	205	8
Engelmann spruce—subalpine fir	206	8
Red fir	207	8
Whitebark pine	208	6
Bristlecone pine	209	n.c.
Middle Elevations, Interior		
Interior Douglas-fir	210	1
White fir	21	8
Larch—Douglas-fir	212	7
Grand fir—larch—Douglas-fir	213	7
Ponderosa pine—larch—Douglas-fir	214	4
Western white pine	215	5
Blue spruce	216	8
Aspen	217	10
Lodgepole pine	218	6
Limber pine	219	6
Rocky Mountain juniper	220	9
North Pacific		
Red alder	221	10
Black cottonwood—willow	222	10
Sitka spruce	223	2
Western hemlock	224	2
Sitka spruce—westen hemlock	225	2
Pacific silver fir—hemlock	226	8
Western redcedar—western hemlock	227	2
Western redcedar	228	1
Pacific Douglas-fir	229	1
Douglas-fir—western hemlock	230	1
Port-Orford-cedar—Douglas-fir	231	1
Redwood	232	3
Oregon white oak	233	10
Oak—madrone	234	10

TABLE 6.2. List of the Forest Cover Types of Western North America Showing Arrangement by Natural Groups." — Continued

COVER TYPE	TYPE NUMBER	U.S. FOREST SURVEY NUMBER
Low Elevations, Interior		
Cottonwood—willow	235	10
Bur oak	236	10
Interior ponderosa pine	237	4
Western juniper	238	9
Pinyon—juniper	239	9
Arizona cypress	240	9
Western live oak	241	n.c.
Mesquite	242	n.c.
South Pacific Except for the High Mountains		
Ponderosa pine—sugar pine—fir	243	4
Pacific ponderosa pine—Douglas-fir	244	4
Pacific ponderosa pine	245	4
California black oak	246	10
Jeffrey pine	247	4
Knobcone pine	248	n.c.
Canyon live oak	249	n.c.
Digger pine—oak	250	n.c.

"These types are arranged in a natural grouping, starting with those characteristic of the coldest region and working toward the warmest. Within each temperature region the types on moist sites precede those on dry sites. Subclimax types come next to the climax into which they pass.

Types are numbered from 201 in order to avoid the use of the same numbers as were used in the eastern classification.

Region of Quebec. In order of increasing productivity, his classes are the following.

Kalmia (sheep laurel)—*Ledum* (Labrador tea)—black spruce swamp of low productive capacity.

Sphagnum (sphagnum moss)—*Oxalis* (wood sorrel)—generally swamp type somewhat higher capacity than Kalmia-Ledum with some hardwood, fir, and northern white-cedar present.

Cornus—steep hillsides (spruce ledge) and on borders of swamps and streams Bunch berry (*Cornus spp.*) predominates with some mountain maple (*Acer spicatum*) and blueberry (*Vaccinium spp.*). Trees are generally red spruce and balsam fir and white birch in minor amounts.

Oxalis-Acer—recognized by predominance of wood sorrel (*Oxalis montana*) with an abundance of mountain maple and striped maple (*Acer pennsylvanicum*) on the better sites. Good sites for balsam fir, and grows spruce, yellow birch, but not much beech and sugar maple.

Viburnum - Oxalis—witch hobble (*Viburnum lantanoides*) present, oxalis present as is mountain maple and striped maple. Good quality sites for hardwoods and softwoods.

Viburnum—thick undergrowth of witch hobble and striped maple. Good quality for yellow birch, sugar maple, and beech, and spruce-fir when present.

A major criticism of this method of classification is that a recent disturbance to a stand may alter the lesser vegetation; also, early successional stages are not represented in the classification. Such species as aspen, black cherry, tamarack, and white pine occur in a variety of types during early stages of secondary succession and the lesser vegetation is not always present at the onset of succession. It is not until the later stages of succession that the indicator species, as they are called, will appear in the understory. Missing from the classification of Ray are the indicator types that are indicative of dry site conditions.

Stanley (1938), working in an area of New Hampshire where sites were drier than those in Quebec and which supported subclimax stands of eastern white pine, recognized the following six types in two groups:

Group A Communities characteristic of the poorer soils of the region and indicating conditions too unfavorable to the production of well-stocked thrifty stands of white pine.
    I. *Cladonia - Andropogon* (beard grass) type
    II. *Vaccinium* (blueberry)—*Gaultheria* (wintergreen) type
Group B Communities characteristic of the better soils of the region, the indicating conditions which favor the production of well-stocked, thrifty stands of white pine.
    III. *Maianthemum* (false lily-of-the-valley) type
    IV. *Cornus - Lycopodium* (club moss) type
    V. Subtype of IV. *Aspidium* (wood fern)—*Dicksonia* (Hay scented fern)
    VI. Subtype of IV. *Mitchella* (partridgeberry) - *Pterii* (bracken fern)

Allen (1965) found that for eastern white pine two species, *Dennstaedtia punctilobula* (hay scented fern) and *Lycopodium complanatum* (club moss), accounted for 86% of the good sites (site index 73 feet or more); *Lycopodium complanatum* and *Viola* pp. accounted for 50–59% of the medium sites (site index 66–72 feet), and *Vacciunum augustifolium* with *Oryzopis asperifolia* (mountain rice), *Gaultheria procumbens*, and *Pteridium aguilinum* accounted for 45–55% of the poor sites (site index 65 feet or less). The indicator species represent those that most frequently occurred on the sites indicated. Allen said that, as a result of his study, the occurrence of an individual plant in a particular environment at a particular time is due to timing and chance rather than to a particular need of the individuals for a common environment. He stated that if enough elements of the environment that affect an organism are considered, then plant behavior can be described. The lack of an ability to define with precision plant associations concerned Allen as it has concerned other plant sociologists.

Students who pursue an interest in site type classification will discover the need to study in more detail the relationships between different forms of dominant and subordinate vegetation occurring in different regions. Research in this area is called plant sociology (phytosociology). Ecologists in the United States have developed neither the number nor the detail of the approaches to phytosociology as have Europeans. Perhaps this is because there is greater diversity of vegetation in the United States and therefore a greater number of plant groupings. In any event, the student who pursues the subject of phytosociology will want to study the different European approaches (see Becking, 1957 and Braun–Blanquet, 1951).

Part Three

By observing differences in growth between stands within a forest, it is not too difficult to distinguish sites of low productivity from those of highest productivity. Such qualitative judgments in some instances may be all that is required to measure site quality for some types of forest management planning. As management intensity increases, however, it becomes quite important that differences in site quality classes be measured with precision. At times it is possible to judge site quality by the growth of the trees on a site. More frequently stand density differences or the absence of desired species makes it difficult to measure site quality from the vegetation growing there. For these reasons it is sometimes necessary to be able to classify the productive potential of different sites by using their permanent features. The dis-

The Forest Site

cussion that follows will explain briefly how different site factors affect tree growth, and later there is a discussion of the ways these factors have been combined into a number of expressions of site quality for several species.

It is important to know the factors and their individual and combined effects upon tree growth. It is often necessary to determine whether one or more of them is limiting growth and if they can be modified through some silvicultural practice, or whether it is possible to improve the environment for one of the major forest tree species. By altering a limiting factor it may be possible to: increase tree growth; substitute one tree species for another and gain in timber production; substitute a genetically improved tree that may be resistant to a particular disease or insect; make conditions more suitable for wildlife habitat.

Up to this point in the discussion, the only site factors listed have been atmosphere, soil, physiographic location, and biota. These now need to be elaborated upon. An outline of the components of each factor is given below in the order in which they will be discussed.

Atmospheric factors (climate)
    Solar radiation
    Air movement
    Atmospheric moisture

Soil factors
    Water relations
    Physical properties
    Mineral cycling
    Biotic properties

Physiographic location
    Aspect
    Degree of slope
    Slope position
    Elevation

Biotic
    Competition between trees
    Effects of higher animals, insects, diseases, and man on growth
    Reaction of trees on site

In the introduction a statement was made to the effect that timber production for a particular stand is the result of the interaction of

---

genetic constitution and the physical and biological environment. The first expression used in that statement was for individual components of the environment. Now it is necessary to put the environmental components together into known relationships of growth. Keep in mind that in order to quantify site quality it is necessary to substitute variables into a factorial form equation for environment, where

Site quality = (climate) + (soil) + (physiography) + (biota)
+ (climate) (soil) + (climate) (physiography)
+ (climate) (biota) + (soil) (physiography) + (soil) (biota)
+ (physiography) (biota) + (climate) (soil) (physiography)
+ (climate) (physiography) (biota) + (soil) (physiography) (biota)
+ (climate) (soil) (physiography) (biota)

It is assumed in the foregoing statement that species composition and stand density and age have in some way been fixed between stands. Not all the factors listed in the equation will affect growth at the same time, but stand growth over the years will be influenced by each of the elements included among the four factors. The influence may be the effect of a single element or of several factors acting together or as the factors influence each other.

In Chapter 5 the concept "law of the minimum" was introduced to explain growth response of species to reduced light, and to explain relative tolerance. It is now necessary to expand the concept to include the range of environmental conditions that are experienced by forest trees growing in other situations. For example, trees grow better on soils that have the capacity to supply water in sufficient amounts to meet transpiration needs. If a soil is either excessively or poorly drained, tree growth may be affected because of a lack of water in the first case, and because of a lack of soil aeration in the second. Tree growth would increase for the excessively drained condition to the situation where moisture was adequate and then decline again on the very wet soils (Figure III-1a). In another situation an inadequate supply of minerals may limit growth (Figure III-1b). More generally, however, both water and mineral supply affect growth and they may interact (Figure III-1c).

The biological function of a plant is to take the constituents of growth available on a site and arrange them into organic compounds of varying form so that the compounds can be used to increase the size of various plant organs and to satisfy reproductive needs. A tree

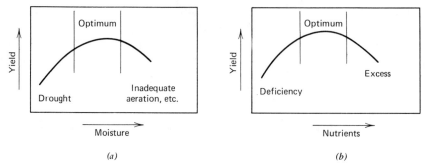

(a)   (b)

FIGURE III-1. (a) and (b) Effect of moisture and nutrient supply as single factors.

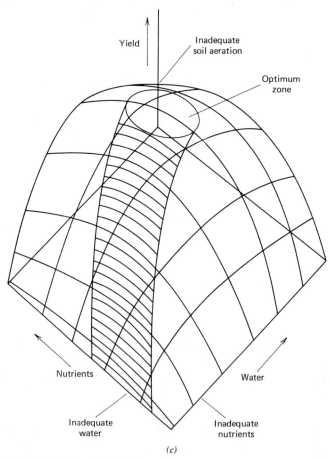

(c)

FIGURE III-1. (c) Effect of nutrient-moisture interactions on tree yields (Hansen, 1976).

is autotrophic, thus capable of fulfilling all of the requirements for growth and reproduction within a single structure, as contrasted with animals and plants, which are heterotrophic and must utilize some or all organic products that have been produced in some form by an autotrophic plant. Trees take carbon dioxide from the air, combine it with hydrogen obtained from soil water, and then synthesizes these into basic carbohydrates that, when combined with other soil mineral elements, can then be used for cell division and enlargement. Soil minerals (calcium, magnesium, iron, phosphorous, potassium, molybdenum, manganese, boron, zinc, copper, and nitrogen) are combined with the organic compounds made by a plant to produce amino acids for protein synthesis, chlorophyll molecules for storing energy in the photosynthetic process, and hormones that regulate growth and reproduction. Other organic compounds derived from the basic carbo-

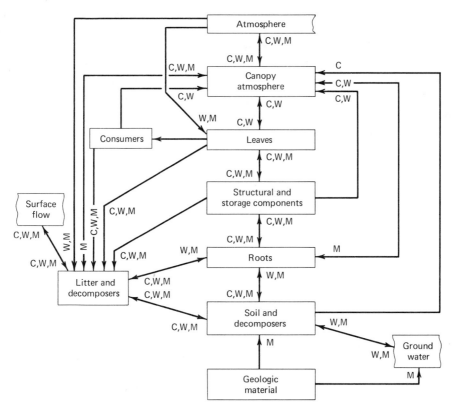

FIGURE III-2. Conceptual compartment and flow model for carbon (C), water (W), and elements (M) in a forest ecosystem (Henderson, 1974).

hydrates produced in photosynthesis are fats, starches, and cellulose which are essential to conservation of energy, cell division, and cell differentiation. Energy released in respiration as well as solar energy is used to carry out growth processes; in fact, starches and fats are materials that can be used by plants to conserve energy for future energy and food needs.

The various physiological responses of trees to the environment must be viewed as a series of cycles where there is a continuum between plant and environment. For example, the water cycle has been characterized as the soil-plant-air continuum. The carbon and mineral cycles are similar (Figure III-2). The trees are not only affected in their growth by the environment, but they in turn affect the environment. The total system is the forest ecosystem; however, in order to be able to understand the system it is necessary to examine separately the energy cycle, the hydrologic cycle, and the mineral cycle.

The processes just described are included in the study of plant physiology. For a more complete explanation of the physiology of trees the student is referred to texts on the subject. Foresters, however, should be able to translate the responses observed and explained by physiologists into the production function of growth response where varying combinations of light, temperature, moisture, and minerals affect tree growth and reproduction. For this reason, a brief discussion of the physiology of tree growth is included.

Chapter 7

The above-ground atmosphere of the forest includes diurnal and seasonal effects of solar radiation, atmospheric moisture, and air movement. These factors vary with region and from one physiographic location to another. Trees, during the one hundred million years or more
since they first appeared on the earth, have evolved species and races
that can take advantage of the different atmospheric conditions found
throughout the world. Tree species that grow in temperate zones have
developed a particular adaptation to the cold winter season that in-
terrupts growth each year. Species in subtropical climates have had
to adapt in some locations to dry periods that occur annually. Trees
have adapted to a variety of situations; however, the essential elements of growth—energy, water, and minerals—must be obtained

The Atmosphere

from a site in order for the trees to continue to exist. If any of these elements is lacking or not supplied in sufficient amount, then some species will be absent, and in their place will be some other tree species or some other form of vegetation.

PRIMARY ENERGY FLOW IN THE FOREST

The electromagnetic spectrum transmitted to the earth from the sun is composed primarily of shortwave radiation with wavelengths from 0.1 to 3.0 μ.* Between these limits is the visible spectrum that is used primarily by green plants for energy in photosynthesis with wavelengths from 0.4 to 0.7 μ. Most of the ultraviolet light in the solar beam with wavelengths 0.001–0.40 μ is filtered out by the ozone layer in the upper stratosphere; the remainder is absorbed by leaf epidermal cells. The ultraviolet wavelengths are harmful to living organisms because of mutagenic properties, and are therefore undesirable in the natural environment as it has evolved on our earth.

The chlorophyll pigments within green plant leaves absorb light within the visible spectra—90% in the violet-blue range (0.41–0.50 μ), secondary in the yellow-red range (0.58–0.7 μ)—and reflect about 75% of the light in the green-yellow range (0.51–0.59 μ) (Figure 7.1).

Radiation from the sun strikes the earth's atmosphere and surface at different angles at different times of the year resulting in different amounts of radiation being received at the surface. During the winter

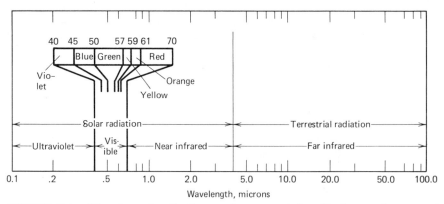

FIGURE 7.1. Wavelength of solar and terrestrial radiation and spectral bands (Reifsnyder and Lull, 1965).

*1 μ = one micron = 0.001 mm = 10^{-6} m = 10^3 nm.

season, the earth, although closer to the sun, has its northern axis pointed away from the sun at about a 23° angle (Figure 7.2). The result of this orientation is that there is less radiation received at northern latitudes during winter (Table 7.1; also see Table 7.5). As the solar year progresses, radiation increases in northern latitudes (with a corresponding decrease at southern latitudes) reaching a peak in intensity during the summer when day lengths and solar angle are at a maximum.

Incident radiation at the surface represents the net accumulation of total incoming radiation less losses due to reflection and absorption by clouds and dust in the atmosphere (representing a loss of up to 53%), and to the earth being a curved rather than a flat surface (a reduction of an additional 25%). So, if 2 ly (langleys) of incident radiation were to enter the earth's atmosphere, about 0.23 ly of photosynthetically active radiation (PsAR, radiant energy between 0.4–0.7 μ) would be received as direct radiation. However, reflected long wave (>0.7 μ) radiation increases light intensity at the surface (Table 7.2).

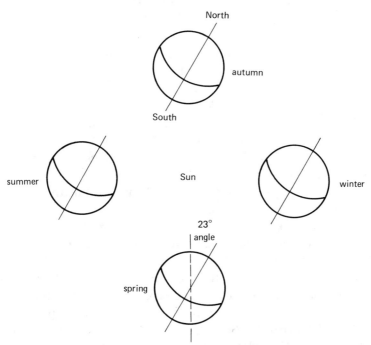

FIGURE 7.2. Relationship of the earth (northern hemisphere) to the sun during the winter season.

TABLE 7.1. Average Daily Solar Radiation for Different Regions of the United States (Adapted from Reifsnyder and Lull, 1965, p. 27)

	ALL-WAVE (ly/day)				
SEASON	NORTH-EAST	SOUTH-EAST	MID-WEST	NORTH-WEST	SOUTH-WEST
winter (Dec.–Feb.)[a]	158	225	216	167	300
spring (Mar.–May)	367	458	443	458	592
summer (June–Aug.)	500	533	566	600	667
autumn (Sept.–Nov.)	242	333	325	300	417

[a]There is a delay of about one month in cooling in winter from the winter solstice and warming in summer from the summer solstice with the result that the seasonal calendar is adjusted for this effect.

In Table 7.2, total all-wave radiation entering the system in spring was 1405 ly, and in the autumn it was 984 ly. These measurements are higher than the corresponding ones in Table 7.1 (compare values for the southeast where the Gay and Knoerr study area was located). These would be the result of measurements in Table 7.2 representing measurements during clear days when the sun was not obstructed by clouds, while those in Table 7.1 represent averages for different sky conditions. It can be expected that on a daily basis incident radiation would be lower than that indicated in Table 7.2.

Some authors suggest that the proportions of total incident radiation used in photosynthesis (PsAR) is about 0.40–0.45 times the all-wave radiation $(0.3–3\,\mu)$. In the Gay–Knoerr study the ratio of shortwave to all-wave was 0.46 = (652/1405) in spring and 0.40 = (390/984) in autumn. However, if only the amount of shortwave energy that is absorbed by the dominant canopy is considered, then in spring (582 + 51)/(1405) = 0.30 is available as PsAR and in autumn (340 + 14)/(984) = 0.36 is available as PsAR.

It is important to recognize that the photosynthesis process uses very little of the PsAR received. Hellmers and Bonner (1959) estimated trees utilize only 2.4% of PsAR. Zavitkovski (1976) indicated the Hellmers–Bonner estimate might be too low. He had found as a result of his work that red alder stands fixed 4.4% of PsAR and western hemlock 2.8%. Other authors report even different results.

Hickok and Hutnik (1966) reported a study of productivity of 43 year-old red pine planted at 6 × 6, 6 × 8, and 10 × 10 foot spacings. Productivity of main stem, branch wood, bark, and needles was meas-

TABLE 7.2. Mean Clear-day Radiation Budget in a 32-year-old Loblolly Pine Plantation (ly/cm/day)[a] (Gay and Knoerr, 1975)

	SPRING			AUTUMN		
	SHORT-WAVE	LONG-WAVE	ALL-WAVE	SHORT-WAVE	LONG-WAVE	ALL-WAVE
Above canopy						
Downward	652	753	1405	390	594	984
Upward	− 70	−960	−1030	− 50	−814	−864
Net	582	−207	375	340	−220	120
Below canopy						
Downward	208	855	1063	64	714	778
Upward	−51	−895	− 946	− 14	−754	−768
Net	157	− 40	117	50	− 40	10

[a]One langley (ly) = 1 calorie/cm² total incident radiation. Shortwave radiation can be estimated by multiplying all-wave by 0.45.

ured both on a dry-weight and energy transfer basis. Dry matter accumulation varied somewhat with spacing, but was not significantly different. About 46% of total production occurred on main stems; 54% was branch production. The same proportions held when the data for energy transfer were analyzed. On the basis of total solar radiation available at the earth's surface during a five month growing season, the red pine converted solar to chemical energy at a ratio of 0.8%; when root production was considered the accumulation reached 1%. When solar energy was computed as the amount of energy contained within PsAR, then the conversion rate was 2%.

Leak (1970) estimated that 120–160 year-old even-aged beech-maple stands utilized about 6.7 to 7.0 million kilogram-calories of net incoming solar energy per acre per growing season (100 days). He estimated net incoming solar energy for the growing season at 1700 million kilogram-calories per acre. Estimated energy utilization for wood, leaf, and seed production was less than ½% of net solar radiation (0.39–0.41%). Leak explains that his estimates are less than those of Hellmers and Bonner because they had based their estimates on visible solar radiation, which is roughly half of the net solar radiation.

Heat Transfer.

The greater part of incident and reflected light energy is transformed into heat energy. The pathways can be characterized as

$$Q_I + Q_M + Q_P + Q_S + Q_H + Q_E = 0$$

in which energy is directed into various segments of the ecosystem. In the statement above each element may be either plus or minus, that is, gaining in energy or losing energy. The elements are:

Q_I = total amount of incident radiation—large positive input.

Q_M = energy released as a result of metabolic activity (small about 1–2% total exchange)

Q_P = heat stored in the plant mass

Q_S = heat stored in the soil

Q_H = sensible heat exchange through conduction and convection

Q_E = latent heat exchange in evaporation—evapo-transpiration

These exchanges can be illustrated in terms of the amount of energy obtained and emitted by plant leaf surfaces:

$$\text{Net absorbed} = \text{Absorbed} - (\text{reflected} + \text{re-radiated} + \text{transmitted})$$
$$- \text{transpiration}$$

Gates (1965) showed that tree leaves modify the thermal environment through their effect upon energy flow as measured by different amounts of energy being dissipated through each pathway. For example, a leaf exposed to an energy source of 1.30 ly and ambient air temperature of 30°C (Gates, 1965) would attain surface temperatures as follows:

Re-radiation only—temperature of a leaf $= 30°C + 55°C = 85°C$
Re-radiation and free convection $\quad = 30°C + 28.8 = 58.8°C$
Re-radiation and
   forced convection (1 mph wind) $\quad = 30°C + 13°C = 43°C$
   forced convection (5 mph wind) $\quad = 30°C + \ \ 6°C = 36°C$
   Re-radiation, forced convection
   and transpiration (1 mph wind) $\quad = 30°C + \ \ 8°C = 38°C$
                  (5 mph wind) $\quad = 30°C + \ \ 4°C = 34°C$

Waggoner and Shaw (1952) showed that leaves perpendicular to the sun received and lost greater amounts of energy than leaves parallel to the sun:

DISPOSITION OF ENERGY	PERPENDICULAR (ly/min)	PARALLEL (ly/min)
Incoming radiation	+1.20	+0.3
Absorbed	−0.60	−0.15
Transpiration	−0.30	−0.075
Heat of leaf	−0.29	−0.072
Photosynthesis	−0.01	−0.003

These vectors are in line with those of Buamgartner (1956) who traced the energy flow in a spruce plantation. He determined that for a 586 ly/day two-thirds of the radiation was lost through evaporation/transpiration, one-third through convection, and less than 1%, about 5 ly was stored in trees and in the ground. He also found that during the course of one day there was nearly a complete turnover of the energy budget. Keep in mind that energy cannot be created or destroyed, but only transformed or directed in its flow.

It is quite apparent that solar energy utilized in photosynthesis represents the smaller portion of available energy; however, it should not be concluded that the larger portion is surplus and is not utilized. In the Waggoner-Shaw example, energy utilized in transpiration, 0.30 ly/min (perpendicular leaves), represents 25% of available energy. This energy is used, in part, to move water through a plant and is responsible for a portion of the energy required by plants to extract water from the soil. (This is discussed in more detail in the next chapter.) Twenty-four percent of available energy in the example, 0.29 ly/min, was absorbed by plant leaves and 50% by the plant biomass and the soil. It can be assumed that part of this energy, 74% of total available energy, was reradiated back into the atmosphere; but a part was used to maintain a suitable temperature level within and around the plants and the soil for metabolic functions. Most temperate zone species reach optimum metabolic levels near 29°C, and early morning air temperatures in the temperate zone are often below this level. Therefore, solar energy is needed to raise air and plant temperatures to a more suitable level for metabolic activity. Soil organisms respond in the same manner to increased temperature. Consequently, solar energy is responsible not only for photosynthetic needs, but for movement of water (and minerals) within the plant and for maintaining a suitable temperature for growth of plants and animals occupying a site.

Gay and Knoerr (1975) showed that the amount of energy entering a system varies with the season. In the spring (Table 7.2) 652–70 ly = 582 ly entered the dominant canopy; 582 − 208 + 51 = 425 ly were absorbed, or 65% = (425/652)(100) of the incident shortwave radiation. In autumn, 390–50 ly = 340 ly entered the dominant canopy; 340 − 64 + 14 = 290 ly were absorbed, or 74% = (290/390)(100) of the shortwave radiation. Differences here are small in terms of energy exchange per minute being 0.48 ly in the summer and 0.41 ly in autumn. The difference is a result of there being only 696 minutes (11.6 hours) of sunlight in the autumn compared with 878 minutes (14.6 hours) in the spring (See Table 7.5). As indicated only a small fraction of this shortwave energy is accumulated in the plant as biomass; however, as will be demonstrated in the next section, there is such a large leaf mass available that large accumulations of energy can occur.

Carbon Accumulation in the System
Energy flow in an ecosystem can be considered from the stand point of carbon production and consumption as it moves through the system

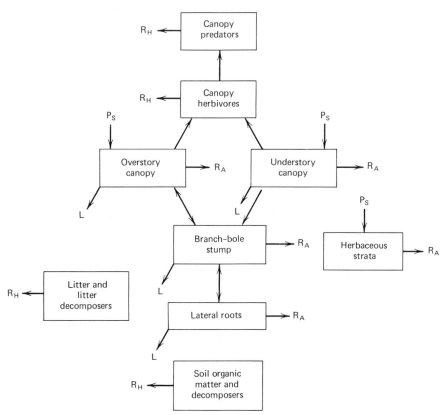

FIGURE 7.3. Diagram of conceptual model of organic matter/carbon storage and flow in a temperate forest ecosystem. Ps = net photosynthesis; R_A = autotrophic respiration; R_H = heterotrophic respiration; L = losses due to litter-fall or root sloughing. Litter and soil decomposers include both microbial and invertebrate organisms. Standing deadwood is included in the branch-bole-stump compartment. Double-headed arrows denote photosynthate translocation pathways (Harris et al., 1975).

(Figures 7.3 and 7.4). In this approach it is necessary to trace each step in the production-consumption process so that all of the system is monitored. The system (Figure 7.3) is represented by two major components: a productive component—the autotrophs—that includes trees, shrubs, and other green plants occupying the site, and a consumer component made up of herbivores that feed on the green plants and predators that feed on the herbivores. The system has two phases: photosynthesis and respiration. Photosynthesis (Ps) is confined to the autotrophic segment of the system, whereas respiration is carried on

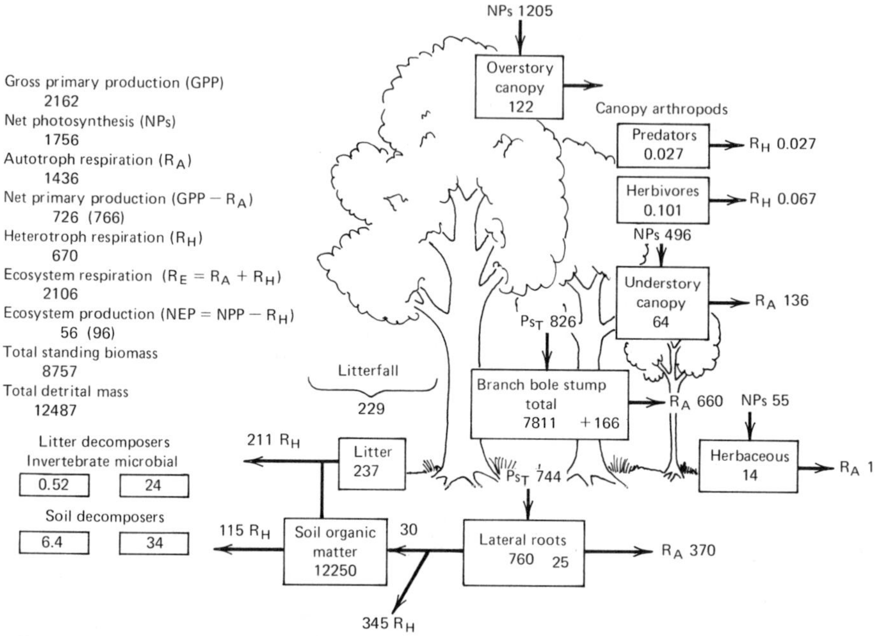

Gross primary production (GPP)
2162
Net photosynthesis (NPs)
1756
Autotroph respiration (RA)
1436
Net primary production (GPP − RA)
726 (766)
Heterotroph respiration (RH)
670
Ecosystem respiration (RE = RA + RH)
2106
Ecosystem production (NEP = NPP − RH)
56 (96)
Total standing biomass
8757
Total detrital mass
12487
NPs 1205
Overstory canopy 122
Canopy arthropods
Predators 0.027
RH 0.027
Herbivores 0.101
RH 0.067
NPs 496
Understory canopy 64
RA 136
PsT 826
Litterfall
229
Branch bole stump total 7811 +166
RA 660
NPs 55
Herbaceous 14
RA 1
Litter decomposers
Invertebrate microbial
0.52
24
211 RH
Litter 237
PsT 744
Soil decomposers
6.4
34
115 RH
Soil organic matter 12250
30
Lateral roots 760 25
RA 370
345 RH

FIGURE 7.4. Annual carbon cycle in a temperate deciduous forest. Major fluxes in the system are illustrated by arrows. Ps_T denotes translocated photosynthate. Mean annual standing crops are summarized in the center of each box; net annual increment is shown in the lower, right corner. Units of measure are g C m^{-2} yr^{-1} and g C m^{-2}. A summary of ecosystem metablism and standing crops are shown to the right of the figure. Values of NPP and NEP in parentheses are based on harvest/allometric methods; other values are based on gasometric analyses (Harris et al., 1975).

by both autotrophics (Ra) and heterotrophics (Rh). During the year some plant materials are lost and deposited as litter on the surface soil (L).

It is necessary at this point to introduce the use of several constants, and to point out that carbon flow through the ecosystem is, perhaps, a simple way to monitor energy exchange. The energy equivalents used here are averages for a number of species. In the illustration (Figure 7-4) energy is expressed in terms of grams of carbon per square meter of surface contained in different segments of the system. For each gram of original dry matter of plant biomass there are approximately 0.45 grams of C.* Each gram of dry matter has an energy equivalent of 18.5 K Joules, and, since 1 ly contains an energy equivalence of about 4.1869 K Joules (see Larcher, 1975), it takes 4.42 ly to produce one gram of dry matter and 0.45 grams of C.* If the gross primary production of carbon (Figure 7.4) is 2162 grams, there is contained in this mass approximately

$$\frac{(2162 \text{ g}) (4.42 \text{ ly})}{0.45 \text{ g}} = 21.23 \text{ K ly}$$

of energy. Because the translation of grams C contained in the dry matter to energy equivalents requires only the multiplication of a constant, it can be ignored. Only estimates of dry matter production and consumption need be considered to obtain an understanding of the relative amount of energy transferred but contained within a system.

Net production of the system = gross primary production − (autotrophic respiration + hetertrophic respiration). Using the data for Figure 7.4, 2162 − (1436 + 670) = 56 g C/m²/yr. In a balanced system it would be expected that the annual increase of carbon would be small unless the equilibrium was disrupted. There would not normally be a net increase or decrease in carbon accumulation in a balanced system.

In two other systems, net primary production was 8.3% for the *Liriodendron* forest and 10% for the *Quercus-Pinus* forest (Table 7.3). These amounts are relatively low compared with the total product of the system. In the two examples 2.11 Kg/m²/yr (24%) in the *Liriodendron* and 1.01 Kg/m²/yr (19%) was lost to respiration. These relationships are similar to that shown by Möller et al. for a beech system (Figure 7.5, *see also Figure 5.5*).

In Möller's system the net increment was greatest in the early life with production decreasing in later years. In forests used primarily for timber production, it is important to know that the standing crop of tree boles produce maximum growth biomass early in the life of the

---

*The proportion of carbon in sucrose ($C_6H_{12}O_6$), using the atomic weight of each element, is

$$\frac{6(12)}{6(12) + 12(1) + 6(16)} = 0.40$$

However, not all organic carbon is in the form of sucrose; some is combined in molecules with greater molecular weight. It is necessary to use a weighted proportion of carbon, in this case 0.45 grams carbon per gram dry weight of organic matter.

TABLE 7.3. Comparison of Metabolism and Structure of Two Terrestrial Ecosystems. Units of Measure are kg C m⁻² and kg C m⁻² yr⁻¹ for the Forests and Fluxes Unless Otherwise Noted (Harris et al., 1975).

PARAMETER	LIRIODENDRON FOREST	QUERCUS-PINUS FOREST
Total standing crop (TSC)	8.76 kg C m^{-2}	5.96 kg C m^{-2}
Net primary production (NPP)	0.73 kg C m^{-2} yr^{-1}	0.60 kg C m^{-2} yr^{-1}
Relative production (NPP/TSC)	8.3%	10%
Autotroph respiration (R_A)	1.44	0.68
R_A/TSC	0.16	0.11
Heterotroph respiration (R_H)	0.67	0.29
Ecosystem respiration ($R_E = R_A + R_H$)	2.11	1.01
Net ecosystem production ($NEP = NPP - R_H$)	0.06	0.28
Annual decay	0.70	0.36

stand, and that as the stand ages its relative biomass potential declines. This does not mean that wood value or aesthetic value also decline; actually, these may increase at a rate faster than the biomass potential. Nevertheless, young vigorous stands, because they are able to utilize the full potential of the site, are capable of producing max-

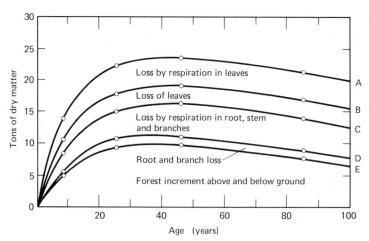

FIGURE 7.5. Dry matter losses due to various causes in European beech trees of varying age. The ordinate gives dry matter increment per hectare per year (from Möller et al., 1954).

imum annual growth up to the point where the less vigorous trees in the stand begin to die, before mortality detracts from the net biomass potential of the stand. In timber production it is important to balance the growth–death process and to attain a net production level that offsets high mortality as it occurs in older stands by reproducing stands before they become too old.

The residence time of carbon in different segments of a system varies with plant parts. Leaves, small roots, bark, and other material that are shed annually break down rapidly—about 1 to 2 years (Table 7.4). Woody stems live much longer and are reincorporated into the organic-mineral cycle only at longer intervals—156 years. Humus in the soil takes nearly as long as stemwood to breakdown. However, because of the large amount of fine litter that is produced each year compared to the amount of stemwood, the average cycle time for carbon in the system shown is at a 10-year interval.

CARBON DIOXIDE EXCHANGE IN LEAVES

Fixation of atmospheric carbon dioxide (CO_2) occurs during photosynthesis while release of CO_2 occurs during respiration. Autotrophic plants take CO_2 from the atmosphere, combine it with water from the soil using the energy stored in the chloroplasts, and produce carbohydrates. Before CO_2 can enter into this simply described but phys-

TABLE 7.4. Turnover Rates and Residence Times for Carbon in the (*Liriodendron*) Forest at Oak Ridge, Tennessee. Turnover Rates (yr^{-1}) Are Calculated from Flux (g C m^{-2} yr^{-1}) Divided by Compartment Size (g C m^{-2}) Harris et al., 1975).

COMPARTMENT	TURNOVER RATE[a] (yr^{-1})	MEAN RESIDENCE TIMES (yr)
Rapidly decomposable litterfall	0.89	1.1
Lateral roots[b]	0.49	2.0
Aboveground woody component	0.0064	156.0
Total wood component	0.049	20.0
Soil organic matter	0.00931	107.0
Forest ecosystem	0.10	10.0

[a]Turnover rate calculated from ratio of R_E (total carbon efflux) to total ecosystem carbon pool.
[b]Lateral roots are all roots except the central stump.

iologically complex process, it must first enter the protoplasm of the leaf mesophyll cells. The rate at which CO_2 enters a leaf determines the rate at which the photosynthetic process proceeds.

Assimilation of CO_2 by a leaf is determined by the concentration of the gas in the atmosphere surrounding a leaf. Normally this concentration is 0.03% of the atmospheric gases surrounding a leaf. Nitrogen makes up about 78% of the mass and oxygen about 21%; other gases and water vapor make up the balance. Concentration of CO_2 will vary daily, or even hourly, being highest usually in early morning when it may have increased during the night from respiration of plants and animals. In early morning CO_2 may reach a concentration of 0.04% in the air within a forest stand. During periods of photosynthetic activity the concentration may be reduced below 0.03%, and deficits in CO_2 can limit plant growth.

The rate at which CO_2 enters a leaf depends on diffusion resistances occurring at several points. These can be defined as follows:

$$A = \frac{\phi}{r_a + r_s + r_i + r_w + r_p + r_x}$$

where A = CO_2 assimilation in g/m²/sec
ϕ = g/m³ CO_2 at the leaf-air interface
r_a = boundary layer resistance (sec/cm)
r_s = stomatal resistance
r_i = intercellular resistance
r_w = cell wall resistance
r_p = protoplasm resistance
r_x = carboxylation resistance

There may be a CO_2 deficiency at the leaf-air interface (the boundary layer). The air may be still and the boundary layer depleted of CO_2. Turbulence at the interface as a result of air movement will result in a stable concentration of CO_2.

A most important source of resistance to diffusion of CO_2 into the interior of a leaf occurs at the stomates. If they are closed obviously no CO_2 can enter the leaf. This occurs during the night and during periods of soil moisture stress. Stomates respond to light during the day when the guard cells that control the stomatal aperture become turgid. When guard cells become turgid their shapes change so that a stoma enlarges. As the cells lose water and become flaccid a stoma closes. The presence of gibberellic acid and cytokin in guard cells promote opening; abscisic acid promotes closing. Humidity of the air surrounding a leaf determines the rate of transpiration, which, in turn, determines how much water is available to maintain guard cell turgor pressure.

Once CO_2 enters a leaf its rate of diffusion into the interior of the mesophyll cells of deciduous leaves, or chlorenchyma cells of conifers is determined by intercellular and cell wall resistance to diffusion. This resistance is associated with the rate at which CO_2 can be dissolved in water and transported across the cell wall and through the protoplasmic membrane. If autorespiration within the protoplasm produces CO_2 at a rate greater than the rate at which it can be fixed by photosynthesis, carboxylation occurs thus reducing diffusion of CO_2 into the mesophyll cells. Carboxylation is the result of excess CO_2 produced by respiration in mesophyll or chlorenchyma cells. The excess Rh CO_2 blocks diffusion of additional CO_2 into the cells.

Under natural conditions CO_2 can be fixed by photosynthesis at a rate of 15–25 mg per gram of leaf dry weight for deciduous leaves and 3–18 mg/g per dry weight in boreal conifer leaves (Larcher, 1975). Leaf respiration can account for 3–4 mg of CO_2 in deciduous leaves and approximately 1 mg in conifer needles. As a result, the net rate of CO_2 fixation is 12–21 mg in deciduous species and 3–17 mg in boreal conifers. The rate of fixation will vary depending upon the age of leaves, soil moisture, available light, air and leaf temperature, soil mineral supply, and genetic differences among individuals within a species.

PHOTOPERIOD

Spectral and thermal radiation are properties of solar radiation used by plants. We are prone to think of plant response to solar radiation only in terms of heat intensity or energy transfer. Solar radiation is the primary source of energy; however, that energy may be absorbed and utilized in several ways. Studies have shown that trees respond to differences in the photoperiod and the intensity of light as well as the quality of light.

Photoperiodic responses (the responses of plants to the length of daily light and dark periods) were first recognized in relation to flowering and fruiting of annual plants. It has since been shown that initiation of reproductive growth by trees in the spring may be related to the length of the light period. Apparently trees in temperate regions absorb radiant energy for some period of time before a sufficient change in chemical structure in the plant causes them to begin their annual period of reproductive growth. Trees of a particular provenance adapt to the photo-temperature regime of the locality.

Whether vegetative growth in the spring is also associated directly with photoperiod can be questioned. Further, increased photoperiod is directly related to increased temperatures, which may be the primary factor for resumption of growth in spring. However, within the balsam fir range, races from the southern extreme of the range when moved north tend to begin growth later in the spring and cease growth later in the fall than northern races that utilize the long light and warm temperature period during the June solstice (Table 7.5). Vegetative growth period of temperate species is associated with the increasing photoperiod, hence warm temperature periods, that occurs during the spring and summer seasons. Along parallels of latitude the photoperiod increases from winter solstice to summer solstice and then decreases again to the winter solstice. Summer in northern lat-

TABLE 7.5. Hours Between Sunrise and Sunset for 30°N, 40°N, and 50°N Latitudes for the Year 1975 (Determined from Tables in the Nautical Almanac, 1975, U.S. Govt. Printing Office)

DATE (1975)	LATITUDE 30°N	40°N (hours and tenths)	50°N	DATE (1975)	LATITUDE 30°N	40°N (hours and tenths)	50°N
Jan. 1	10.27	9.40	8.18	July 1	14.05	14.98	16.32
15	10.40	9.60	8.50	15	13.88	14.73	15.93
Feb. 1	10.80	10.15	9.30	Aug. 1	13.53	14.22	15.17
15	11.15	10.72	10.15	15	13.25	13.78	14.53
Mar. 1	11.58	11.35	11.05	Sept. 1	12.75	13.05	13.48
15	11.95	11.88	11.82	15	12.42	12.55	12.75
21	12.13	12.15	12.02	21	12.13	12.15	12.02
Apr. 1	12.50	12.67	12.93	Oct. 1	11.87	11.76	11.63
15	12.93	13.32	13.85	15	11.42	11.12	10.70
May 1	13.33	13.92	14.72	Nov. 1	11.02	10.52	9.83
15	13.68	14.42	15.48	15	10.65	9.98	9.05
June 1	13.95	14.82	16.07	Dec. 1	10.37	9.57	8.43
15	14.08	15.00	16.35	15	10.23	9.35	8.10
21	14.08	15.02	16.37	21	10.22	9.33	8.08
				Jan. 1, 1976	10.25	9.37	8.15

itudes exposes plants to high energy compared with the winter season when solar energy availability is much lower (Table 7.1). These differences are due not only to the shorter photoperiod in winter, but also to the lower angle of the sun that results in even less radiation reaching the earth's surface (Figure 7.2). Cessation of growth in autumn is associated with a decrease in photoperiod.

Gregory and Wilson (1968) reported that white spruce in Alaska of the same size and vigor as trees growing in Massachusetts produced during three growing seasons the same number of tracheid cells. The Alaska trees accomplished this in a 95 day growth period compared with 145 days in Massachusetts. Growth began (noted as time of first mitosis) after trees in each area had experienced 11 growing-degree-days based on days when mean temperature exceeded 6°C. The Alaska trees were able to accomplish their greater relative growth as a result of having a faster rate of tracheid production. The rates for each area where:

$$Y \text{ (Alaska)} = 0.02173(x) - 0.00002(x^2)$$
$$Y \text{ (Massachusetts)} = 0.01245(x) - 0.00003(x^2)$$

where Y = rate of tracheid production (cells per day)
x = tree vigor (annual rate of tracheid increment)

The reproductive response of trees to a varying photoperiod is not documented as well as it is for other plants, but one striking example does exist with witch-hazel. This species initiates flowering during the equinox of autumn and completes the cycle in early spring. It represents what appears to be a short-day species. American elm and red maple are probably examples of short-day plants since they both flower after the spring equinox; however, basswood, tulip tree, and perhaps locust can be cited as possible examples of long-day species. Although it should be noted that flowers of these species develop on newly formed growth rather than from separately formed buds as with early flowering species. Pines flower in the intermediate period between January and June and so do not appear to favor either short or long days.

LIGHT INTENSITY

Light intensity is important to tree growth because relatively slow growth of all species is associated with reduced light. However, most species appear to make maximum growth at light intensities less than full sun (Figure 7.6). Few species can survive when light intensity is less than 1% of full sun (approx. 100 f.c.). Under controlled conditions where moisture and temperature are maintained at constant levels, the growth curve of a seedling of one tree species shows that a maximum is attained rapidly, as with Species A. Other growth patterns are also possible. Species A in Figure 7.6 could represent an intolerant species, while Species B would represent a relatively more tolerant one. Note that Douglas-fir, intermediate in tolerance, has a compensation point (photosynthesis = respiration) at 100 ft/candles and maximum photosynthesis occurs at about 2400 ft/candles (Figure 7.7).

AIR TEMPERATURE

Growth responses to variations in air temperature are more completely documented than are the spectral and net radiational light relationships because temperature is an easily recognized energy relationship. Also, plant growth is not solely related to the current incident solar energy level since energy is continually released within the plant as a result of the respiration process, although in this last case temperature increase is not always associated with the exchange of energy, and the amounts of energy involved are quite small. In addition, photosynthetic responses are also related to air and plant temperature as are the more basic responses of cell division and cell elongation. All enzyme controlled reactions in plants are affected by temperature change.

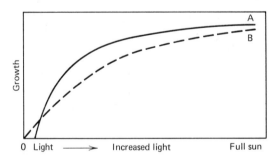

FIGURE 7.6. Relationship between light intensity and growth of two hypothetical species, Species A an intolerant species and Species B a tolerant species.

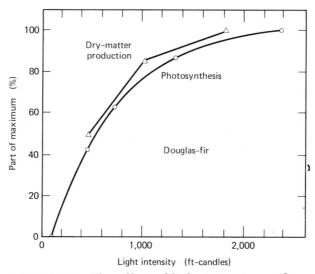

FIGURE 7.7. The effect of light intensity on dry-matter production and on rate of net photosynthesis of Douglas-fir seedlings at 18°C, expressed as the percentage of maximum observed (Brix, 1967).

Tree growth occurs when there is a net accumulation of photosynthate. If by reason of too high or too low temperature no accumulation of photosynthate occurs, then growth as represented by net increase in size does not occur (Figure 7.8). However, cell division can take place without an increase in dry weight. Growth here is represented as a net accumulation of dry matter. In the illustration (Figure 7.8) net growth occurs between 32° and 120°F. It is important then to recognize that dry weight accumulation varies as a result of differences in net photosynthetic activity. In Figure 7.8 net photosynthesis is maximum between 70° and 90°F and decreases with changes in temperature above and below this range. At 120°F photosynthesis and respiration are equal and above 120°F respiration exceeds photosynthesis and "growth" is negative; more photosynthate is used in respiration and, therefore, is not available for cell division and cell differentiation.

Difference in photosynthesis also results when light intensity is increased but temperature is maintained at a constant level (Figure 7.9). From this illustration it could be concluded that a continuous day-night temperature of 70°F would be most effective of those used.

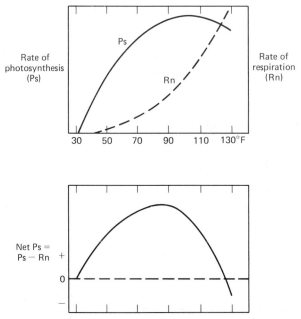

FIGURE 7.8. Response of photosynthesis and respiration to increase in temperature. Net photosynthesis is illustrated in the lower diagram and was derived as the difference between the Pn and Rn curves in the upper diagram.

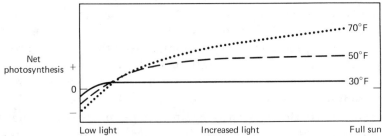

FIGURE 7.9. Response of net photosynthesis to increased light intensity for a range of constant temperatures (from Muller, 1928).

However, studies by Kramer (1957) and Hellmers and Sundahe (1959) of thermoperiodicity have shown that a continuous high temperature does not result in maximum growth of loblolly pine, Douglas-fir, and redwood. Results of the Hellmers and Sundahe study are illustrated in Figure 7.10. These studies show that with Douglas-fir, in particular, a warm day temperature should be followed by 10°C reduction of night temperature to obtain best growth; with redwood a 6°C reduction in night temperature was as effective as a 16°C reduction.

The fact that forest tree seed can be moved from south to north within the range of species with the expectation that the trees will put on more growth than local trees of the species, can be attributable in part to the fact that the southern trees, when moved, accumulate more photosynthate during the day and conserve larger amounts than the northern trees as a result of cooler nights. When seed is moved from north to south, trees of a number of species do not show an apparent benefit. Photoperiod is no doubt involved here. Although southern races flush later in the spring, they continue growth longer into the autumn. However, when a southern source is entirely un-adapted to a more northern location, the trees die because of cold damage in the winter or from early autumn frosts.

Air temperature of a site depends not only upon latitudinal location (Figure 7.11), but also upon elevation and whether the site is on a north facing or south facing slope. Temperature decreases at a rate of approximately 3°F per thousand feet increase in elevation (1°C per 100 m). Such a decrease in temperature is equivalent to a change of

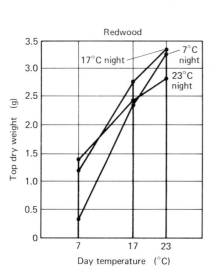
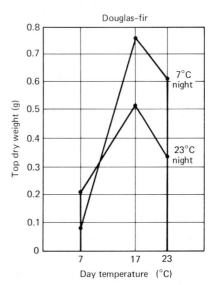

FIGURE 7.10. Growth response to tree seedlings to diurnal differences in temperature (Hellmers and Sundahe, 1959).

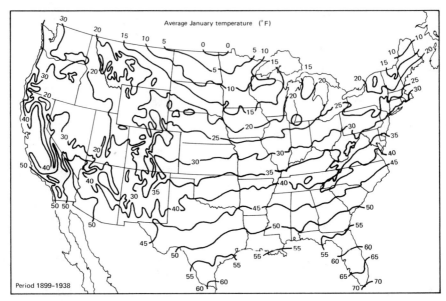

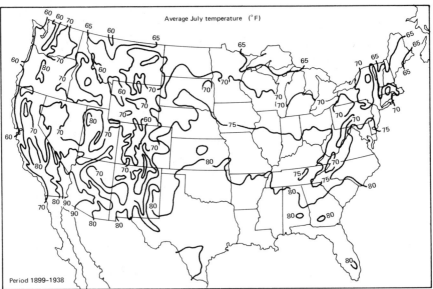

FIGURE 7.11. Average air temperature in the United States for the months of January and July.

approximately 800 miles in latitude. Assuming that latitude change is from south to north, the result is an average annual decrease in mean daily temperature of 3°F. The Rocky Mountains provide an excellent example of how vegetation changes with elevation and the resulting change in temperature and aspect (Figure 7.12). In this illustration vegetation at high elevations is adapted to an arctic climate, and as a result cannot grow as well on the warmer south slopes. Vegetation at the lower elevations is adapted to near desert climate and as a result grows better on warm south slopes, but cannot grow to as high an elevation on north slopes. The mid-elevation vegetation (spruce, fir, and pine) is adapted to temperate climate and therefore occupies a position between desert and arctic vegetation.

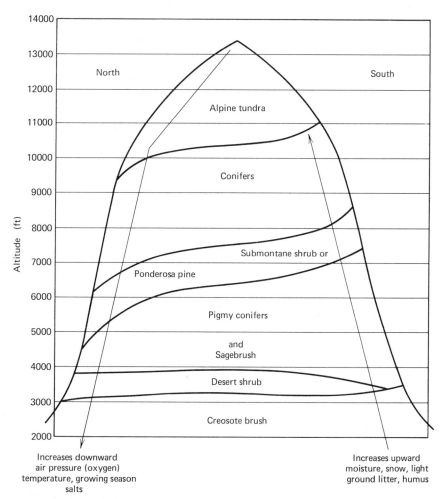

FIGURE 7.12. Vegetation zones in the Rocky Mountains in Utah and northwestern Arizona (from Woodbury, 1947).

TRANSPIRATION

As air temperature increases, the capacity of the air to hold water increases. This is a geometric progression (Table 7.6). The rate of change in vapor pressure for each 18°F change in temperature is roughly equivalent to a doubling in the vapor pressure with increase in temperature, or halving of vapor pressure when temperature decreases (Q_{10} approx. $2x$). It can be expected then that for each 18°F (10°C) increase in temperature, transpiration potential of the air will nearly double, provided that soil moisture is near field capacity, air movement around the leaves is sufficient to maintain a moisture gradient, and leaf temperatures do not increase much above air temperature (for a more detailed discussion of transpiration, see Kramer, 1969).

During the day, as transpiration drain increases the demand for water in leaf and stem tissues, it can be expected there will be a delay in replacing water lost from the leaves because conduction of water within the tree will lag behind the water lost through the stomates, and roots fall behind in absorbing water from the soil. As a result, leaf stomates close partly or completely and the vapor pressure gradient between leaves and the surrounding air will diminish. Hodges (1967) found that the pattern of photosynthesis in several tree species was controlled by leaf water potential, which reacts mainly to stomatal movement and to the resistance of the mesophyll cells to absorb CO_2.

TABLE 7.6. Saturation Vapor Pressure for Increasing Air Temperatures.

TEMPERATURE		SATURATION VAPOR PRESSURE
°F	°C	(mm Hg)
32	0	4.58
41	5	6.54
50	10	9.21
59	15	12.79
68	20	17.54
77	25	23.76
86	30	31.82
95	35	42.18
104	40	55.32
113	45	71.88
122	50	92.51

Daily variation in leaf water potential occurred primarily in response to exchange in atmospheric moisture, primarily to vapor pressure gradient. When excessive water loss occurs, water loss by transpiration diminishes as a result of a lag in water absorption by the roots. In fact, during drought periods it is possible that no water is lost through the stomates because there may not be sufficient water to maintain the turgor pressure in the stomate guard cells thus closing off the stomatal opening. This reaction can be viewed as an adaptive mechanism preventing plants from losing water essential to their survival.

Transpiration demand in cool climates will be less than in warm ones because the vapor pressure gradient will be less and transpiration water loss reduced. Stated another way, if a locality experiences a large number of warm, dry days, a high rainfall is needed to compensate for the higher transpiration losses as compared to localities having lower temperatures. If a high temperature region does not also have a relatively high precipitation, then tree growth will be diminished.

EFFECTS OF TEMPERATURE EXTREMES

Climatic-induced abnormalities in trees and tree growth are termed noninfectious diseases because they are not pathogens, although injury may expose a plant to attack by a pathogen. Noninfectious diseases are the result of high or low temperature, drought, wind, lightning, snow, ice, hail, and air pollutants. Pathogens that cause injury are viruses, bacteria, fungi, and higher plants; these will be discussed in Chapter 11. Tree response to temperature extremes are considered here; response to other climatic factors will be treated later in this chapter.

Low Temperatures

Winter injury is common among conifer species growing in climates where soils freeze and daytime temperatures frequently rise above freezing. The result is a type of foliar dessication. The needles lose water during the day, and are unable to replenish it because soil water is frozen. If a tree is exposed to this condition for too long a period, it will be killed or severely defoliated. Winter injury can be classed as a form of drought because trees may die from a lack of water; however, the primary cause is low soil temperature, and not because soil water is in low supply as occurs in summer months. If the soil were not frozen, then injury might not occur, or if it did it would be slight.

Winter injury, also termed "winter killing" or "winter drying," is common in the mountainous regions of the western United States after warm weather with a "chinook" wind follows a period of freezing weather. Boyce (1961), citing the reports of others, indicated that ponderosa pine, lodgepole pine, Douglas-fir, and blue spruce are frequently injured. In the east, eastern white pine will show the effects of winter injury. Damage is particularly acute where cold weather precedes snow accumulation. Once a snow pack has formed, the occurrence of winter injury diminishes. Where damage is extensive so that foliage of whole stands turn color, the terms "red belt" or "parch blight" are used to characterize the condition (Boyce, 1961). In some instances entire trees are killed; in others only the foliage is killed and new growth develops from undamaged buds.

Frost cracks are the result of unequal shrinkage of the outer and inner portions of the bole. As air temperature drops, the outer wood shrinks faster than the inner wood, and if the elastic limit of the wood is exceeded a rupture can occur. This can be either a radial or tangential tearing of the wood. Fergus (1956) reported that oaks are damaged as a result of frost cracks. If the injury occurs several years in succession, a protruding callus growth ("frost rib") can develop, particularly in hardwoods (Boyce, 1961). This injury may appear as a tangential break in the outer bark, or it may not be visible externally. Boyce (1961), summarizing the work of several workers, indicated that, in addition to oak, species affected by frost crack are lodgepole pine, red and white fir, Japanese larch, the northern hardwoods (beech, birch, maple), and yellow poplar.

Sunscald occurs during the winter, but can also occur occasionally in the summer. Particularly susceptible are thin-barked young trees of aspen, sugar maple, and eastern white pine (Boyce, 1961). During the winter the bark on a tree whose bole is exposed to direct sunlight on the south side may undergo extreme changes in temperature. Consider a tree growing in a climate where night air temperature reaches −20°F and +40°F the following day. During the night the bark and phloem of the tree attain an equilibrium temperature with the air (−20°F), but during the day because of its dark color and because it receives direct radiation since it is in an upright position, the bark may reach a temperature in excess of 80°–90°F. Bark and phloem of the tree would have to undergo a change of 100°F in temperature within a period of a few hours. Such extreme changes may cause damage to the phloem and in some cases to the cambium as well, and stem lesions often result exposing the inner wood of a tree to attack by insects and disease. If not so affected a stem will probably exhibit abnormal growth in the damaged area. During the summer, broad-leafed plants in the understory afford some shade to tree boles. Then too, temperature changes are not so extreme, and so it can be expected that sunscald may not be as frequent as during the winter. In summer, sunscald could occur where bole temperatures excessed 120°F, the temperature at which living plant tissue is injured by heat.

Sunscald, although the result of high temperatures primarily occurring in the winter, is probably most severe during periods of rapid diurnal fluctuation of stem temperatures rather than to high temperature alone (Huberman, 1943; Daubenmire, 1959).

Frost heaving occurs because tree seedlings are pushed out of the ground as a result of the soil water expanding on freezing and later retracting on thawing (Heidmann, 1976). First, when the soil freezes the tree is thrust up with a thin layer of soil as the soil water expands, and possibly some roots which extend below the frost line are broken (Figure 7.13). Later when the soil thaws, the mass separates into small aggregates that return to the previous level, leaving the tree with its roots exposed to the air where they dry. As a result the tree dies. Fine-textured soils, silt, and clay loams are most prone to frost

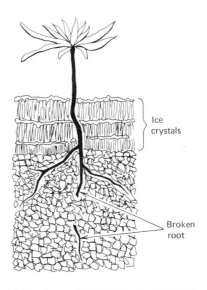

FIGURE 7.13. Frost heaving of soil due to the formation of ice crystals beneath the surface layer (from *Principles of Silviculture*, F. Baker, McGraw-Hill, New York).

Ice crystals

Broken root

heaving. Fall planting on fine-textured soils can result in almost 100% loss of planted trees.

PHYSIOLOGICAL CHANGE IN TREES IN AUTUMN

In the autumn as the daylight period shortens, the physical and physiological condition of the active cells within a tree change. When cell division occurs the cells laid down by the cambium are smaller than during the mid-summer period. Viscosity of the cytoplasm of active cells decreases and their permeability to water increases. Osmotic pressure and sugar supply increases and there is a reduction of cell water content. It is extremely important to a cell's ability to resist frost damage that it is able to exchange water rapidly with the intercellular spaces. If a cell cannot undergo rapid hydration and dehydration, the cellular membranes can rupture and the cell will die. Apparently the enzymes that transform sugars remain active to very low temperatures although their activity decreases with a decrease in temperature.

Conifer species adapted to cold climates can carry on photosynthesis when they have been exposed to night temperatures below freezing. Douglas-fir and ponderosa pine exposed to night temperatures of $-3°$ to $-4°C$ had photosynthetic rates of 55% in Douglas-fir and 75% in ponderosa as compared to other trees that were held at night temperatures above freezing (Pharis and Hellmers, 1964). Trees exposed to five days of sub-freezing night temperatures decreased their photosynthesis rate to a stable level and no further reduction was noted. Zimmerman (1964) found that sap flow in trees is interrupted when temperature of the stem tissues reaches $-1°$ to $-2°C$ and that ice formed in the tissues at these temperatures.

The annual cycle of growth of temperate region trees is interrupted by the cold winter period. Trees in late summer and early autumn reduce growth and set buds. Upon close examination the buds are really compressed stems upon which are leaf initials, flower initials, and branch initials. After a suitable cooling period these buds will expand and grow, exposing the new terminal growth. It can be expected that even during cold winter days some cell division may occur within the buds when temperatures are near but above freezing after a sufficient cooling period has occurred, but before bud break in spring.

LATE AND EARLY FROST

Late frost injury occurs in the spring, early in the growing season. It is the result of a period of cold weather that kills back the new growth of trees. Particularly susceptible are ash, beech, black locust, sassafras, walnut, sycamore, white oak, yellow poplar, horsechestnut, Norway spruce, red and slash pine, balsam fir, Douglas-fir, and Sitka spruce. Conifers in the southwestern United States are subject to late frost injury when growing in locations where there is excessive radiational cooling, as occurs in mountain valleys, especially "frost pockets." In these situations seedling stands are particularly susceptible to damage (Hough, 1945; Byram, 1948). In fact, frost damage may occur at such frequent intervals in these locations as to prevent the establishment of new stands after clearcutting, which is not recommended in "frost pocket" locations (see Figures 7.20 and 7.21). Campbell (1955) reported that in 1955 spring frost injury to shortleaf pine flowers in Louisiana was so extensive as to preclude a seed crop of that species in 1956. Reports of seed crop failure are not uncommon after late frost.

Boyce (1961), reporting on the work of others, indicated that cambium injury from late spring frost occurs in Douglas-fir and Sitka spruce. In these instances the entire cambium may be killed, which eventually causes the tree to die, or only part of the cambium may die, in which case a scar may appear exposing the tree to fungus attack.

Reports of early frost injury are not so extensive as those of late frost. Early frost injury occurs near the end of the growing season before new growth and winter buds have had a chance to harden-off to prevent cold injury. Boyce (1961) reports that some conifers, black locust, basswood, sycamore, and elm have been reported as being injured by early frost. In instances where southern races of a species are moved too far north early frost injury will occur.

HIGH TEMPERATURES

Stem lesions and heat cankers occur on exposed seedlings and young trees as a result of excessively high temperatures. Such damage is particularly noticeable to seedlings growing in nursery seedbeds where they are exposed to direct solar radiation (Boyce, 1961; Smith, 1970). Damage to small seedlings can be severe enough to girdle and kill them. On exposed natural seedbeds the stems of young seedlings may not reach excessively high temperatures as a result of direct radiation, but the soil around them may reach lethal temperatures, and as a result the trees are killed. The temperature at which growing cells are killed and the protoplasm destroyed varies in intensity and duration and with the type of cellular material involved.

Jack pine cones can be exposed to a temperature of 700°F for as long as three minutes before the enclosed seeds are damaged, while roots of this species are killed in five minutes when exposed to temperatures of 129°F; however, a two-hour exposure at 118°F is required to kill the roots (Beaufait, 1970). Needles may be killed in only three seconds when exposed to 147°F temperatures, but at 126°F an 11-minute exposure is required to cause death. Short exposures to high temperatures will deactivate the enzyme systems in a plant; but a longer exposure to a lower temperature may be just as effective.

High air temperature may be a contributing cause of death rather than a direct cause. Periods of extreme high temperature cause plants to be "starved" to death because of the high respiration rate resulting in a negative net photosynthetic rate. High winds and high temperature cause excessive transpiration with the result that trees die from lack of water.

AIR MOVEMENT

Air movement is important to tree growth and the distribution of forests in a number of ways. It is essential to plants because it stabilizes the carbon dioxide and oxygen environment, preventing a saturation or a deficit of either of these gases from occurring. Air movement also dissipates heat radiated from the plant surface and tends to equalize temperatures and vapor pressure over large areas so that the moisture in an air mass is more uniformly distributed. Large amounts of water vapor are transported to the land masses of the world from the oceans on streams of air. Air movement is important to the distribution of tree seeds, and wind can influence the form of trees. Forest fires are greatly influenced by wind.

If there was no diurnal heating and cooling of the earth's surface, the atmospheric pressure around the earth would decrease at a rate of 0.1 inch in pressure for every 100 feet increase in elevation; from sea level to 100 feet elevation the pressure would decrease from 29.9 to 29.8 inches of mercury. Because of the continual change in the receipt of energy as a result of seasonal and diurnal changes in heating of the atmosphere and the earth's surface, pressure areas develop resulting in regional pressure gradients. Deflection of air movement occurs because of the earth's rotation and the friction from the earth's surface as the air moves across it. This is known as Coriolis force. Above 2000 feet, loss in velocity to friction is negligible.

Air Mass Movement.

If a mass of air settles as a result of its being cooled, a high pressure anticyclonic cell develops and the air movement around the center of pressure will be clockwise and away from the center as a result of Coriolis force (Figure 7.14a). When there is change in pressure caused from convective air movement as a result of heating, a low pressure cell or cyclonic cell develops. Air movement around a low cell will be counterclockwise, and air moves into the center (Figure 7.14b). In the northern hemisphere large low and high pressure systems move from west to east as a result of vector forces from the earth's rotation and the movement of air within the system.

In the northern hemisphere high altitude jet-stream wind directions are predominantly from the west because of the lack of friction and because of the earth's rotation. In reality, the earth rotates within an outer covering of air that does not rotate at the same speed as the earth. The maximum velocity of these jet-stream winds occur near 40,000 feet and tend to be greatest along the 30°N parallel (Figure 7.15). The jet stream winds do not blow in a uniform westerly direction but shift their direction in southerly and northerly waves as a result of regional high and low pressure systems. Although there is a general poleward movement of the jet-stream during the winter

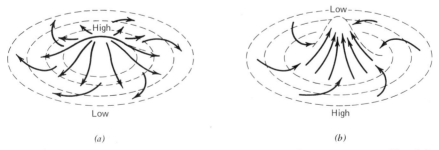

FIGURE 7.14. Air circulation around high and low pressure cells. (a) Anticyclonic sinking. (b) Cyclonic lifting.

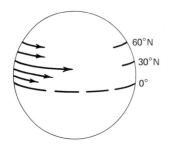

FIGURE 7.15. High altitude wind direction and relative velocity in the northern hemisphere.

60°N
30°N
0°

(35°–25°N), in the summer the jet stream is closer to the equator (20°–25°N). The north–south movement will tend to modify the movement of large polar and maritime high and low pressure systems that affect the climate of North America.

Because there is less heating of the polar region as compared with the equatorial region, there is generalized high pressure air mass movement from north to south, and a low pressure warm air mass movement from south to north. Jet-stream movement interferes with high and low pressure cell movement. Also, differences in heating of land mass as compared with the ocean masses, and the difference in frictional loss between land and ocean air, cause movement patterns of pressure systems to vary. As a result, there are several generalized patterns of air movement around the earth (Figure 7.16).

Wind Velocity
The velocity at which air moves depends upon the steepness of the pressure gradient in a high or low pressure system. The steeper the pressure gradient the greater the wind velocity. This can be thought of as analogous to water moving down an incline; as the angle of inclination doubles, velocity of water will nearly triple. There are, however, generally prevailing low velocity winds in a particular locality. The winds tend to be westerly in the mid-latitutdes of the northern hemisphere, and the velocity is greater in winter than in summer. Upon approaching the equator wind direction shifts to the east. Surface winds average between 2–4 mph in winter and diminish to 1–2 mph during summer. During passage of a cold front (an incursion of cold arctic air) or the occurrence of air mass thundershowers, surface wind speeds may increase to 20–30 mph. Hurricanes are accompanied by winds of 75 mph or higher, and tornadoes have wind velocities of much higher magnitude.

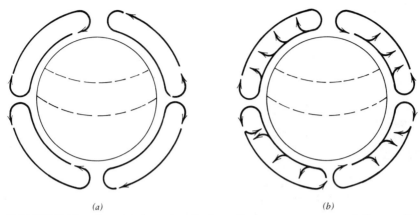

(a)　　　　　　　　　　　　　　　　*(b)*

FIGURE 7.16. Convective circulation of air masses around the earth. (*a*) Ideal without rotation forces or ocean-land difference. (*b*) Generalized movement of air from north to south in northern hemisphere and the reverse in the southern hemisphere.

Effects of Wind at High Elevation

With increase in elevation, friction from the earth's surface diminishes and wind velocity increases. At high elevations vegetation is exposed to a continuous pressure from high velocity winds. Wind velocities of 125 mph are observed fairly often on the top of Mt. Washington (located in the White Mountains of New Hampshire; 6000 feet elevation). In fact, the highest wind speed ever recorded on this mountain was measured at over 200 mph. Wind pressure tends to stunt trees and other high altitude vegetation (Figures 7.17a and 7.17b). Along major mountain passes where the wind is funneled and velocities increased, tree crowns are shaped like pennants streamed in the wind (Figure 7.18).

High elevations bring not only high wind velocities, but also cold air temperatures and large amounts of snow. These combine to create a harsh environment for all except the hardiest of vegetation. Adaptation of alpine vegetation is in the form of small-leaved, dwarf plants. In fact, the alpine environment has been classed as a cold climate desert because of the extreme, relatively dry conditions that prevail in these regions. Tree vegetation at higher elevations, because it is continually damaged by ice, snow, and wind, has an appearance termed *Krumholz* (broken trees).

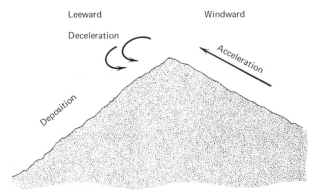

Leeward   Windward

Deceleration

Acceleration

Deposition

FIGURE 7.17. (*a*) Air as it is lifted against a wind-ward slope, its velocity accelerates as it approaches a leeward slope its velocity decelerates. Snow accumulates to greatest depths on lee slopes because of wind action (from Parland and Martinelli, 1976).

FIGURE 7.17. (*b*) Limber pine at upper timberline on Long's Peak, Colorado. Wind pressure has caused the tree to develop in a prostrate position from the beginning. Blasting ice particles have eroded the bark from a large area at the base of the trunk (from Daubenmire, 1947).

Seasonal and Diurnal Winds

In the summer during daylight hours land masses accumulate heat faster than water bodies because water has a higher specific heat than

FIGURE 7.18. Wind-trained ponderosa pine and Oregon white oak along the Columbia River gorge in Oregon. In *Pinus* strong and persistent wind pressure has bent the branches on the windward side around until they point permanently in a leeward direction. Toward the base of the canopy secondary growth has deeply buried the curved bases of these branches (from Daubenmire, 1947).

soil and vegetation. As a result, during daylight land masses have lower air pressure than water bodies and there is a movement of air from the water to the land. At night the air moves from land to water. During the winter, diurnal movement of air is reversed because the land mass is generally cooler than the more temperature stable water mass. The velocity of winds that accompany these air movements usually does not exceed 10–15 mph.

Major winds with velocities of hurricane force (greater than 75 mph) occur in connection with deep low pressure systems, hurricanes, or tornadoes (Figure 7.19). High velocity winds may occur in connection with high pressure systems, but not as frequently as with low

FIGURE 7.19. Breakage and uprooting by the 1938 hurricane, on Mt. Washington, New Hampshire. During the hurricane wind velocity exceeded 160 mph on the summit of this mountain (from Daubenmire, 1947).

pressure systems, nor will they usually attain as high a velocity. Daily air movement occurs mostly as a result of differential heating of local land and water masses that produce localized high and low pressure systems. Along coastal areas—either lakes or oceans—air will tend to move onto the land during the day and off of the land during the night.

In mountain terrain, during the day air will move up the valleys as a result of convective flow of air and during the night air will flow down the valleys as a result of radiational cooling at low elevations, when the influence of a larger air mass pressure system does not interrupt the air flow (Figure 7.20). Also, at night in hilly terrain a warm blanket of air may form above the valley floor preventing further upward movement (Figure 7.21). This inversion of hot air above cool air results in "smog" where there is industrial and automobile air pollution.

If an upper air mass subsides and it contacts warmer air its pressure will increase. As the air becomes warmer its vapor pressure deficit will also increase and it becomes both dry and warm. In the western United States air mass subsidence is the cause of the *Chinooks*

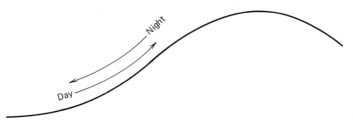

FIGURE 7.20. Diurnal movement of air along a ridge and valley.

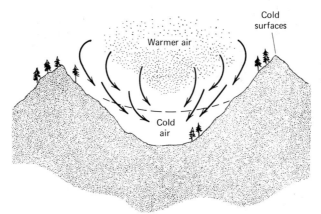

FIGURE 7.21. Temperature inversion in a mountain canyon.

(or foehn) in Montana, Wyoming, and Colorado (Figure 7.22); the *Santa Ana* winds in southern California; the *Mono* winds of central California; and the strong east winds of Washington and Oregon. There are instances where a single large subsiding air mass over the western plateau region of Utah and Nevada that was accompanied by low pressures on the opposite sides of the mountains resulted in a Santa Ana in southern California, a Mono in central California, and a Chinook in Montana, all blowing at the same time (Brown and Davis, 1973). These extremely dry winds are capable of driving wild fires at great speed both up and down hill, and there is a little change in burning conditions between day and night. Usually, at night, winds will diminish accompanied by increased humidity so burning conditions are moderated, but this does not occur during a Chinook. Also, trees experience extreme moisture stress when exposed to these hot dry winds because they accelerate transpiration rates.

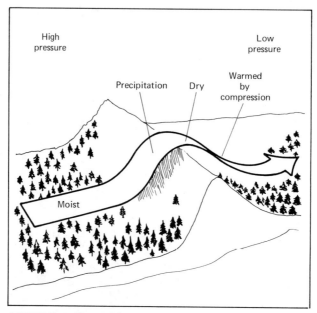

FIGURE 7.22. Mechanism of chinook winds. Although precipitation is not required, it greatly accentuates the chinook effect (from Perla and Martinelli, 1976).

AIR POLLUTION

In the geological past, it is believed that there was no free oxygen in the earth's atmosphere. Not until green plants evolved and could transform CO_2 and H_2O into carbohydrates with the evolution of O_2 did oxygen begin to occur in the atmosphere. As human populations began to use large amounts of fossil fuels, the amount of oxygen in the atmosphere decreased and the CO_2 content increased. Some ecologists feel that the change in amounts of these gases in the air poses a threat to the human environment. As CO_2 increases in the atmosphere, infrared radiation is prevented from escaping, which results in an increase in air temperature. However, as air temperature increases the atmosphere can hold more water. This will tend to reduce the amount of incoming radiation, with a resulting decrease in air and land temperature. Whether the two effects will balance, or whether accumulation of CO_2 will cause air temperature to continue to increase, is not certain. Also, the oceans can dissolve large amounts of CO_2 and green plants may be able to fix increased amounts, so there is a chance that no real problem exists; but there is reason for concern that increased CO_2 in the atmosphere may have a long-term effect on global climate.

As a result of burning fossil fuel in automobiles, in electrical generating plants, and in copper and iron smelting plants, there has been an increase in the amount of ozone and sulphur dioxide in the air. These two gases and the effect of their synergistic interaction endanger tree growth; when present in excess they kill trees. In white pine trees short yellow needles and dwarfing are the result of exposure to SO_2 and O_3 at levels of 6–10 parts per hundred million (pphm) (Dochinger and Selisker, 1970). Not all white pine trees prove susceptible to sulphur dioxide and ozone; quite a few appear to have some degree of resistance to these air pollutants. When sulphur dioxide combines with water in the form of precipitation, it falls as "acid" rain that accelerates the soil weathering process as it percolates through the soil. Acid rain has been evident for many years in the accelerated weathering effects it has had on stone buildings and statuaries.

Reduction in forest growth in Sweden has been attributed to acid rainfall causing an inbalance in available Ca^{++}, Mg^{++}, and K^+ in shallow soils (Miller and McBride, 1975). Normal precipitation has a pH 5.8; acid precipitation is pH 4.03–4.19.

Smog, a combination of smoke and fog, occurred in London until it was reduced through control of smoke emission. In the United States smog occurs regularly in and near large cities where there is concentrated industrial activity and automobile traffic. Smog usually develops when there is a temperature inversion with the result that smoke containing a large volume of pollutants is trapped near the ground under the layer of cool air.

One of the products of combustion of fossil fuels is nitrogen dioxide, a major contaminant in smog. Upon reacting with sunlight and other compounds in the air, a series of changes in NO_2 occurs produces a group of compounds named peroxyacyl nitrates. (The reader is referred to Treshow, 1970, for more detailed treatment of air pollution.) Three of the compounds, *peroxyacetyl nitrate* (PAN), *Peroxypropionyl nitrate* (PPN), and *peroxybutyryl nitrate* (PBN) have been used in fumigation studies to determine their effects on plants. Of the three, PAN appears to be the primary constituent of smog that causes injury to plants. Concentrations of 5 pphm of PAN occur on smoggy days; however, as little as 2.5–3.0 pphm in the air can damage sensitive plants, including trees, causing foliage to turn yellow and die.

Another product of incomplete combustion is *ethylene*. It is also found in illuminating gas and is a natural plant product. Injury to plants from long exposure to 1 pphm ethylene has been noted. Near Los Angeles ethylene concentration in the air averages 20 pphm, with peaks of 142 pphm (Stephens and Burleson, 1967).

Under special circumstances other gases and dusts have been shown to be injurious to trees. Such gases as fluorides, ammonia, and chlorine, when produced in excess or even in low concentration can injure vegetation.

Atmospheric Moisture

If any compound can be considered to be a universal part of all plants, it is water. Water is important to plants in the following ways (Kramer, 1949):

1. Reagent in hydrolytic reactions
2. Solvent for the translocation of products within the plant
3. Medium within which reactions take place
4. Maintain turgidity
5. Major constituent of protoplasm and represents 52–60% of the fresh plant weight
6. Pathway between plant and the soil environment

The major source of water for precipitation is from the ocean masses that border the land and occupy over 70% of the global surface. Water evaporates from ocean surfaces and is carried over land where it is precipitated (Figure 7.23).

Trees lose water through transpiration, but the water budget of the forest is more detailed than shown in the hydrologic cycle. It is best illustrated by Figure 7.24 as the water and chemical budget because it is not possible to separate the two. In fact, the term biogeochemical relationships implies that the plant water and chemical budgets are closely interrelated. The chemical budget will be discussed under mineral cycling in Chapter 8, and the hydrologic budget will be discussed in Chapter 12.

Precipitation of atmospheric water occurs when an air mass cools sufficiently so that it becomes saturated and the water changes from a vapor to a liquid or solid state, and individual droplets or granules attain sufficient weight to fall to the earth. Cooling of an air mass occurs as a result of one of the following processes:

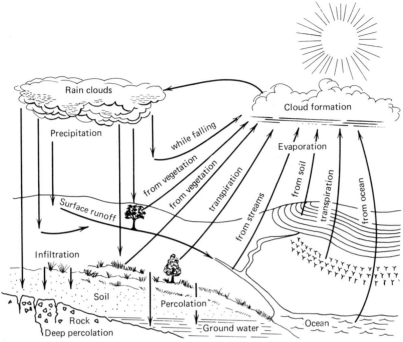

FIGURE 7.23. The hydrologic cycle (from the 1955 Yearbook of Agriculture, USDA.)

(a) Convection disturbances that occur when warm air is forced aloft by cool air underrunning it. The warm air cools as it rises; its relative vapor pressure increases to the point where saturation occurs thus causing a release of latent heat, which results in additional convection movement upward. If the convective currents thus created contain sufficient energy, the rising air creates a flow that can pull air in for some distance; towering cumulus clouds are formed as a result. When air rises its temperature decreases at a normal adiabatic lapse rate of −3°F per one thousand feet. The temperature lapse rate decreases as the air approaches saturation, and when water changes from a liquid to a solid energy is released further reducing the rate of temperature decrease; however, instability may occur. Continued circulation of air results in further cooling, rain, and possibly hail. Usually convective disturbances are local in nature, and the surface precipitation pattern that results is quite irregular with some areas receiving considerable amounts of precipitation while others nearby receive none.

(b) Cyclonic disturbances of massive size occur when two air masses meet. One air mass, if it is colder than the other, can move under the warm air

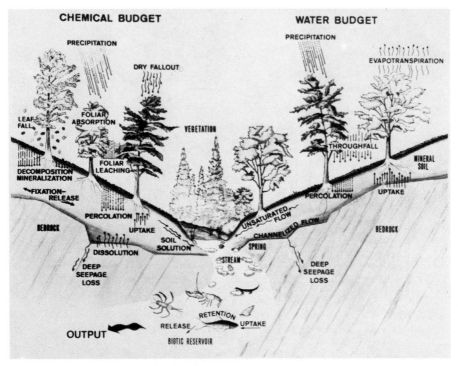

FIGURE 7.24. The biogeochemical relationship of the chemical and hydrologic cycles of watershed ecosystems (Curlin, 1970).

and push it aloft, or if the warm air has greater momentum it may override the cold air. In either case, the warm air will be cooled and it will reach saturation. If it contains sufficient water, precipitation as rain, snow, or sleet results. Precipitation from cyclonic disturbances is more regional than local convective disturbances and the surface amounts of precipitation received are more uniformly distributed over the land areas along the front that separates the air masses.

(c) Moving air when it encounters an inclined land mass such as a mountain range is carried aloft. Cooling occurs and precipitation can result. Such precipitation occurrences are termed "orographic" and the surface precipitation, although it is less wide-spread than with cyclonic storms, covers larger areas than with local convection disturbances.

Not all air masses that move across the earth's surface cause precipitation. When a relatively cold, dry polar air mass is heated as it moves over warmer land surfaces it takes up water from the atmosphere. Such an air mass is capable of holding increasing amounts of water as it becomes warmer (its capacity to hold water doubles with a temperature increase of +18°F). For example, if a dry polar air mass crossed the United States–Canadian border with 0.44 inches of rainfall equivalent, it could, by the time it reached the Gulf Coast, contain a 0.96 inch rainfall equivalent, an increase of 0.52 inches of water (an increase in temperature of approximately +21°F). If it took three days for the air to traverse the United States, the air would have accumulated 5,909,000 cubic feet of water per second or a volume equivalent to nine times the flow of the Mississippi River.

Land regions close to the moisture of ocean air masses receive greater amounts of precipitation than inland areas that are close to dry land air masses (Figure 7.25). Tropical air masses hold more water than arctic air masses so areas near warm, tropical oceans receive more precipitation than these near arctic polar regions. Windward sides of mountains generally receive more precipitation than regions on the leeward sides. High elevation positions along mountains receive more precipitation than low elevation locations.

Forms of Precipitation

Most any form of precipitation can provide moisture for plant use; however, some forms are more readily available than others. *Rain* is the most readily available form since it penetrates directly into the soil; and it seems that small amounts of rain that wet leaf surfaces can be directly absorbed through the leaves. *Fog* can be almost as effective as rain if it occurs regularly. The effectiveness of fog as a

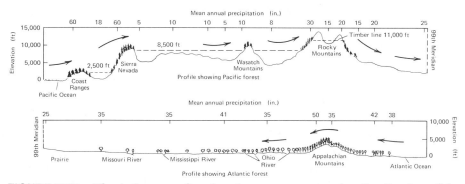

FIGURE 7.25. The influence of moist air currents upon the distribution of forests in the west and east along the thirty-ninth parallel of north latitude (Zon, 1941).

source of soil moisture will be discussed in a later chapter. *Dew*, although not as capable of supplying as much water as fog or rain, may nevertheless supply moisture when it is most critically needed—during periods of low rainfall. Survival of ponderosa pine in areas of low summer rainfall may depend upon the amount of dew that is precipitated. *Snow* comes during the season of minimum plant growth; however, rain and snow during the winter in temperate regions are responsible for providing water for soil moisture recharge. Soil moisture, therefore, is at its annual maximum capacity when plant growth begins in the spring.

Sleet and *hail* supply water in lesser amounts. These forms of precipitation can also cause considerable physical damage to trees, as can freezing rain. Large hailstones can completely defoliate trees that are in full leaf, or may damage the bole of the tree if they strike with sufficient force. *Freezing rain* (ice) is responsible for considerable damage in sections of the Appalachian range. Frequent ice storms in this region result in heavy deposits of glaze forming on trees, and if relatively high winds accompany glaze or if ice weight is too great, numerous tree limbs and boles are broken. Less frequent glaze storms occur throughout many sections of the eastern United States, which, in winter, are subject to incursions of warm, moist tropical air that coincides with the arrival of subfreezing polar air. The result is a freezing rain that spreads a coat of ice over vegetation and other features of the landscape.

Frost can be an unwelcome form of precipitation if it occurs later than usual in the spring or earlier than usual in the autumn. In these instances new growth, or growth not fully hardened, is killed, and the precipitation is of no benefit. Otherwise, frost should be considered in the same category as dew—contributing to the survival of plants—but probably not providing water for growth.

Drought

It is difficult to define drought in absolute terms since it occurs in varying degrees. The transient condition when a tree does not continue height growth because soil moisture tension reaches three atmospheres (3 atms) is a form of drought. A more severe example might be where 60% or more of the seedling trees planted as part of a regeneration program die because of lack of rain. Drought has been defined variously as follows:

1. Periods during which not more than 0.10 inches of precipitation occurred during any consecutive 48-hour period.
2. A period of 14 days or more in which there is not 0.25 inches of precipitation in one 24-hour period.
3. Drought occurs when plants can no longer obtain the moisture required to sustain growth.

Perhaps the last definition would more truly define drought as it is normally visualized. However, growth losses can occur at soil moisture levels higher than the wilting point and these are, in effect, drought-induced. In addition to the amount of precipitation that a locality receives, the availability of water in the soil or the loss of water through evaporation-transpiration depends upon the following (Thornthwaite and Hare, 1955):

1. The amount of solar radiation (energy received)
2. Wind velocity (water removed from the evaporating surface)
3. Type of vegetation and depth of rooting
4. Amount of available water in the rooting zone

Earlier, Thornthwaite (1948) found that mean air temperature could be used in estimating evaporation-transpiration (E/T). Nelson (1959), using the Thornthwaite approach and other published information, developed a method for estimating soil moisture depletion for pine forests in the southeastern United States (Table 7.7). By accumulating the daily moisture loss through evaporation-transpiration from Table 7.7, and by comparing the daily or weekly E/T loss with the amount of precipitation and soil moisture, one is able to quantify different degrees of drought.

Effects of Drought on Forest Trees

Drought in various forms causes greater injury to trees than perhaps any other environmental agent. Trees are subjected to annual and sometimes daily moisture stress from the very moment a seed germinates. Availability of soil moisture will determine to a great extent the amount of height and diameter growth of a tree in a particular year. Also when trees are subjected to moisture stress for an extended period, such as one year, they will generally experience a loss in height and diameter growth the following year.

TABLE 7.7 Estimated Soil Moisture Loss (Evapo-transpiration) According to Daily Mean Temperature (From Nelson, 1959)

DAILY MEAN TEMP (°F)	EVAPO-TRANSPIRATION LOSS (in.)	DAILY MEAN TEMP. (°F)	EVAPO-TRANSPIRATION LOSS (in.)
50	0.02	68	0.13
51	0.03	69	0.14
52	0.03	70	0.14
53	0.04	71	0.15
54	0.04	72	0.16
55	0.04	73	0.17
56	0.04	74	0.18
57	0.05	75	0.19
58	0.06	76	0.20
59	0.06	77	0.21
60	0.07	78	0.23
61	0.08	79	0.24
62	0.09	80	0.25
63	0.09	81	0.26
64	0.10	82	0.27
65	0.10	83	0.28
66	0.11	84	0.30
67	0.12	85	0.32

During the winter season, trees suffer injury as a result of drought primarily because of low soil temperature. During the growing season trees may experience, when subjected to drought of long duration, early leaf drop or may show scorch along the margins. In cases where trees are growing on southern exposures and have experienced drought over several years, top killing of the larger ones may be evident. Maple decline and birch dieback are injuries resulting from drought. Birch, in particular, after it has been exposed after logging to a series of winters without snow may exhibit growth decline (Hepting, 1971). In general, species found on mesic sites are most likely to suffer drought injury, particularly when they have moved onto sites subject to drought at infrequent intervals. Smith (1970) and Boyce (1961) indicate that maples, firs, hemlocks, cedars (*Thuja* spp) are sensitive to drought; while oak, pine (particularly the hard pines), and cedar (*Juniperous* spp) are resistant to drought.

TEMPERATURE AND PRECIPITATION

Precipitation and temperature interact in their effects upon tree growth. If a high temperature is not counteracted by a correspondingly high rate of precipitation, a grassland vegetation or a tropical desert vegetation condition will prevail. Similarly, a cold climate is one of low precipitation, and this condition results in a desert of another kind, which may be termed an arctic desert. The vegetation in such regions is called tundra. An effective method of presenting a graphic illustration of the progress of temperature and precipitation is the climagraph. The temperature-precipitation climates for the boreal and deciduous forests, and the tall and shortgrass prairies are shown in Figure 7.26.

Using published monthly records of temperature and precipitation, Hocker (1956) correlated the region of distribution of loblolly pine to the amount and frequency of precipitation during the summer season and to the minimum temperature in winter. Bethune (1960), using the techniques developed by Hocker, found that slash pine distribution was confined to a rather uniform temperature zone, and its distribu-

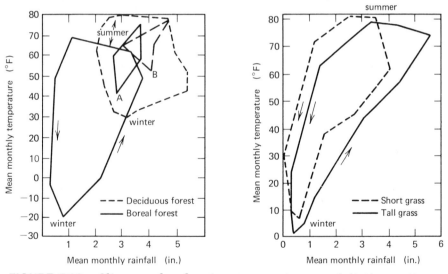

FIGURE 7.26. Climagraphs showing temperature-precipitation patterns for several vegetable formations in the United States (from Smith, 1940). Superimposed on the deciduous forest are the patterns for (A), the zone adjacent loblolly pine distribution, and (B), the zone within loblolly pine distribution.

tion within this zone is limited by frequency of heavy precipitation during each of the four seasons of the year. Shoulders and Terry (1968) found that survival of planted longleaf pine could be related to mean annual temperature at the seed source and that 10-year height growth was related to the early season rainfall (January–April) at the seed source. Seed from locations with a mean annual temperature near 63°F appeared to produce the highest survival rates. A rainfall amount of 17.4 in. from January to April at a source location resulted in the tallest seedlings. Sources having less rainfall produced shorter seedlings.

Precipitation Effectiveness

Not all of the precipitation that reaches the earth's surface is effective in replenishing soil moisture. Even the meagerest amount of vegetative cover will intercept some fraction of the precipitation, preventing it from infiltrating the soil. A two- or three-storied forest stand presents many opportunities for precipitation to be intercepted and later evaporated directly into the air. As the amount of precipitation received in any particular area increases, however, the relative amount of precipitation reaching the soil surface increases. Leonard (1961) found that in a beech-birch-maple forest, of the total annual precipitation, approximately 85% (from storms with more than 0.5 in.) reached the ground as stemflow and throughfall, with 15% being intercepted. The relationship between precipitation and throughfall for a loblolly pine plantation is illustrated in Figure 7.27.

Stemflow is precipitation that is detained by the crowns and boles of trees in the stand, but that finally reaches the soil surface by flowing down the branches and bole. Throughfall may be precipitation that falls directly through the branches of the crown, or reaches the ground after it is detained temporarily by the crown but later drips from the leaves and branches onto the soil.

Some of the water from precipitation that reaches the soil runs off as surface flow and some infiltrates the soil. The relative amounts of each depend on the physical condition of the ground cover and the soil layers below it. However, a portion of the water of infiltration will be detailed for only a short period and will later be released as subsurface flow. The remaining water is retained against the force of gravity by the soil. The relative amounts of detention and retention will depend upon the physical properties of the soil.

A number of indices have been derived to indicate effectiveness of

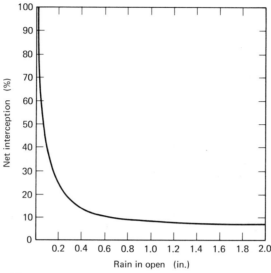

FIGURE 7.27. Net interception as a percentage of rainfall in the open (Hoover, 1953).

precipitation for an area in relation to its temperature. Transeau's index uses the ratio between precipitation and evaporation and is most useful; however, lack of sufficient data on evaporation prevents its general use in the United States. Two different approaches to indexing precipitation effectiveness are:

Transeau (1905)

$$R = 10,000 \frac{\text{Precipitation (in.)}}{\text{Evaporation (mm)}}$$

Thornthwaite (1948)

$$R = \frac{\text{Precipitation (in.)}}{11.5 \text{ (temp.)}}$$

In addition to temperature and precipitation, transpiration depends upon wind velocity, exposure, and the seasonal nature of precipitation. The ratios (indices) of precipitation and temperature can be used either for annual or for monthly data. The latter provide a more definitive picture of changes in the environment.

Paterson (1956) developed a sophisticated expression of climate that is termed the CVP (climate, vegetation, productivity) index. This index combines several expressions of temperature, precipitation, and solar radiation to develop a climate index, which can be related to tree growth in different climatic zones. The CVP index is as follows:

$$\text{CVP index} = (\text{Tv/Ta}) \ (\text{P}) \ (\text{G/12}) \ (\text{E/100})$$

where

Tv = average temperature °C of the warmest month

Ta = temperature difference in °C between average temperature of the warmest month and the average temperature of the coldest month

P = total annual precipitation in mm

G = number of months in the growing season

    (a) In the boreal and cold temperature regions this is the number of months when the average temperature equals or is above 3°C.

    (b) For warm temperate and sub-tropical climates Paterson uses de Martonne's aridity index as G.

$$I = \frac{P}{T + 10}$$

E = 100 $Rp \cdot R$, where Rp = gm cal/cm² radiation at the pole for an entire year and R = gm cal/cm² at the latitude of the place under consideration

Paterson used the CVP index as a variable against which he correlated forest stand growth. The formula derived is: mean annual growth (m³/hectare) = −7.25 + 5.20 (log CVP).

Lemieux (1961) derived CVP index values for 152 Canadian stations in Quebec and New Brunswick and showed a definite relationship between forest regions and CVP index values. Comparison of volume growth data from a number of field plots and the CVP index for the locality showed a fair comparability, but Lemieux indicated there was a need to modify the CVP index and to improve the estimate of growth.

GENERAL REFERENCE

Gaertner, E.E. 1964. Tree growth in relation to the environment. *Botanical Review* 30 (2):393–436 (11 pages of references).

Kozlowski, T.T. 1962. *Tree Growth*. Ronald Press, New York.

Kramer, P.J. and T.T. Kozlowski. 1960. *Physiology of Trees*. McGraw-Hill, New York.

Larcher, W. 1975. *Physiological Plant Ecology* (trans. by M.A. Biederman-Thorson). Springer-Verlag, New York.

Treshow, M. 1970. *Environment and Plant Response*. McGraw-Hill, New York.

Trewartha, G.T. 1954. *An Introduction to Climate*. McGraw-Hill, New York.

Scientific American, 1970. *The Biosphere*. Freeman, San Francisco.

Schroeder, M.D. and C.C. Buck. 1970. *Fire Weather*. U.S.D.A. Forest Service Agric. Handbook 360.

Chapter 8

Trees obtain water and minerals from the soil through their roots. The soil provides a foundation for the roots to support the bole and crown. If a soil lacks the ability to meet any of these needs, tree growth will suffer. Forest soils, in general, are deficient in water each year during part of the growing season, and they contain only minimum amounts of the minerals essential for plant growth. Yet trees are able to make acceptable growth on most soils because their perennial nature permits them to extend their roots into all parts of the soil mass capable of supplying water and minerals essential for tree growth. At the same time the extensive root system is able to support the above-ground mass against storm winds, except under exceptional conditions.

Jenny (1941) defined soil symbolically as:

$$soil = f(cl, o, r, p, t)$$

Forest Soil

where cl represents climate, p represents the parent material, o represents organisms in the soil, t represents time, and r represents topography. This function is similar to the one used to define a forest stand. It would appear that either one system exists within another system or that there are two systems that are interdependent. If the soil is taken as a relatively permanent system that is only slightly modified by the presence of vegetation, then one can say that forest ecosystems are systems that vary within the soil ecosystem. The ecology of the soil system is not greatly different in its organization from the forest; the organisms are of different species, and the composition (parent material) is of different substance, but the principles that underlie an understanding of the two ecosystems are the same, and the two are essential to the soil-plant-animal ecosystem.

Soil, compared with climatic factors, is more local in its influence on tree growth. The effects of parent material, topography, and time can change within relatively short distances and these changes are reflected in the vegetation. The soil system is made up of four major physical parts: solids (organic and inorganic), liquids, gases, and biotic.

Soil solids are the "backbone" of the soil and are represented by varying proportions of stones, sand, silt and clay, and organic residue. Attached to the clay fraction are various mineral ions essential to plant growth. The liquid phase of the soil is predominantly water that may contain various suspended minerals and various gases. A portion of the liquid phase is intimately associated with the solid phase since soil water is bound to soil particles with such force that the water molecules can be removed only by large inputs of energy. Most of the water that sustains growth is retained in the capillary pores. Between the soil particles are, in addition, noncapillary pores that are too large to hold water against the force of gravity. These voids become filled with air or other gases following the drainage of free water. The biotic soil fraction is made up of microscopic organisms of both plants and animals, as well as the macroscopic plants and animals that live in the soil. These organisms live within and may subsist on the soil by utilizing the solids and liquids for their life needs.

From the standpoint of tree growth, in addition to serving as a rooting medium the soil must be capable of supplying water and minerals essential for growth. The relative availability of these factors in a particular soil determines the rate at which trees grow. Since the soil is influenced by the same general climate as the forest, the rhythm of growth of plants and animals within the soil itself will nearly coincide with that of the forest plants and animals that live above ground.

SOIL WATER RELATIONS

Generalizations concerning tree growth are difficult to make, particularly those that have few exceptions. One certain generalization is that water is the most important substance which the tree derives from the soil. It is from the soil that trees obtain the water needed to satisfy transpiration and growth needs. Small amounts of water are no doubt absorbed directly through leaves and other above-ground portion of trees, but the total amount gained in this way is very small in comparison with the amount of water that enters through the roots.

Soil water is retained or detained within the soil with varying degrees of force and for different lengths of time. One of the forces that retains water in the soil is the molecular force of chemical bonds derived from the negative surface charges occurring on the surface of clay particles. This force attracts water molecules making them a part of individual soil particles. Another force is the capillary attraction between the water molecules themselves. This force also develops between soil particles and water molecules and is great enough to retain water against gravitational force. The noncapillary channels in the soil are of such size that the capillary attraction is not sufficient to retain water for any period. The volume of water that transiently occupies the noncapillary pores is called detention storage. There are even larger channels in the soil formed by animals and dead roots that can possibly be considered as ultra-macroscopic. Here water moves as it would in a large pipe, being only ever so slightly detained.

While a portion of the water that infiltrates the soil is retained within it, some of the remainder moves through and replenishes subterranean aquifers. Gravitational water than leaves behind open pore spaces of noncapillary size that become filled with air or some other gas. Some of the water retained by the soil occupies the capillary pore spaces; the rest is attached to the surface of soil particles.

Water within the soil can be classified as follows:

1. Noncapillary—detention water
2. Capillary—retention water
3. Adsorbed
4. Hydration

Water of hydration and water adsorbed on soil particles is probably not available for plant growth. A portion of the detention water is available, but the greater part is not detained long enough to be used for growth and to satisfy the transpiration requirements of plants. However, there are topographic situations where a water table may be sufficiently close to the soil surface that tree roots can obtain water from noncapillary pores.

Capillary soil water capacity of a soil depends upon the relative proportions of sand, silt, and clay, the texture of the soil, and its structure and density. A coarse textured soil such as sand will have very low capillary potential. By contrast, a fine textured soil—clay—may have a high capillary potential but may be poorly aerated (Figure 8.1). A deep loamy soil, having proportions of solids, liquids and gases approaching 50% solid, 30% water, and 20% air space is one which appears to be a suitable environment for tree growth, provided there are sufficient amounts of sand, silt and clay to produce a loamy texture and the soil is sufficiently deep to supply adequate water. The greater the proportion of solid volume in the soil the less space there is for air and water.

The volume to weight ratio of a soil is a measure of apparent specific gravity and is referred to as bulk density or volume-weight. Bulk density is related to the moisture-holding capacity of the soil. High density soils have more solids per unit volume and, therefore, are not

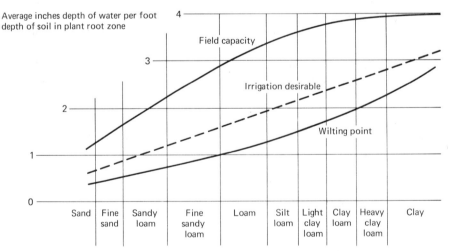

FIGURE 8.1. Typical water-holding characteristics of different textured soil (from USDA, 1955).

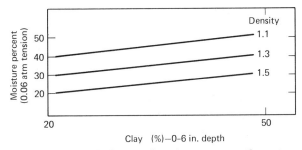

FIGURE 8.2. Relationship between soil moisture and bulk density and percent clay (for soil types Sharkey, Loring, Lakeland, Lynchburg, Ruston) (data from Doss and Broadfoot, 1956).

capable of holding as much available water per unit volume as soils with the same texture but lower density. Figure 8.2 shows the results of an analysis of the relationship between capillary water capacity and texture and bulk density of a number of soils. The relationship can be expressed as: percent moisture (0–6 in. depth) = 83.213 − 43.371 (bulk density) + 0.218 (percent clay). As the percent of clay increased, capillary water increased; capillary water decreased as the soil bulk density increased.

DETERMINING PROPORTIONS OF SOLID, LIQUID, AND AIR VOLUME OF SOILS

Solid volume (volume of sand, silt, and clay) of a soil can be computed by dividing bulk density of the solids by the specific gravity of the rock material from which the soil was derived. Specific gravity (particle density) of rock material ranges from 2.60–2.70: 2.65 is considered average. To determine the solid volume of a soil with 1.5 bulk density thus requires that 1.50/2.65 = 0.57; for a bulk density of 1.10 the proportion solids as a ratio of total soil volume would be 1.10/2.65 = 0.41. Bulk density of a soil can be measured by removing a sample of known volume. The soil sample is dried and weighed. The volume-weight ratio (dry weight) volume or bulk density of the soil can then be computed.*

To determine the water retention of a soil it is necessary to measure the capillary water-holding capacity, which can be estimated by saturating it and allowing free drainage to occur. After the soil has been brought to saturation and allowed to drain until the gravitational

*Weight is expressed in gram (g) units and volume on cubic centimeters (c³).

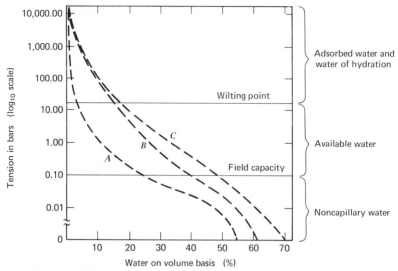

FIGURE 8.3. Moisture retention curves for three soils: A, fine sandy loam; B, loam; C, clay.

water is removed, the amount of water remaining is capillary water. Under field conditions, 24 to 48 hours is the time generally allowed for noncapillary water (gravitational) to drain. Under laboratory conditions a saturated soil is placed in a situation where tension (0.10 bars)* can be applied to the water in the soil to simulate field conditions. After a period of drainage under tension when no more water is removed from the soil, it is assumed that water remaining is capillary water. Whether the field capacity (total capillary capacity) of the soil water is estimated in the field or in the laboratory, it represents an estimate of the capillary water capacity, which is the upper limit of soil water available for plant growth (Figure 8.3).

All of the capillary water in the soil is not available for plant growth. In fact, it has been shown that when water tension in the capillary pores becomes equivalent to 2.5 atm, growth is affected. It is generally assumed, however, that plants are capable of removing water from the soil up to a tension equivalence of 15 atm (Figure 8.3). At this tension it has been shown that plants reach permanent wilting and do not recover until water is added to the soil. Moisture content of a soil at permanent wilting can be estimated in the field or in the laboratory.

*1 atm at normal air pressure = 14.7 lb./in.² 15 atm tension = 220.5 lb/in.² suction pressure. 1 bar = 0.987 atm = 10^6 erg/cm³ = 1000 mbar.

When field capacity, permanent wilting, and bulk density are known for a particular soil, then the proportion of solid, liquid, and gas can be estimated, and the important parameter—available water—can be determined. Examples follow that present physical data for field capacity, wilting percent, and bulk density for three different soils. Soil A represents a fine sandy loam; soil B a loam; and soil C a clay. The constants obtained were used to construct the curves illustrated in Figure 8.3

Since the values for solids, capillary water, and air shown in the following table represent relative amounts of soil volume, it is possible to determine the volume of solid, liquid, or gas for any soil thickness. The amount of available capillary water for each inch of soil depth is: A—0.24 in.; B—0.40 in.; and C—0.48 in. A soil depth of 12 in. would then have a total capillary capacity equivalent to 2.88 in. for A, 4.80 in. for B, and 5.76 in. for C.

	SOIL A	SOIL B	SOIL C
Field capacity (0.10 bars)[a]	20%	40%	60%
Wilting percent (15.0 bars)[a]	2%	15%	20%
Bulk density	1.2	1.0	0.8
$\dfrac{\text{Proportion}}{\text{Solids}} =$	$\dfrac{1.2}{2.65} = 0.45$	$\dfrac{1.0}{2.65} = 0.38$	$\dfrac{0.8}{2.65} = 0.30$
Capillary water =	$(0.20)(1.2) = 0.24$	$(0.40)(1.0) = 0.40$	$(0.60)(0.8) = 0.48$
Air volume =	$1.00 - (0.45 + 0.24) = 0.31$	$1.00 - (0.38 + 0.40) = 0.22$	$1.00 - (0.30 + 0.48) = 0.22$

[a]Determined on a dry weight basis.

To determine the amount of available soil water requires the amount of water held at the wilting percent be deducted from the field capacity. The following table shows estimates of available water.

	SOIL A	SOIL B	SOIL C
% available water by weight	$0.20 - 0.02 = 0.18$	$0.40 - 0.15 = 0.25$	$0.60 - 0.20 = 0.40$
Available by volume	$0.18\,(1.2) = 0.22$	$0.25\,(1.0) = 0.25$	$0.40\,(0.8) = 0.32$
Equivalent volume per 12 in.	$(0.22)(12) = 2.64$ in.	$(0.25)(12) = 3.00$ in.	$(0.32)(12) = 3.84$ in.

Periodically during the year, generally in winter and early spring, the entire solum of soil will be saturated with water. When this occurs both the capillary and the noncapillary pores are filled with water. The soil atmosphere changes to one in which oxygen becomes limited, and soil temperature may be depressed because of the high specific heat of the water that is present. Root growth, as a result, is inhibited by lack of oxygen and low temperature. During the growing season, completely saturated soils do not occur regularly except in topographic situations where water runoff is hindered as a result of depressions or flat terrain. In these instances the soil may be saturated for long periods, thus depressing tree growth.

There are instances when additional amounts of water will increase tree growth. Broadfoot (1967) observed that where shallow water impoundments were created from February to July in each of four years, soil moisture increased in pole and timber-sized stands of cottonwood, red gum, and green ash. This resulted in a 52% increase in mean annual diameter growth. Oxygen supply in the water was depleted 15 days after flooding, but because the water was shallow, oxygen was replenished quickly after each rain. New seedlings appeared each year soon after the water was released from the impoundments; established seedlings made rapid growth. On the other hand, tree growth may be severely retarded and trees killed where water depths in excess of those reported in the study occur, or when an impoundment is permanent; except for species such as baldcypress, water tupelo and other trees capable of growing on swampy soils.

SOIL DENSITY

Forest soils generally have surface horizons lower in density than agricultural soils (Table 8.1). The difference between open and forest soils in the 0–3 in. depth class shown in the table is attributed to the accumulation of humus in the forested soil. Differences between open and forest soils in the 3–6 in. depth class is attributed to the fact that the open area had been used for a garden and had been plowed to about a 9 in. depth, thus incorporating the organic matter throughout the 9 in. of surface soil.

Coile (1940) did not find any difference in bulk density for soils that supported loblolly pine stands of 10, 20, and 70 years, and there was no difference between soils under pine and oak stands (Table 8.2).

Broadfoot and Burke (1958) reported that the bulk density for sur-

TABLE 8.1. Mean Bulk Density of Soils in Two Locations (Sartz, 1961)

DEPTH CLASS (in.)	OPEN	FOREST
0–3	1.08	0.70[a]
3–6	1.07	1.22[a]
6–9	1.26	1.36
9–12	1.40	1.37
12–15	1.44	1.43
15–18	1.48	1.46
18–21	1.54	1.49[a]

[a]Difference is significant.

face soil layers under forest conditions averaged 1.12, under pasture conditions, 1.27, and under old field conditions, 1.31. To predict the bulk density for surface and subsoils combined, they found the percent sand and percent organic matter in the soil were the most efficient of a number of predictors: bulk density = 1.47 + 0.001 (percent sand) − 0.100 (percent organic matter). This equation accounted for about 80% of the variation in the data.

TABLE 8.2. Bulk Density of Forest Soils of the Lower Piedmont of North Carolina (Coile, 1940)

SOIL DEPTH CLASS	LOBLOLLY PINE AGE (YEARS)			RED OAK–WHITE OAK– BLACK OAK (uneven-aged)
	10	20	70	
	BULK DENSITY			
A 1–3	1.41	1.33	1.35	1.24
B₁ 7–9	1.46	1.41	1.32	1.36
B₂ 20–22	1.28	1.30	1.28	1.21

From research reported it appears that forest soil, because of the accumulation of litter and formation of humus, has lower densities of the A layer (particularly the A_1) than agricultural soils; however, soil density depends more upon the physical structure of the soil than upon land use.

SOIL MOISTURE AND TREE GROWTH

The amount of available water in the soil at any time is directly related to the amount of precipitation and the amount of water lost through transpiration and evaporation. The ability of a soil to supply water to trees determines the rate at which photosynthesis will progress. Photosynthesis takes place when the stomates in the leaves are open to permit the passage of CO_2 into the interior of the leaves. The guard cells control the opening and closing of the stomates and they must be able to develop full turgor pressure in order for the stomates to remain open. If soil moisture is limiting, leaf stomates will either not open fully or remain closed, thus limiting the amount of CO_2 exchange; however, when the stomates fail to open or only partially open, the amount of water lost in transpiration is reduced, which in a sense protects trees from excessive water loss.

Zahner and Stage (1966) introduced a method of using daily soil moisture stress to estimate tree growth. They proposed that soil moisture stress greater than 2 atm reduces growth, and that an inventory of soil moisture by budgeting daily loss through transpiration and daily soil moisture increase through precipitation can be correlated to tree growth. For red pine growing in Michigan, their analysis showed the following relationship: deviation from average height growth = a_0 + a_1 (water deficit of current year shoot-elongation period) + a_2 (sum of deficits of preceding year needle-elongation plus food storage period) + a_3 (square sum of deficits for preceding year). Previous research showed that red pine shoots cease growth each year, regardless of soil moisture conditions, in early July (May 1–July 15). Buds for current year shoot growth are set by early July of the previous year (June 15–July 15) and needle elongation occurs from mid-June to early September (June 15–August 31). The food storage period begins in September and continues until the end of October (September–October 25). The analysis showed that water deficits during the current shoot elongation period and food storage period of the previous year and the squared sum of the total deficits of the previous year contributed significantly (72% of the variation in the data) to deviation of annual height growth of red pine from the average for the 1950–1960 period.

Moehring (1967) determined that diameter growth of sawtimber stands in Louisiana was more closely related to the rate of soil moisture loss (transpiration-evaporation depletion) than to the amount of available water in the soil. In loess soils he found that for transpiration-evaporation rates of 0.25 in. per day, diameter growth ceased when available soil moisture reached 13%; a 0.35 in. per day E/T rate caused growth to cease when available water was 26%. Apparently both height and diameter growth respond not only to the total amount of available soil water but to the temporary deficits that occur during the current and previous growing seasons. Stransky and Wilson (1964) found during spring growth flush, height growth of loblolly and short-leaf pine seedling was inhibited by soil moisture tensions greater than two atmospheres and that it stopped completely as tension increased to 3.5 atm. Incipient wilting occurred on new growth at 5 atm and all plants were dead at 15 atm. Luftus (1975) noted that height growth of yellow-poplar seedlings was completely inhibited by soil moisture tensions at less than 15 bars. Root development and height growth were drastically inhibited and seedlings wilted at a soil moisture tension of 4 bars.

A fact that needs to be emphasized is that in pure or nearly pure stands the trees do not always function as individuals with respect to soil moisture depletion. Bormann (1966) explains that because of the widespread occurrence of natural root grafts in white pine (these occur with other species), a pure stand or nearly pure stand of trees should be considered as being a single unit. Because of numerous root grafts, water and food are exchanged from dominant to intermediate and suppressed trees. Dominant trees exchange only food between each other. When this occurs the trees do not suffer a loss on the exchange as they do when dominant trees prolong the existence of intermediate and suppressed trees by contributing food and water to them. The fact that grafts exist and that they provide a pathway between trees, particularly from the more vigorous to the least vigorous, indicates that overtopped trees probably survive longer than if they had to depend entirely upon the amount of food and water that they could supply for themselves. It would seem that the photosynthetic efficiency (use of solar energy) is reduced as a result of prolonged survival of decadent suppressed trees.

SOIL MOISTURE DEMAND

The ability of a soil to retain water depends upon its capillary pore space, which in turn is dependent upon the texture and density of the soil. The amount of water that a soil contains in the capillary pores at a particular time depends on the demands made upon the water supply by transpiration and evaporation. When there is no vegetation on the soil, water is lost from the surface of a bare soil through evaporation. When there is vegetation growing on the soil, water is lost through transpiration and evaporation. The relative and total amounts of this lost water are difficult to determine separately, and total loss is taken to indicate water demand. Thus the term evapotranspiration is most generally used to describe loss of soil capillary water under natural conditions.

Studies of water depletion rates in forest stands indicate that water is removed from all portions of the soil occupied by roots. Annual depletion of the soil water supply in temperate climates begins in early spring with a drawdown of the water accumulated during the previous autumn and winter. Depletion proceeds at a fairly rapid rate, particularly following leafing out of deciduous trees. The rate of depletion of water by trees can equal or exceed the rate for agronomic crops. During the growing season and after a protracted period of drawdown, the lower soil horizons are nearly exhausted of water, that is, they are near wilting percentage. The upper soil layers, particularly the organic layers, play an important part in supplying water to plants during the remainder of the growing season.

Curtis (1960) found that under full sun conditions both pine and hardwood litter retained the same percentage of water. For storms less than 0.5 in., two-thirds or more of the water received was retained in the litter and for all storms the litter retained one-third of the total precipitation (Table 8.3).

TABLE 8.3. Precipitation Stored in Leaf Litter During a Three-year Period at LaCrosse, Wisconsin (Curtis, 1960)

PRECIPITATION CLASS (in.)	NUMBER OF STORMS	TOTAL PRECIP. (in.)	MOISTURE STORED IN HARDWOOD LITTER (in.)	%	SCOTCH PINE LITTER (in.)	%
0.0 –0.25	56	5.71	5.36	93.7	5.39	93.7
0.26–0.50	25	9.51	6.31	66.4	6.12	64.4
0.51–1.00	35	25.03	10.38	41.5	9.87	39.4
1.100	19	35.71	3.96	11.1	3.86	10.8
Total or Average	136	75.96	26.00	34.2	25.24	33.2

Trimble and Lull (1956) report that the retention storage is the same for both humus types but varies according to the thickness of the humus layer (Table 8.4). Mader and Lull (1968) found that in white pine stands in Massachusetts partial wetting of the forest floor took place readily, but complete resaturation of the organic layer required an extended period of heavy rainfall. Only a small portion of many rains is absorbed even though the forest floor was not completely saturated when rains occurred. Absorption of 0.10–0.25 in. of rain occurred during rainy periods in the summer, but resaturation to 0.4–0.5 in. of total capacity did not occur even with rainfall amounts of 1–2 in. per storm. The authors concluded that either a good deal of high intensity rain moves through the layers too rapidly to be absorbed—the moisture does not have easy access to all absorbing surfaces—or some soil materials do not re-wet easily once dried.

It can be expected that tree roots will deplete the water in the solum during spring and early summer. During mid- and late summer, water for growth will come primarily from the surface soil layers in which moisture is replenished by periodic precipitation. Occasional heavy storms (rainfall amounts in excess of 1.0 in.) will supply enough water to fill the capillary pores of the surface soil and perhaps a surplus to reach the subsoil. However, rainfall of less than 1.0 in. is not enough water to wet other than the organic surface layer and part of the upper A horizon of most soils.

Metz and Douglas (1959) report that after a depletion period of 40 days, barren fields had more available water than soils supporting grasses or pines. Soil moisture losses during this period were 2.89, 3.99, and 5.85 in. for barren field, broomsedge, and pine plantation.

TABLE 8.4. Retention and Detention Water in Stands of Different Species Composition and Humus Depth (Trimble and Lull, 1956)

FOREST STAND MED. TEXTURED SOIL	HUMUS DEPTH (in.)	WATER STORAGE RETENTION (in.)	CAPACITY DETENTION (in.)
Aspen, pin cherry, gray birch, and paper birch	3.5	0.91	0.98
Beech, birch, maple	7.3	2.19	2.19
Hemlock, spruce, balsam fir	7.6	2.51	2.43

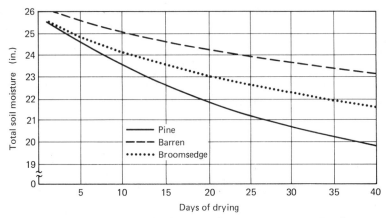

FIGURE 8.4. Moisture depletion of 0–66 in. zone (Piedmont, South Carolina) over a 40-day period (Metz and Douglas, 1959).

Average daily rate of loss was 0.07, 0.10, and 0.14 in. for barren field, broomsedge, and pine plantation. Effective rooting depth for the broomsedge was about 36 in., and twice that for the pine plantation (Figure 8.4).

The rate of water depletion depends not only upon the plant species concerned, but also upon the demands of the atmosphere. Holkias, Weihmeyer, and Hendrickson (1955) recommend that the difference between the amount of water lost through black and white atmometers be used in predicting transpiration-evaporation of field crops. The relationship is linear and permits easy extrapolation. Monthly evapotranspiration rates for fruit trees such as walnuts, apricots, peaches, and prunes growing under irrigation ranged from 4 to 8 in. per month during the growing season. Zahner (1956) reports that in southern Arkansas, water needed by forests during June, July, and August is 8 in. per month based on Thornthwaite's method of estimating evapotranspiration. He calculated that for May and September monthly water loss was 5.0 in. and for October was 2.5 in. These rates conformed to measured rates of depletion. It would appear that the evapotranspiration rate for a forest stand during the growing season would range between 0.13 and 0.27 in. of water per day, and from bare soil it would average about 0.07 in.

Urie (1959) reported water depletion under pine began in April and May and later under oak. Both pine and oak stands reached maximum

---

depletion August 18, depleting the soil by an equal amount of water (3.9 in. of stored water). This would support the argument that different plant species remove water in equal amounts from the soil, and any difference between them is the result of a difference in rooting depth. Trees and other perennials, however, have the capacity for their roots to penetrate the soil to greater depths than annual and biennial herbaceous plants. So it is that trees tend to use more soil water than annual plants and some perennial grasses where rooting depth is not limiting.

SOIL TEMPERATURE

Soil temperature has a pronounced effect on the rate at which water enters a plant. If soil temperature is above freezing, water can move as a liquid and is relatively easily obtained; however, if soil temperature is below freezing, then soil water is immobile and unavailable to plants.

Root growth and development are also affected by soil temperature. If soil temperature is too high or too low (115°F or 40°F), root growth will be curtailed. Root growth is important to absorption of water since roots must come into contact with soil having available water in order to continue absorbing it. Bilan (1966) observed that root growth of loblolly pine seedlings ceased when soil temperature reached freezing in late November and early December, and growth resumed in early March when the minimum soil temperature rose to 30°F. The beginning and end of the seasonal cycle of root growth is controlled largely by temperature. Soil temperature requirements vary among species, however. The minimum soil temperatures for most species ranges from 32° to 41°F, the optimum 50° to 77°F, and the maximum 77° to 85°F.

ABSORPTION OF WATER BY TREE ROOTS

Absorption of water through tree roots occurs as a result of either passive or active transport of water molecules from soil water to the root xylem elements. Passive transport of water involves a plant that is transpiring water from its leaves, resulting in a large negative water potential in the root cells. The water potential created is sufficient to overcome resistance to water movement by root cell walls. When water reaches the xylem elements, it is transported rapidly to the leaves. The direction of flow is toward decreased water potential. Kramer (1969) expressed the relative water potential of the different segments that affect flow as

$$\Psi_w = \Psi_m + \Psi_p + \Psi_s$$

where
Ψ_w = water potential of root cells, soil, or xylem elements (expressed as bars of tension (negative) or pressure (positive)

Ψ_m = matric pressure, which is the surface potential in the soil, cell walls, protoplasm, and other substances that can bind water (this is a negative force occurring both in the tree and the soil)

Ψ_p = cell wall pressure, which occurs *only* within the plant. Wall pressure may be negative during periods of transpiration and positive during guttation.

Ψ_s = solute potential of the soil or cell solution (a negative force)

As an example, when passive water movement takes place, the water potential of soil water could be

$$\Psi_w = \Psi_m + \Psi_p + \Psi_s$$
$$-0.3 \text{ bars} = -0.2 + 0 - 0.1$$

(Soil water in this case would be less than field capacity.) The water potential of the xylem sap could be

$$-5.5 \text{ bars} = 0 - 5 - 0.5$$

(Transpiration has applied tension to water within the xylem elements and across the root cells.)

Water in the system will move from the soil into the xylem elements and up the stem. It will then replace water lost through the stomates from the mesophyll cells. Mass flow of water during passive transport primarily occurs along the cell walls of epidermal and cortical cells surrounding the central vascular cylinder (Figure 3.11). Therefore, it can be expected that the influence of the protoplasm of living cells would not greatly affect rate of transport.

Active transport of water occurs during periods when transpiration demand is low. There is some doubt about the controlling mechanism or mechanisms that result in differences in water potential between soil and xylem. Kramer (1969) indicates a need to consider more than one possible explanation of active transport. He suggests, however, the most probable reason for water to move from soil solution into root cells, and then into the xylem, is as a result of difference in osmotic potential between root cells and soil water. In short, there needs to be a greater solute potential in root cells. Using the water potential equation:

$$\Psi_w = \Psi_m + \Psi_p + \Psi_s$$

and the water potential of soil as before:

$$-0.3 \text{ bars} = -0.2 + 0 -0.1$$

while that of the xylem sap could now be

$$-1.5 \text{ bars} = 0 + 0.5 -2.0$$

(Here osmotic potential of cell sap is sufficient to cause water to move from the soil into the root.)

Whether water is moving into root cells either actively or passively, as matric pressure of soil water increases less water will be available to be absorbed. As was indicated earlier in this chapter, when matric pressure of soil water increases to -2.0 to -3.5 bars, tree growth is greatly affected. In general, it can be assumed that only through transpiration and passive absorption can trees develop a root water potential sufficiently high to remove water from soils with water potentials up to -15 bars.

SOIL PROFILE
The soil under a forest is not of the same texture and density throughout its entire depth. It also contains different amounts of mineral and

organic fractions as well as different biotic elements at different depths. Soil can be divided into strata, or horizons, on the basis of differences in physical, chemical, and biotic characteristics. A well-developed forest soil will have a layer of litter on the surface, below which is a mineral soil layer with an accumulation of well-decomposed litter (humus); below this are layers of mineral soil exhibiting difference in color and texture.

Zone of eluviation or leaching	—undecomposed organic material—litter	L	A horizon*
	—partially decomposed—fermentation	F	
	—decomposed—humus	H	
	—horizon of incorporated humus	A_1	
	—horizon of maximum leaching	A_2	
	—horizon of transition	A_3	
Zone of illuviation or accumulation	—horizon of transition	B_1	B horizon
	—horizon of maximum accumulation	B_2	
	—horizon of transition to C	B_3	
	—weathered parent material	—	C horizon

A soil profile may have fewer layers in a particular horizon than shown, or a horizon may be absent. Immature soils do not exhibit nearly the number of horizons indicated in the example, which represents a soil that had weathered in place and had reached a matured state with a minimum of erosional disturbance. Alluvial, water-borne soils show less differentiation as would loess soils that are wind deposited. In glaciated regions the underlying substratum might bear no relation to the soil layers above it. From the standpoint of water relations, it is important to be able to recognize and to measure the thickness and physical properties of the different horizons. Bunting (1965) presents in detail the factors that influence soil formation and the manner in which soils can be classified.

---

*The diagnostic horizons are the A_2 and B_2 layers. In the example the L and F compare with O_1 and H to the O_2 horizon in standard terminology.

SOIL MINERAL ELEMENTS

Of the 16 known essential elements for plant growth, the soil must supply all the plant needs of 10 of these elements (P, K, Ca, Mg, Cu, Zn, B, Fe, Mn, Mo) and except for leguminous plants all of the needs for an eleventh (N). The other five elements (C, H, O, Cl, S) are at least partially and sometimes primarily supplied by the atmosphere (Fried and Broeshat, 1967), Part III, p. 115).

The soil elements needed for plant growth are derived from weathered rock material. The rocks that form soil can be classified into three groups: igneous, sedimentary, and metamorphic.

Igneous rocks are formed by magma when it cools. Magma is hot molten material from the earth's mantle and core. Common igneous rocks are granite and basalt.

Sedimentary rocks are formed from bits of rock that come from other rocks. Shale, sandstone, and limestone are examples. Some sedimentary rocks have fossil imprints of shellfish within their structure, indicating that they were formed in shallow seas or in lake bottoms.

Metamorphic rocks are igneous or sedimentary rocks that have been changed as a result of heat, pressure, and chemical action. Examples are slate (formed from shale), schist (from shale and igneous rock), gneiss (from shale and igneous rock), quartzite (from sandstone), and marble (from limestone).

A *rock cycle* is recognized as a sequence beginning with magma that cools thus becoming igneous rock (Judson, 1968). Igneous rocks disintegrate as a result of the action of weather, and the individual particles are later hardened to form sedimentary rock. The hardening process occurs when large deposits of unassorted eroded material build up to rather great depth (perhaps 2000–4000 ft), and the material in the lower part of the "pile" is compressed into a sedimentary form. Individual particles of the material may also be cemented by chemical action. Igneous and sedimentary rock can be metamorphosed through heat and pressure. In time, any of the three kinds of rocks will be melted to form magma.

The solid phase, or coarse skeleton, of soil is made up of weathered rock material of various sizes and from which most of the mineral elements, in ionic form, are obtained by tree roots. Soil clay minerals are most important in the ion exchange process. Even though a particular soil has an ionic exchange potential for supplying certain amounts of mineral ions, past land use alters the amounts of the different minerals that are actually present in the soil. The cation exchange capacity of a particular soil may be 17.7 milli-equivalents*; however, the availability of particular types of ions depends upon the kinds of ions that saturate the clay valences.

The surface area of organic and inorganic colloids is quite large and the amount of surface, hence proportion of clay, in a soil determines exchange capacity. The degree of saturation of the surface of the clay particles in a particular soil depends upon the rainfall and temperature of the locality as well as its past use. In general, H^+ ions predominate on the surface of clay particles where rainfall is high, as in eastern U. S. forest regions. In dry western United States forests, soils are lower in colloids and the mineral ions Na^+, K^+, Mg^{++}, and Ca^{++} can saturate the clay surfaces.

ABSORPTION OF IONS BY TREE ROOTS

To be absorbed ions must come in contact with a root surface. This requires they move within the soil water solution. Movement within this solution can occur either as mass flow or as diffusion. Mass flow would take place during soil moisture recharge; diffusion could occur during those periods when soil water was not moving—periods when soil water was at or below field capacity.

Ions enter roots and are transmitted across the cortical zone to the xylem elements against a concentration gradient. For this to be possible an expenditure of metabolic energy may be required. Kramer (1969) proposed that to accomplish the transfer of anions and cations from soil solution to the xylem, passive absorption occurs and ions are carried along in water solution in the transpiration stream. As a result, no biological energy would be involved, only the physical forces that moved water from the soil through the plant and out of the leaves. A

*

CATION EXCHANGE CAPACITY (mg/100 g SOIL)						
	TOTAL	Ca^{++}	Mg^{++}	K^+	Na^+	H^+
Spodosol	17.7 mg	2.0	4.2	0.1	0	11.4 (a northeastern U. S. soil group)
Ultisol	8.0 mg	2.0	1.0	0.1	0	4.9 (a southeastern U. S. soil group)

Organic matter between pH 4.5–5.5 has an exchange capacity of 50–120 mg/100 g.

second process, which does involve biological energy, may occur in activated diffusion. Active ion absorption may occur in a receptor system where organic molecules act as carriers of ions from the root cell membrane, then deposit them into a root cell vacuole. Apparently organic acid metabolism is involved in maintaining ionic balance between root cells and between the root and soil solution. Extension of roots into soil areas where ions are readily available also requires an expenditure of energy. In addition, mycorrhizal activity can increase the effectiveness of a root to absorb ions. Nitrate ions after they are absorbed are incorporated into organic compounds in the roots.

Tree species differ in their ability to absorb ions (See Table 8.7). In addition, absorption is affected by concentration of soil ions, and it has been demonstrated that one type of ion may affect the absorption of other types. There may be a differential absorption of ions. In some cases the presences of one type of ion may enhance the absorption of another type; on the other hand, there can be competition between ions where some types interfere with the absorption of other types (Fried and Broeshart, 1967). Soil temperature, moisture, and aeration affect root metabolic activity and growth, and therefore will affect ion absorption.

MINERAL CYCLING

The process by which mineral ions enter trees and are returned to the soil to be later recycled is called mineral cycling. Although the process is analogous to the hydrologic cycle, there are many differences; the process of mineral cycling is in some ways more complex (See Figure 7.17). If the movement of one element was to be followed along the pathways of the mineral cycle, the element could follow any of a number of different paths. The cycle can be considered as having three stages: *uptake* or absorption; *retention*—annual accumulation in the biomass (roots, boles, branches of trees and other vegetation); and *restitution*—annual return of litter (leaves, other organic debris, and wash from the air and from the vegetation) (Duvigneaud, and Denaeyer–De Smet, 1970). A fourth stage can be added where an ion may be *leached* from the soil and is lost to a particular site; however, leached ions may later be captured and recycled at another site, either terrestrial or aquatic. It can be speculated that eventually all mineral ions find their way to the oceans where they are incorporated into rock masses, to be recycled at some later time as new land masses are formed. However, unlike water that is recycled from the oceans in a relatively short time, minerals would be recycled from ocean depths at very long and infrequent intervals. Hence, once minerals are "lost" in the ocean, they can only be "reclaimed" through mining operations if they are to be used in the lifespan of the present human populations. The soil mineral reserve, therefore, must not be squandered.

In the retention stage, minerals are held for varying periods depending upon where they are finally lodged. If an ion is incorpoated into the leaf tissue of a deciduous species, it will be returned to the litter within the year and perhaps re-absorbed the following year if litter breakdown is quite rapid. On the other hand, if an ion is immobilized in the xylem cell wall tissue of the tree bole, it can be expected that it will not be recycled until the tree dies and that section of the bole decays at the point where the ion was released. If the ion was not made a part of the cell wall tissue but a part of the cell protoplasm, it could be translocated; later it might become a part of the tree where it could be retained for an additional period, or it could be restituted to the soil litter if incorporated into root tissue that later died. In any event, the ion will eventually return to the soil mineral reserve.

Ovington and Madgwick (1959) found that fairly large proportions of the total mineral supply is retained in the litter layer and the boles of standing trees (Table 8.5).

The mineral cycle can be separated into two general states: one where minerals are part of a retention pool, either organic or inorganic; the other defines minerals in flux—in the process of uptake,

TABLE 8.5. Distribution (Percentage) of Plant Nutrients in a Scots Pine Plantation (Ovington and Madgwick, 1959)

| | PORTION OF THE FOREST ECOSYSTEM | |
ELEMENT	LITTER AND DEAD PLANTS	BOLES OF TREES
K	42	22
Ca	59	21
Mg	39	29
P	66	10
N	82	5

retention, restitution, or leaching. A study of a mineral cycle in a particular situation must take into account the disposition of each element at each stage. In Figure 8.5, nitrogen is traced as it moves through a deciduous forest system in the southern Appalachians. Each year there is added to the system from the atmosphere 13 kg/ha/year of nitrogen and there is a loss of about 3 kg/ha/year, indicating that there is an annual increase of nitrogen in the system. Each year about 312 kg/ha of nitrogen are in the flux stage where nitrogen in organic and inorganic form is in transit within the system. The pool of nitrogen contains about 5850 kg/ha of nitrogen in organic and inorganic form and represents the base pool. It would appear that each year about 5% of the nitrogen in the system is in flux.

When a forest stand is harvested, it can be expected that a significant amount of the mineral reserve of the site will be removed. If greater utilization is made of the total wood from forest stands, then additional losses of the soil minerals will occur. In forest nursery practice it is quite important that a constant check be made of the amounts of the various soil nutrients, because when a crop of tree seedlings is lifted a large amount of the soil mineral supply is lost.

Of the major soil elements, nitrogen (N), phosphorus (P), potassium (K), and calcium (Ca) are retained in the greatest total amounts by

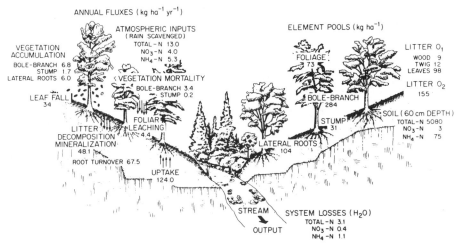

FIGURE 8.5. Representation of the nitrogen cycle on Walter Branch Watershed. Amounts of nitrogen in ecosystem components are shown on the right (kg/ha) while annual transfers are on the left (kg/ha/year) (from Henderson and Harris, 1975).

the standing forest crop (Table 8.5) (Figure 8.6). In the case of Douglas-fir the amount of K in the standing tree crop nearly equals the amount of K in the soil; with Ca, about one-half as much is retained in the standing crop as there is in the soil. About one-sixth to one-ninth of the amount of nitrogen found in the soil is present in the standing crop for Douglas-fir and loblolly pine.

Although the soil appears to have large reserves of all elements, the annual depletion of N, P, K, and Ca in rapidly growing juvenile stands is at the expense of these reserves (Table 8.6). If such high rates of depletion indicated in this table were to continue throughout the life of a stand, serious mineral deficiencies could occur. It can be expected, however, that the depletion rate will decrease as the stand approaches maturity and that there will be no serious deficiencies. Switzer, Nelson, and Smith (1968) found that for loblolly pine 30–40 years old, the demand for N, P, and K stabilized and that by age 45 nutrition exchange is mostly a matter of recycling. However, the addition to forest soils of mineral fertilizers containing N, P, and K in several instances has resulted in increased stand growth. Apparently

TABLE 8.5. Distribution of N, P, K, and Ca (kg/ha) in the Major Components of Several Forest Stands

	SCOTS PINE[a]				DOUGLAS-FIR[b]				LOB-LOLLY PINE[c]
	N	P	K	Ca	N	P	K	Ca	N
Leaves	89	9	43	36	102	29	62	73	89
Branches	79	9	43	43	61	12	38	106	
Boles	97	12	84	115	125	19	96	117	85
Roots	81	11	54	33	32	6	24	37	135
Total	346	41	224	227	320	66	220	333	309
Subveget.	37	6	4	17	6	1	7	9	0
Forest floor	1594[d]	76[d]	162[d]	311[d]	175	26	32	137	124
Soil					2809[e]	3878[e]	234[f]	741[f]	1900[g]

[a]Scots pine 33 year old plantation (from Ovington and Madgwick, 1959).
[b]Douglas-fir 36 year old natural stand (Cole, Gessel, and Dice, 1967).
[c]Loblolly pine 20 year old plantation (from Switzer, Nelson, and Smith, 1968). To convert kg/ha to lb/ac, multiply kg/ha by 0.892.
[d]F and H layers only.
[e]Totals.
[f]Exchangeable with pH ammonium acetate.
[g]Six inches of clay loam.

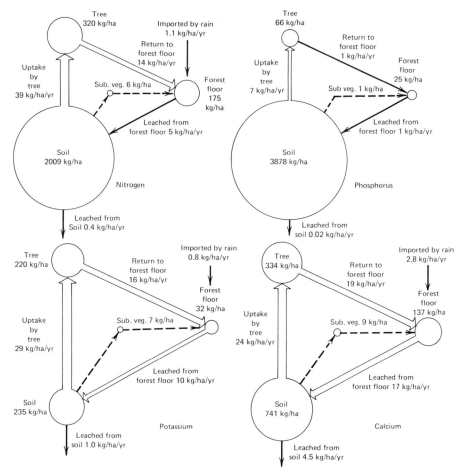

FIGURE 8.6. Distribution and annual biogeochemical cycle of N, P, K, and Ca in a second-growth Douglas-fir ecosystem (after Cole et al., 1967).

these elements are limiting growth on some sites and are not supplied in sufficient amounts to permit forest stands to reach optimum growth.

Jorgensen, Wells, and Metz (1975) found that at age 16 a loblolly pine plantation had an annual uptake of 104 lb N, 19 lb P, and 51 lb of K per acre. Studies have shown that slash pine plantations in the 15–20 year age group respond to application of nitrogen fertilization; annual growth increases of 18% were measured (Malac, 1968). It is quite important with slash pine that there be a sufficient number of trees (300–400 stems per acre) on the site to permit the stand to show

TABLE 8.6. Annual Accumulation of N, P, K, and Ca (kg/ha) Within the Major Components of the Total Ecosystem (Cole, Gessel, and Dice, 1967)

COMPONENT	N	P	K	Ca
Forest	23.5	6.6	14.4	8.7
Forest Floor	11.6	−0.4	5.3	1.1
Soil	−34.6	−6.3	−19.9	−11.5

maximum response. Other studies have shown that sycamore, tulip tree, cottonwood, and sweetgum respond to N fertilizer. Douglas-fir responds to the additon of nitrogen in the early stages of stand development. In natural Douglas-fir stands the best response occurred when the stands were thinned at the time of fertilizer application. Soils on which response to nitrogen treatment was best were those with high moisture-holding capacity. Nitrogen has a better chance to increase tree growth there than on sites where water is limited. Stated another way, where moisture is not limited, a soil mineral deficiency may be the reason tree growth is restricted. Only recently has it been shown that forest trees exhibit a deficiency in mineral supply. Apparently forest trees are able to mobilize the scarce mineral reserves of the soils on which they grow and in time attain an equilibrium between uptake and restitution.

Slash pine seedlings on poorly drained soils grow better when fertilized with phosphorous (P) fertilizer. The additional P appears to stimulate root growth, which results in seedlings becoming established soon after planting and contributes to greater growth in subsequent years over seedlings not receiving the treatment. Slash pine planted on deep sands does not appear to benefit from P fertilizer. Wells and Crutchfield (1969) found that one-year loblolly pine seedlings with foliar P greater than 0.11% did not respond to additional amounts of P; however, when foliar P was less than 0.11%, trees did respond. Those with very low amounts of foliar P grew significantly better. Douglas-fir did not respond to P when it was applied alone; but when it was applied with large amounts of N a response was detected (Steinbrenner, 1968).

Tree response to additions of potassium has been noted on acid, sandy to loamy soils that are low in organic matter and total cation exchange capacity. These soil properties are characteristic of alluvial and glacial outwash soils of the northeastern United States. Heiberg and Leaf (1960) reported the response of pine and spruce plantations to potassium fertilization.

Zinc, which is necessary in only minute quantities but is essential for sustained growth, is deficient in some soils in western Australia, and growth of Monterey pine is inhibited unless zinc is applied to the soil (Stoate, 1951). A number of other instances are cited by White and Leaf (1956) where addition of trace elements increased growth of forest stands. However, some amounts of these elements are obtained from ions present in the air and carried to the soil by rain and snow particles and thus may prevent deficiencies of these elements in some areas. Readers should review proceedings of various forest soil symposia that are held from time to time. Particularly helpful are reports such as those of T.V.A. (1968) and Youngberg and Davey (1970).

As progress is made in the use of fertilizers to increase tree growth, it has become apparent that all trees in a stand do not respond in the same manner. Studies that have been made with selected genotypes show that some genotypes respond to fertilization while others show depressed growth; or in other cases there is no response.

When increase growth of forest stands by nitrogen fertilization is obtained, it appears that soil moisture is not a limiting factor. Phosphorous deficiency may be found on wet sites where loblolly pine and slash pine are to be planted. Potassium may be limiting on soils of low inherent fertility. In general, where soil moisture is not limiting growth, application of nitrogen to stands in the sapling and pole stages of development may result in a growth response. In other situations soil moisture may be the factor most limiting to growth and the greatest growth response will be to the addition of water. Schultz (1969) showed that after nine years a 13% response in height growth of selected clones of plantation slash pine occurred when the trees were both fertilized and irrigated, and that only a 5% response occurred when trees received only fertilizer. Schultz also observed, as have others, that use of nitrogen fertilizer increased female flower production; in the case of slash pine, in one year there was a threefold increase in female flowers the year following treatment.

Gemmer (1932) and Allen (1953) both observed increased cone production on longleaf pine that had received fertilizer; there was, as well, a positive response to increased soil moisture and to crown release. Steinbrenner, Suffield, and Campbell (1960) reported a tenfold increase in cone production on Douglas-fir trees that had received both nitrogen and phosphorous treatment.

SOIL ORGANIC MATTER

The addition of dead plant and animal residue to the soil surface, which occurs each year in the forest, has profound physical and chemical effects on the forest environment. These materials support the life processes of other plants and animals, alter the rate and pathways of water into the soil, and produce chemical compounds that result in the formation of additional soil material and which in turn promote the soil forming process. The antiseptic and mineral exchange properties of humus, and humus-derived compounds, is only beginning to be revealed. It has been proposed that about 50% of the exchange capacity of humid region surface soils is due to organic matter (Fried and Broeshart, 1967). The fact that organic matter is the primary source of nitrogen is generally accepted. Examination of the organic soil-mineral soil boundary (the H and A_1 soil horizons) shows a presence of large numbers of fine tree roots, indicating these layers, containing large amounts of organic matter, are particularly favorable for tree root growth.

In the upper layers of the litter (L and F) and in the deeper soil mineral horizons (B and C), root development declines rapidly from that observed in the H and A horizons. Within the H and A soil horizons maximum "feeder" root concentration occurs. During the growing season, it is in these zones that water and minerals are available in greatest amounts.

The water holding capacity of sandy soils (sands, loamy sands) can be improved considerably by adding humus. The acid residues produced by leaching of humus and from dissolved carbon dioxide in the soil water are effective in causing soil "skeletal" material to break down (weather) thus releasing new soil material.

Humus is then responsible for not only contributing to the mineral cycling, but to soil water relationships and to soil weathering.

Most of the soil biotic activity is associated with the organic surface (L, F, and H horizons) on top of the mineral soil and in the organic-mineral interface (A_1 horizon). The thickness of the organic layer reflects not only the productivity of the sites for tree growth, but also the amount of microorganism activity.

The annual accumulation of organic matter deposited on the forest floor varies with age, species, stand density, site quality, and climate (Figure 8.7). Annual leaf-fall can be considered to begin in later summer when some foliage accumulates during dry weather. Later, in early autumn, leaf-fall becomes progressively greater and by early

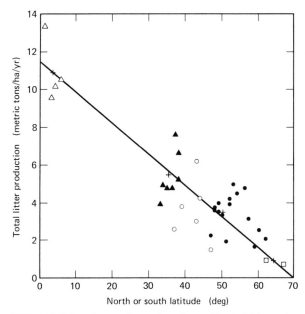

FIGURE 8.7. Annual production of total litter in
relation to latitude. Open triangles—equatorial,
solid triangles—warm temperate, circles—cool
temperate North American (open) and European
(closed), squares—Arctic-Alpine. Line fitted vis-
ually to means for climatic zones, shown by large
crosses. One alpine Californian stand is excluded
(Bray and Gorham, 1964).

winter most deciduous species are bare and older needles of conifers
(two and three year needles) have fallen. Other tree debris adds to the
accumulation of organic litter: twigs, branches, dead roots, bark, and
flowers accumulate from living trees, and the boles and roots of dead
trees add to the accumulation. Organic matter is deposited continually
throughout the year, there is really no particular season of deposition,
rather, there are relative differences in the amount of deposition between seasons.

Willston (1966) found he could relate the weight of forest floor to
stand age and basal area for young loblolly pine plantation: the relationship in tons per acre (forest floor) = 1.38 + 0.246 (age) + 0.0218
(basal area/acre sq. ft.). Weight of organic residue varied between
stands, and the type and amount of mineral material varied as well.

Lyford (1941) reported that red maple was more efficient than white pine in accumulating minerals (Table 8.7). Metz (1954) reported litter fall in southeastern Piedmont stands averaged 4400 lb per year, 3500 lb of which was leaves, the remainder being twigs, bark, and fruit. In pine stands, annual leaf fall contained 18 lb of calcium and 13 lb of nitrogen per acre. Leaf fall in hardwood stands contained 85 lb of calcium and 26 lb of nitrogen, while in mixed stands (pine-hardwood) leaf fall contained 43 lb of calcium and 23 lb of nitrogen. The amount of the annual accumulation does not indicate how much organic and mineral matter is contained in the forest floor. Raw litter takes several years to decay with the result that the forest floor at any one time contains organic material in various stages of decay (Table 8.8).

Lunt (1932) reported the weight of the forest floor in a New Hampshire birch-maple-spruce stand was 262,791 lb per acre, and 118,388 lb in a sugar maple-white pine hemlock stand in Connecticut. Mader and Lull (1968) reported the weight of organic matter (L, F, H) in white pine stands in Massachusetts ranged in weight from 10,200–85,800 lb per acre and was related to stand and site factors as follows:

Organic matter (lb per acre) = 43.104 + 0.156 (stand age)
  − 0.0184 (stand basal area) − 0.335 (site index) − 0.0051 (elevation in feet) + 0.38 (drainage class).

Codes for drainage classes were:

0 = very poorly drained   1 = poorly drained
2 = somewhat poorly drained 3 = moderately drained
4 = well drained   5 = somewhat excessively drained
6 = excessively drained

TABLE 8.7. Percent Mineral Content of Freshly Fallen Leaves (From Lyford, 1941)

	NITROGEN		PHOS-PHORUS		POTASSIUM		CALCIUM	
	RM[a]	P[a]	RM	P	RM	P	RM	P
Brookfield loam	0.59	0.42	0.16	0.06	0.56	0.01	1.40	0.67
Gloucester silt	0.60	0.53	0.44	0.12	0.64	0.24	1.73	0.91

[a]RM = red maple; P = white pine.

TABLE 8.8. Weight of Forest Floor (lb per acre) in Some Piedmont Stands (From Metz, 1954)

STAND	ORGANIC SURFACE HORIZON		OVEN-DRY MATERIAL	NITROGEN
Loblolly pine 12 yr	L		4640	22
(plantation)	F		8290	59
		Total	12930	81
Shortleaf pine 25 yr	L		4110	19
	F		12600	107
		Total	16710	126
Shortleaf pine 40 yr	L		4260	24
	F		7260	61
	H		11750	80
		Total	23270	165
Pine-hardwood[a]	L		4870	45
	F		8360	77
		Total	13230	122
Pine-hardwood[a]	L		3530	39
	F		4460	56
	H		9040	89
		Total	17030	184
Pine-hardwood[a]	L		2980	30
	F		5140	59
	H		9930	67
		Total	18050	156
Yellow poplar 45 yr	L		11940	145
Oak 50 yr	L		2780	34
	F		3780	48
	H		6470	52
			13030	134
Hickory 50 yr	L		8000	85

[a]Mixed stands with trees up to 50 years old consisting of shortleaf pine and upland hardwoods.

An approximate ratio of 100 parts dry matter to 1 part of nitrogen is found in the soils listed in Table 8.8. In the mineral soil layers below the organic layer this ratio will be much lower. However, where the ratio between organic dry matter and nitrogen is lower, the litter breakdown by microorganisms is more rapid than where it is higher. For example, the data shown in Table 8.7 indicate that decay rate of red maple, other conditions being the same, would be more rapid than that of white pine.

The difference in character of the surface litter layers has resulted in their being classed in one of two groups: mor or mull. *Mor humus* types result in a litter layer that is unincorporated and distinct from the mineral soil below. The surface organic layer is usually matted or compact, or both, and is distinctly delineated from the mineral soil except as the mineral layer may be blackened by washing in of the humus material. *Mull humus* types consist of an organic and mineral layer so mixed that the transition between them is not sharp and the A_1 horizon is well developed. On mor soils the A_1 may be absent or weakly developed. (A more detailed classification of mor and mull is given in the section "Function of Residue Decomposition.")

Williams and Dyrness (1967) measured depth and weight of the forest floor in Pacific silver fir–Noble fir–California red fir–mountain hemlock stands, and found those measurements to be 1.8 in. with a mean weight of 56,754 lb per acre. Analysis of the nutrient level of the forest floor (L, F, and H layers) and the underlying mineral soil showed that usually less than one-quarter of the total available nutrient supply was contained in the forest floor material.

Gessel and Balci (1965) found that weight of forest litter in a 80–150 year old Douglas-fir stand in eastern Washington averaged 25,670 lb per acre; while in western Washington forest floor weight in stands of similar composition and age was only 12,880 lb per acre. A comparison between mull and mor forest floor weights in five old-growth forests in the Cascade and Olympic mountains showed that there was no difference in average L, F, and H total weight. For the mor type, dry weight of the forest floor was 28,992 lb per acre, and for the mull, 18,978 lb per acre. Although the difference between means is large, they are not significantly so because the errors of estimate of mean litter weight were affected by large differences in stand density, topographic irregularities, occurrence of logs and other woody portions of trees and shrubs, and variation in age among trees.

Weights of organic matter of the forest floor in Piedmont stands are lower than weights of organic matter in stands in the northeast and northwest, reflecting a more rapid rate of decomposition in the warm climate of the southeast. Witkamp (1966) made counts of bacteria and fungi in litter from different species in oak, pine, and maple stands in Tennessee. He found microbial density, microbial respiration, and annual weight loss of litter were controlled by temperature, moisture, and age of litter, as well as bacterial density.

Metz (1954) found a relationship between the organic and nitrogen content of the Piedmont surface soils. The relationship was expressed as:

% total nitrogen (surface 12 in.)
    = 0.0049 + 0.0281 (percent organic matter surface 12 in.)

Although the original source of soil nitrogen is from the atmosphere, the amount of nitrogen available to trees depends to a great extent upon the organic production of the site. A site that produces a large volume of organic material will in time be able to support a large bacterial population which in turn will cycle nitrogen through the system (Figure 8.8).

Stone and Fisher (1968) found that grasses, forbs, and tree seedlings which grew under and adjacent to canopies of a 10–14 year old larch and pine plantation contained more nitrogen and phosphorus than the same species growing at three to four meters in a bordering old field. Nitrogen and phosphorus were also higher in the upper soil levels within the plantation. Apparently conifers cause both nitrogen and phosphorus to accumulate in the litter and upper soil layers. The authors suggest that such an accumulation may influence the growth of understory hardwood species. Also, it should be noted that tree crowns intercept nutrient laden air and the nutrients are washed off the leaves and stems by rain and snow and are deposited on the soil surface where they become a part of the soil mineral reserve.

BIOTIC SOIL FACTORS

Biotic soil factors affect both the relatively complex relationships between the micro and macro fauna and flora that grow in the soil and, in the case of forest biology, the effects these factors have on tree growth.

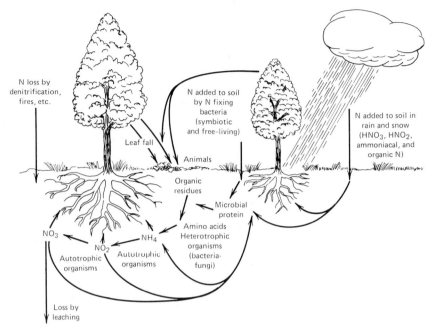

FIGURE 8.8. Diagrammatic representation of the nitrogen cycle (Lutz and Chandler, 1946).

Symbiotic Relationships

Tree roots are very likely to become infected with fungi, and some species support bacteria. These associations result from trees and the endophytic organism. Roots of locust, a legume, may develop nodules as a result of bacterial growth from species of *Rhizobium*.

Species of non-legumes such as alder (*Alnus*), snowbrush (*Ceanothus*), and bitterbrush (*Purshia*) also can develop root nodules. There is some question as to the endophytes responsible for nodulation of non-legumes; however, Youngberg and Hu (1972) were able to isolate a *Streptomyces*, a member of the line *Actinomycetes*, from roots of mountain mahogany (*Cercocarpus ledifolius*). They indicated that these isolates were similar to those from snowbrush. There was a direct association between trees with nodulated roots and increased foliar nitrogen and tree vigor.

It has been shown that when alder is grown in mixture with spruce and pine, or when spruce and pine plots were treated with alder leaf litter, soil nitrogen content was raised and increased leader growth occurred. There were, as well, increased amounts of nitrogen and chlorophyll in the pine and spruce needles (Schalin, 1966). In a heavily thinned Douglas-fir stand, a heavy understory of red alder in about 20 years was responsible for adding 780 pounds of nitrogen to the site (Berg and Doerksen, 1975). In addition, the presence of tree species that have nitrogen root nodules proved beneficial to the formation of humus, soil moisture, and soil nitrogen level.

The relationship between trees and mycorrhizal fungi is not completely understood. Hatch and Doak (1933) and Doak (1934) recognized that the effect of mycorrhizae upon the host was not entirely pathogenic, that a mutually beneficial relationship appeared to exist between tree and fungus. It is now generally recognized that trees whose roots are infected with mycorrhizal fungi appear to benefit from the infection. The initial hypothesis was that the mycelial strands penetrated the soil and extended the effective root zone of trees. More complex relationships appear to exist; however, these relationships do not diminish the importance of the physical effect mycelial strands have upon increasing the absorption area of roots. There is some feeling that in exchange for the carbohydrates needed for growth, mycorrhizal fungi supply growth hormones and amino acids that, in turn, stimulate growth and longevity of the host feeder roots. In addition to increasing tolerance of trees to drought, high soil temperature, and soil toxins, ectomycorrhizal fungi deter infection of feeder roots by root pathogens, especially from species of *Phythium* and *Phytophthora* (Marx, 1973).

Trappe (1962) reported there are about 528 species of mycorrhizal fungi, primarily *Basidomycetes*. These fungi represent 100 orders, 30 families, and 81 genera. In soils under natural stands, the "club"-like mycorrhizae are easily found (Figure 8.9*a*). Few studies have been conducted where a specific fungus has been used, primarily because of the difficulty in handling a causal fungus in pure culture. Marx (1976) reported the potential benefits of using inoculum of *Pisolithus tinctorius*, an ectomycorrhizal fungus, to improve growth of planted conifers and oaks on adverse as well as normal planting sites. Commercial inoculum of this and other fungi could improve growth of trees on a variety of sites.

The initial extension of myceliae strands of the different types of mycorrhizal fungi affect different parts of the tree root cortex. Growth of one group of fungi is confined to the intercellular space of the root cortex producing ectomycorrhizae (Figure 8.9*b*); another group has mycelium that grows within the cells of the root cortex, resulting in

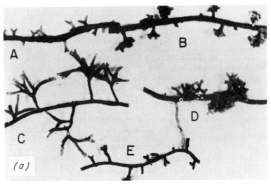

FIGURE 8.9. (a) Infected short roots of shortleaf pine (*P. echinata* Mill.) seedlings grown in shortleaf pine humus for six months in the greenhouse. Five distinct morphological associations are often evident on individual seedlings (A, B, C, D, and E) (from Marx and Davis, 1969).

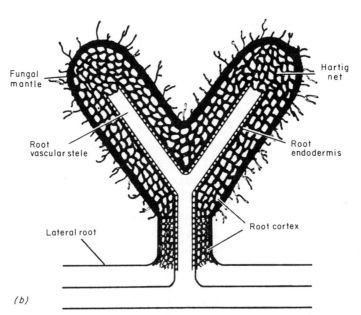

FIGURE 8.9. (b) Diagram of an ectomycorrhizal association with the short shoots of wood plants (courtesy of D.H. Marx, U.S.F.S., Athens, GA.).

endomycorrhizae. A third group has been noted in which the mycelium occupy both the intercellular space and the interior of the cortex cells; these produce ectoendomycorrhizae.

The reader is referred to the summary paper by Marx (1976) for more detail on the role of mycorrhizae on tree growth. Ectomycorrhizae affect all members of pine, spruce, fir, larch, and hemlock as well as certain species of beech, hickory, oak, aspen, and willow. Endomycorrhizae are important to maple, elm, sweetgum, ash, sycamore, and black walnut. Ectendomycorrhizae have the features of both ecto and endomycorrhizae; however, they appear limited in their distribution and are found primarily on roots of trees normally affected by ectomycorrhizae. Fungi primarily responsible for ectomycorrhizae are the *Basidiomycetes,* mushrooms, and puff balls; however, some *Ascomycetes* are also involved. Fungi responsible for producing endomycorrhizae are mainly *Phycomycetes,* which do not produce large, above-ground fruiting bodies. Spore dissemination in this group is primarily below ground from feeder root to feeder root, or by water or soil insects and animals.

The symbiotic relationship between mycorrhizal fungi and trees, and between root nodule bacteria and trees, is evidence that plants have been able to evolve life systems that make them adaptable to environments which may be deficient in one or more elements essential to survival. It seems that not only do mycorrhizae contribute to a tree's enzyme system, but there is a relationship involving mineral nutrition as well. Bowen (1973) reported that N, P, and K uptake per gram dry root weight of mycorrizal roots was 1.8, 3.2, and 2.1 times that of nonmycorrhizal roots. Also uptake of Ca^{++}, Rb^+, Cl^-, $SO_4^=$, Na^+, Mg^{++}, Fe^{++}, and Zn^{++} was affected by the presence of mycorrhizae. Mycorrhizae are able to absorb N and P in an organic form.

It has been demonstrated that forest trees whose roots are not mycorrhizal cannot survive on soils that lack a natural population of fungi; failure of forest plantings on prairie soils has been attributed to such lack. The species of mycorrhizal fungi and their numbers will depend upon a complex set of physical, chemical, and biological balances within the soil. Included within the biological complex is the need for a suitable host or within or upon which a fungus species can obtain the necessary nutrients for growth. Subsequent development of a fungus will then depend upon having temperatures suitable for growth, adequate water, and a soil that is moderately deficient of mineral elements (Harvey, Jurgensen, and Larsen, 1976). In general, it can be assumed that environmental conditions favoring growth of a host species will also favor growth of a mycorrhizal fungus that infects the roots of that species. If the host species suffers poor growth, it can be expected there will probably not be excess of nutrients for mycorrhizal growth. Although the symbiotic relationship between trees and mycorrhizal fungi and root nodual bacteria appears to permit trees to grow on sites that are inherently low in fertility, these root inhabiting microorganisms permit trees also to mobilize the available soil nutrients on most sites to best advantage. The symbiotic relationship permits plants to invade poor or degraded sites and in time to improve the capacity of the sites to support other plants that are more demanding in their growth processes. In effect, some degree of site improvement can be expected as a result of plant growth on disturbed sites; however, ultimate production of each site will be fixed, there being a limit to the amount of improvement that plants can render to a particular physical setting.

Free Living Soil Biota
The free living soil biota can be classified as follows:

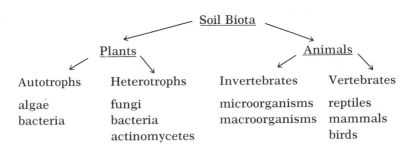

The groups can be arranged in other ways, but this approach does not compromise a functional classification. The main function of each group, other than to obtain some or all of the life needs from the soil, is their abilty to reduce the large amount of organic debris that is deposited upon the soil surface each year. Animal invertebrates attack this raw material reducing it directly, or they expose it to invasion by other organisms that in turn reduce it. Bacteria and fungi are capable of obtaining their life needs directly from the plant and animal residue. They utilize the carbohydrates as a source of energy and other organic compounds furnish basic material for cell division and growth. Some animals are able to benefit directly from plant residues because their digestive systems can obtain energy from the carbohydrates contained in the residue. Other animals and some bacteria and fungi depend upon the primary decomposers to concentrate the energy and minerals before they are able to utilize the potential energy available. These organisms, called secondary decomposers, feed on detrital remains. Next in line are plants and animals that consume the primary consumers directly; these are considered predators. In effect, the soil organisms represent a biotic pyramid where algae, fungi, and bacteria provide nourishment for mites that in turn are consumed by other arthropods, which in turn serve as food for vertebrates (Table 8.9).

The effect of these organisms on tree growth is in many instances not entirely known. Some pathogens such as the damping-off fungi cause considerable loss in germinating seedlings. Other plants and insects such as protozoa and mites are responsible for causing breakdown of fine plant and animal residue, which aids in mineral cycling and humus production. Earthworms are effective in bringing soil material together, causing particles to stick together to form aggregates that improve soil structure. Nonsegmented worms such as the nematodes reduce tree growth by destroying tree roots; other nematodes appear to benefit tree growth by feeding on bark beetles, white grubs, and other insects capable of damaging trees. Large animals burrow in the soil, creating holes that enhance the infiltration of water into the soil, thus reducing the potential for surface water runoff.

TABLE 8.9. Biomass of Groups of Soil Animals under Grassland and Forest Cover[a]

GROUP OF ORGANISMS	BIOMASS (g/m²)		
	GRASSLAND MEADOW	FOREST	
		OAK	SPRUCE
Herbivores	17.4	11.2	11.3
Detritivores			
Large	137.5	66.0	1.0
Small	25.0	1.8	1.6
Predators	9.6	0.9	1.2
Total	189.5	79.9	15.1

[a]Data from Macfadyen, 1963.

Davey (1970) arranged the soil microorganisms into the following groups:

Types of Organisms

I. Plant Kingdom
   A. *Protobacteria*—Very small (less than 1 μ in diameter); not well understood. All appear to be parasitic on bacteria.
   B. *Bacteria*—Very diverse group in their activities. Almost always present in the highest numbers in the soil but usually not the greatest biomass. Some are aerobic and some anaerobic. They average about 1 μ in diameter and are very rapid in reproduction. For example, one cell may multiply to one billion cells in 15 hours. A few produce spores but only one spore per cell (see Figure 8.10a).
   C. *Actinomycetes*—Small, mostly filamentous organisms (average about 1 μ in diameter). Some are aerobic and some are anaerobic.

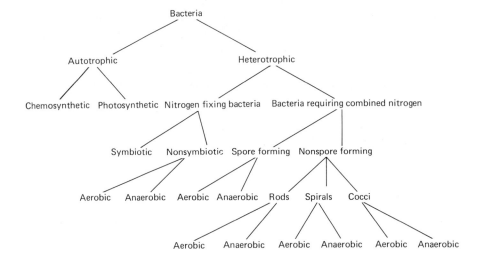

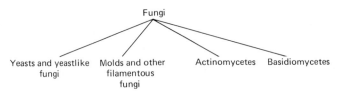

FIGURE 8.10. (a) The classification of soil bacteria and fungi by Waksman (1952).

It has been said that actinomycetes look like fungi but behave like bacteria, that is, they are filamentous in form like the fungi but are small in diameter and have growth requirements like the bacteria. Most of them produce spores like the fungi either singly or in long chains.

D. *Fungi*—Except for the yeasts, fungi are mostly filamentous (hyphae). They are larger than the bacteria or actinomycetes, averaging greater than 5 μ in diameter. They often represent the greatest biomass in forest soil. Nearly all produce spores and many produce more than one kind of spore. As far as is known, there are no truly anaerobic fungi. (See Figure 8.10b.)

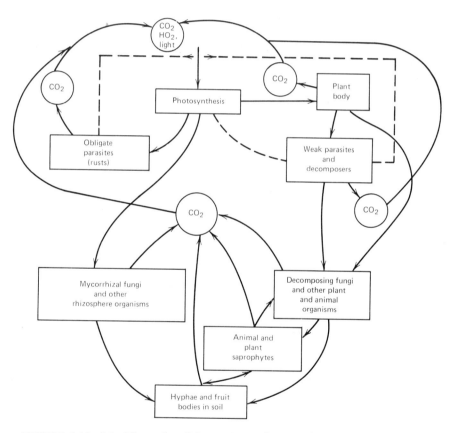

FIGURE 8.10. (b) The role of fungi in cycling carbon within an ecosystem. On the left are rust fungi that affect standing trees obtaining their life needs directly from living tissue. The decomposition cycle is at the right where weak parasitics and decomposers break down dead tissue. In the lower part is represented the symbiotic fungi (after Harley, 1971).

E. *Algae*—Several types of algae exist but most are single-celled chlorophyll-bearing organisms. Nearly all carry on photosynthesis when light is available. All are aerobic. Many produce a resistant resting form but not spores like the fungi. In schemes of classification involving more than two kingdoms, the blue-green algae are usually classified in a different kingdom from the other algae because of their cellular organization.

II. *Animal Kingdom* (all aerobic; no spore-formers) (see Table 8.10 and Figures 8.11a and 8.11b).

A. *Protozoa*—The protozoa are the true micro-animals. They are single-celled and come in three basic forms: amoeboid, cilliate, and flagellate. These designations are based on their type of mobility.

B. *Rotifers*—The rotifers represent a group of small multi-celled animals. They are not true microorganisms but are active and important to forest soil development in several of the same processes as the true microbes.

C. *Nematodes*—The nematodes are also multi-celled animals that are active and important in forest soil.

D. *Acarina*—The acarina include the mites and ticks. They are very large in comparison with the true microbes, and yet many of them are sufficiently small that they cannot be seen by the unaided eye. There are several other groups of small animals in forest soil, but the acarina are usually the most active and represent the types of functions common to several of the groups.

TABLE 8.10. Classification of the Most Important Soil Animals

1. Unicellular organisms (Protozoa)	Amoeba, naked and shell-bearing varieties Flagellates Ciliates
2. Worms	Flat worms (Turbellaria) Rotifers Roundworms (Nematodes) Earthworms and enchytracid worms (oligochacte annelids)
3. Tardigrades (Bear animalcules)	
4. Onychophora (*Peripatus*)	
5. Mollusks	Snails and slugs

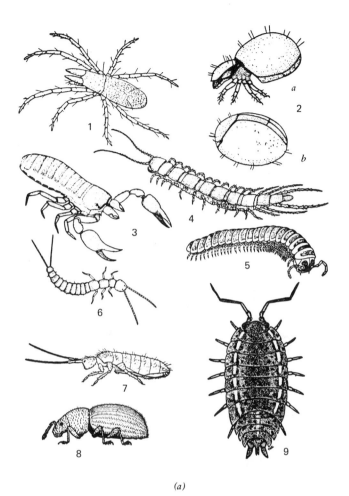

(a)

FIGURE 8.11. (a) Animals inhabiting forest soils (courtesy Eaton and Chandler, 1942. *1*. Rhagidia, a predatory mite, length 1 mm. *2*. Hoploderma, a saprophagous mite, length 0.5 mm; (*a*) walking; (*b*) rolled up for defense. *3*. Neobisium, a false scorpion, length 3 mm. *4*. *Lithobius forficatus*, a common red centipede, length 35 mm. *5*. *Fontaria coriacia*, a mull-forming millipede, brown with yellow bands, length 40 mm. *6*. *Campodia staphylinus*, a bristle-tail, length 4 mm. *7*. Tomocerus, a spring-tail, length 4 mm. *8*. ?, a brown short-bristle, length 3.6 mm. *9*. *Tra-*

6. Arthropods   Crustaceans: Terrestrial copepods
                                 Isopods (wood lice)
                                 Arachnids: Scorpions
                                                Pseudoscorpions
                                              Harvestmen
                                              Soil spiders
                                              Mites
                                 Myriopods: Pauropods
                                              Symphyla

Crustaceans: Terrestrial copepods
Isopods (wood lice)
Arachnids: Scorpions
            Pseudoscorpions
            Harvestmen
            Soil spiders
            Mites
Myriopods: Pauropods
            Symphyla
            Diplopods (millipedes):
                Julids (snake millipedes)
                Glomerids (pill millipedes)
                Pselaphognatha (tufted millipedes)
            Chilopods (centipedes):
                Geophilids
                Lithobids
                Scolopenders
Insects: Primitive insects:
              Protura
              Springtails
              Japygids
              Campodeids
              Machilids
              Silverfish
              Winged insects
              (Pterygota):
              The larvae of many insects and various
          adults forms such as earwigs, crickets,
          beetles, cockroaches, etc.

7. Vertebrates
Amphibians: Gymnophiona (limbless burrowing
                 caecilians)
Reptiles: Amphisbaeridae (wormlike reptiles)
           Skinks
           Typhlopidae (snakes)
Mammals: Moles
          Pocket gophers
          Shrews
          Voles

THE FUNCTION OF RESIDUE DECOMPOSITION

There is a direct association between the various plant-animal groups that inhabit the forest floor. For example, free-living nitrogen-fixing bacteria provide nitrogen to trees and other plants, but nitrogen so

---

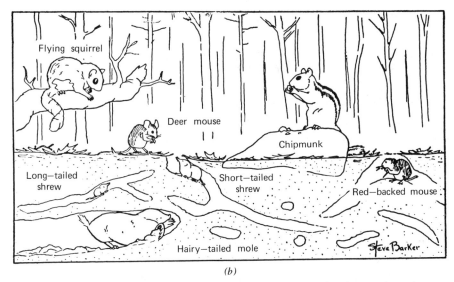

FIGURE 8.11. (b) The relation of small mammals to the forest floor (courtesy Hamilton and Cook, 1940).

produced is also available to fungi and other organisms growing in the soil. Bacteria that live in or are ingested into the gut of an earthworm affect the condition of plant residues as they utilize dead organic matter. Each organism is influenced in turn by the activity of other organisms; however, all will be affected by the micro-climate that prevails on a particular site and by the physical and chemical conditions of its soil. The decomposition process permits carbon, nitrogen, and other minerals that were incorporated into plant and animal bodies to be re-mineralized and in turn to be made available first for plant use and later for animal use. Simplistically, in the process of decomposition, the products and by-products of photosynthesis are decomposed into their original form—water, carbon dioxide, and minerals. The process of decomposition is essential to meet the cyclical need of the ecosystem and it has evolved into a complex ecological network of interrelationships because all organisms are not capable of growing under all climate and soil conditions.

The activities of the plants and animals reduce the several tons or more of organic debris deposited each year on the soil surface of a temperate region forest stand to a small amount of semi-amorphorus dark-colored material called humus. This material is composed primarily of lignin and is essential to the mineral cycling process. In the process of humus formation, different plants and animals will have been responsible for the reduction. Mull humus is found under deciduous stands growing in moderate climates and on mesic sites where the soils are not too acid. Mor humus tends to form on colder and dryer sites, or hotter and dryer sites where the vegetation can be coniferous and the soils less capable of retaining water. At one time it was believed that the mull humus type represented a soil building process and that the mor type a degrading process. It should be recognized that mull and mor types are a result, not a cause, of the conditions that prevail upon a particular site. It is true that microorganism activity is lower on mor sites, but this is brought about by the prevailing climate-soil conditions. On a hot-dry site over a coarse-textured soil, one cannot expect to find many earthworms because these animals grow primarily on mesic sites. Their place in the decomposition process on dry sites may be occupied by arthropods (mites, millipedes, or spring tails), animals that can survive where temperature and moisture are not favorable for mull forming animals. It should not be concluded, however, that arthropods do not populate mesic sites, nor that earthworms may not be found on some xeric sites. Bornebuch (1930) showed that on mull sites earthworms constituted a biomass of 30.4 g/m^2 and on mor sites 1.8 g/m^2, and arthropods on mull had a biomass of 10.2 and on mor 9.4. Wilde (1958) indicated mull humus types were produced by both earthworms and arthropods and mor formed primarily as a result of arthropods and fungi; he later (Wilde, 1976) suggested a system of humus classification that included eight general groups under either of two classes: (A) endoorganic, or mull humus types and (B) ectoorganic, or the mor types (Figure 8.12).

Although the aggregate biomass of the heterotrophic organisms that occupy a forest site is relatively small compared to the primary production plant biomass, the heterotrophs have large energy demands. For example, from a case reported net primary production was 1756 $gC/m^2/yr$ (2162 GPP-1436 Autotrophic Rn), and heterotrophic respiration (Rn) utilized 670 $gC/m^2/yr$ (Harris et al., 1975).

SOIL POLLUTION FOLLOWING FERTILIZATION, BURNING, AND CLEARCUTTING

Actually, what is involved here is the pollution of soil water as a result of excessive accumulation of mineral ions in the soil; the soil is unable to retain the ions. A large increase in free ions causes them to be lost to ground water where they can cause damage to biological populations because of toxic amounts, or from their effect on the balance in numbers of different biological populations. As a example, the bursts of algael growth following increased P concentration in streams can result in the reduction of the water oxygen level, which will limit growth of fish and other water life.

In a managed forest there should be very little increase in mineral ion level of soil water following fertilization. The soil organic layer provides an efficient "sump" where temporary excesses of various minerals can be "stored." The humic clays and the tremendous number of soil microorganisms present in the litter can hold, or use, any excess nitrogen and phosphorus applied to the soil. In fact, the soil microorganisms compete with the trees for the minerals applied to the soil. This competition provides two beneficial effects: first, the temporary excess of minerals can be stored; second, the minerals are released to the trees over a longer period of time (approximately five year in the case of nitrogen) than if microorganisms are absent. Then, too, the amounts of a fertilizer added to the soil are generally less than the amount the soil can retain.

Prescribed burning will have the effect of temporarily decreasing the nitrogen level of the soil because organic nitrogen is volatile and will pass off as gas during combustion. Other elements are released in large amounts and may be leached from the soil. Soil pH will probably increase. There are, nonetheless, heavy clay soils on steep slopes that are quite subject to erosion. Prescribed burning on these should not be considered, or, if burning is done, only the lightest possible burns should be used so that a layer of litter is retained. There have been no studies to date which indicate that burning is detrimental to physical soil structure; however, it can be assumed that on coarse-textured soil, particularly, there could be a significant increase of minerals in the soil water because there would not be enough clay particles in the soil to retain the excess of mineral ions. (The effects of fire on forest soil are discussed in Chapter 11.)

Large, uncontrolled wildfires are capable of reducing the mineral reserve of a site because large amounts of ash that contain various minerals are carried from the site by the accelerated air mass movement created by the heat of the fire. Prescribed fires, being less violent, do not create large air mass movement. Some particulate matter is created by prescribed fire, but most of the smoke is actually water vapor. Some nitrogen is lost but the site is not depleted of nitrogen or

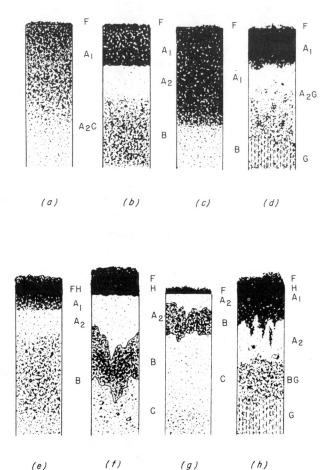

FIGURE 8.12. Types of humus layers of forest soils. (*a*) *Parvital*, microbiotic endohumus. (*b*) *Vermiol*, earthworm mull. (*c*) *Rhizol*, sward of prairie-forest soils. (*d*) *Sapronel*, muck-like fen mull of wet soils. (*e*) *Leptor*, arthropod paramull. (*f*) *Lentar*, lignomycelial raw humus. (*g*) *Crustar*, lichen crust. (*h*) *Uliginor*, swamp saprogenous raw humus (from Wilde, 1976).

A. ENDORGANIC LAYERS—Figure 8.12(*a*),(*b*),(*c*),(*d*)

Parvital (microbiotic or micro-aggregated mull: term from the Latin *parvus* for small and *vita* for life). Dark minero-organic layers inhabited by floristic microorganisms and often by nematodes: characterized by sparse, mycelia-covered litter and fine, bacteria-aggregated structure.

Vermiol (earthworm mull: term from the Latin *vermis* for worms). Dark, crumb-like castings of *Lumbricus* and other worms, occasionally including *Enchytraeidae*.

other minerals as a result of prescribed fire, nor is the important organic reserve depleted. Soil microorganisms are able to survive in the lower organic layers (F,H) that are not consumed in a prescribed fire.

On a watershed where timber had been clearcut and the vegetation sprayed each year for three years with herbicides to prevent sprout growth and to kill annual weeds, Bormann et al. (1968) found the loss of cations was 3 to 20 times greater than on comparable undisturbed systems. The most notable increase occurred in the loss of nitrate nitrogen from the area that was cut. The study reported by Bormann and his co-workers has caused a number of scientists to question the advisability of clearcutting stands. However, it should be pointed out that in this study the treatment following cutting represented an ex-

---

Rhizol (root mull, sward: term from the Greek *rhiza* for root). Dark layers containing residue of decomposing roots: this type is largely confined to soils of the prairie-forest regions.

Sapronel (fen mull, hydro-mull: term from the Greek *sapros* for putrefaction or anaerobic decay). Muck-like layers of hydromorphic, base-enriched soils formed under the influence of hydrolysis and hydrophilic organisms including pot-worms, algae, protozoa, and anaerobic bacteria.

B. ECTORGANIC LAYERS—Figure 8.12(*e*),(*f*),(*g*),(*h*)

Leptor (mull-mor, arthropod paramull: term from the Greek *leptos* for small or fine, as well as for certain mites). Lacerated, but not decayed litter of bran-like appearance underlaid by 1 to 8 inches thick dark brown organic matter resembling fine weathered sawdust: this layer is formed by castings and exoskeletons of mites, ticks, insects, and other members of *Arthropod* phylum. It is largely confined to soils of calcareous deposits.

Lentar (matted mor, lignomycelial mor, raw humus: term from the Latin *lentus* for tough, tenacious). Firmly consolidated holorganic layer penetrated by fungus mycelia, at times over 8 inches thick: as a rule, sharply delineated from the bleached layer of podzol soils.

Crustar (crust mor, lichen crust; term from the Latin *crusta* for crust). Firmly consolidated, light grey to snow-white layer derived from tissues of *Cladonia* lichens.

Uiginor (bog mor, swamp mor: term from the Latin *uliginosum* for swamp). Saprogenous layers of raw and finely dispersed organic matter, usually containing remains of *Hypnum, Polytrichum,* and *Sphagnum* mosses and the lowland-inhabiting *Ericaceae:* confined to gley podzols and other swamp-border soils. A large number of soils exhibit amphimorphic, two-layered varieties, such as *leptor-parvital, leptor-vermiel,* and *uliginor-sapronel.*

---

treme condition that would not be used as a regular silvicultural practice. Research has shown that the leaching of soil minerals can be associated with the amount of water that percolates through the soil. With increased amounts of water moving through the soil, there is a corresponding increase in leaching of soil minerals. It would appear that by maintaining the organic soil layers intact and by not permitting soils to remain bare of plant growth for long periods, the rate of mineral leaching could be reduced.

As a means of reducing ice and snow on roads during the winter throughout the northern United States, highway departments apply "road" salt, usually sodium chloride, although some calcium chloride may be used. During melt periods the salt is carried off from the roads in the melt water and is often deposited in large amounts on the surface soil under trees. After several years of exposure to excessive salt application, trees die. Particularly susceptible to salt injury are eastern hemlock, sugar maple, and eastern white pine.

GENERAL REFERENCES

Black, C.A. 1968. *Soil-Plant Relationships*, 2nd ed. Wiley, New York.
Brady, N.C. 1974. *The Nature and Properties of Soils*, 8th ed. Macmillan, New York.
Bunting, B.T. 1965. *The Geography of Soils*. Aldine, Chicago.
Kramer, P.J. 1969. *Plant and Soil Water Relationships: A Modern Synthesis*. McGraw-Hill, New York.
Lutz, H.J. and R.F. Chandler, Jr. 1946. *Forest Soils*. Wiley, New York.
Marks, G.C. and T.T. Kozlowski (eds.) 1973. *Ectomycorrhizae*. Academic, New York.
Wilde, S.A. 1958. *Forest Soils*. Ronald Press, New York.

Chapter 9

Physiographic location,* sometimes referred to as situation, is used to
define variations in site resulting from topography, direction of slope
(aspect), slope position, degree of inclination of slope, and slope configuration. Physiographic position in itself does not produce a growth
response in trees; the changes in growth that can be distinguished are
the result of the effects the situation has on air and soil temperature,
precipitation, soil moisture, the interaction of temperature and moisture, and other physical and biological site factors.

*See Jones, J.R. 1971. "An Experiment in Modeling Rocky Mountain Forest Ecosys-
tems." USDA Forest Service Res. Pap. RM-75 (19 pp.).

Physiographic
Location

AIR TEMPERATURE

Air temperature varies with physiographic location of a site. For example, the local climate of a level site could have an average air temperature of 62°F for a period during the early growing season. A forest stand located at an elevation 500 ft higher could experience an entirely different temperature condition. On the average, the temperature of the forest site that experienced a similar exposure, except for the difference of 500 ft in elevation, can be expected to have a temperature one and one-half degrees lower than 62°F; however, if the stand were located on a north facing slope, the temperature would be even lower. On the other hand, a stand located on a south facing slope would have a temperature higher than 62°F because it would be exposed more directly to solar radiation than the level site. Whittaker and Niering (1965) found that in Arizona the lower boundary of Englemann spruce-subalpine fir averaged about 900 ft higher on south slopes than on north slopes; this corresponds to about a +4.5°F mean annual temperature difference.

Minckler (1961) reported that the total amount of sun received in forest openings of different size for day-long periods varies with the aspect and time of day and season of year. The center of an opening equal to the height of surrounding trees receive in June about 45% full day-long sunlight on both northerly and southerly aspects. In September, north slope openings of this size receive only 10% full sun, while south slopes receive nearly 60%. In June and July, available soil moisture in both small and large openings is high, often 15–20% greater than under the canopy, but in late summer the soil approaches wilting point throughout the forest. The coincidence of good light and soil moisture conditions of north slopes, and high radiation and low moisture on south slopes, explain the observed differences in composition and the behavior of reproduction. The reader's attention is directed to the section on air temperature, Chapter 12, for further consideration of physiographic effect upon temperature.

Swift (1976) has developed an algorithm to compute potential solar radiation on any sloping surface at any latitude. The required inputs are Julian dates and the latitude, inclination, and aspect of the slope.

Byram and Jemison (1943) found that during the summer solstice at latitude 34°N there was some difference in the intensity of radiation between south and north aspects; however, during the winter solstice steep north aspects received less than one-half the radiation they received in the summer (Tables 9.1 and 9.2). Both in June and December south slopes received more radiation than north slopes.

Radiational cooling is greatest on the floor of mountain valleys and as a result frosts occur with greater frequency there than on slopes or on elevated ridges. In some localities the tops of ridges experience temperatures that are higher than valley sites because the cool air flows downhill and forces the warmer valley air aloft (Figure 7.21). The result is a "blanket" of warm air over a cold air layer—an inversion. Most severe frosts occur when there is a minimum of air movement (wind speed 3 mph or less) and it is during such periods that temperature inversions also occur.

SOIL MOISTURE

Soil moisture is greatly influenced by physiographic location. South facing slopes, termed southern aspects, experience higher temperatures than northern aspects and bottoms, locations at the base or bottom of slopes. Therefore, evaporation potential is greater on south slopes. Water runoff is faster on steep slopes than on shallow ones. Higher elevations lose water to sites farther down slope—lower slopes, flats, and bottoms. This augments the water received by these sites.

If topography is flat, runoff can be prevented to such an extent that water accumulates and fills the noncapillary pores in the soil. If this occurs frequently, the soil is poorly drained and a swamp condition develops. Some soils that exhibit poor drainage, however, may have a subsurface horizon that prevents the water from penetrating, causing it to accumulate in the noncapillary pores. Subsurface layers that inhibit drainage are the result of impermeable parent material, or

TABLE 9.1. Intensity of Radiation Received at Different Seasons on 20% and 40% North Facing and South Facing Slopes (Adapted from Byram and Jemison, 1943)

SLOPE PERCENT	PERCENT MAXIMUM INTENSITY			
	SOUTH SLOPE		NORTH SLOPE	
	June	December	June	December
20	67	33	63	28
40	69	39	56	18

other rock strata, or an impermeable soil layer. A soil layer that is high in percentage of silt and clay and that has a bulk density near 1.3 will not only impede the downward movement of water but will also prevent the penetration of plant roots. Pore space in high density soils is primarily capillary so percolation of water through them is slow. To have rapid infiltration of water, a soil must have a large proportion of noncapillary pores.

In glaciated terrain, silt and clay settle in shallow lakes resulting in poorly drained soils at the base of slopes, along some streams, and on flats. In other geographic locations where glaciers have not influenced soil development, residual soils have been eroded as a result of past agricultural use. Soils on lower slopes are often deeper because of material washed down from sites higher up the slopes. Forest stands exhibit growth differences between sites where surface soils have been eroded. Growth is poorer on the shallow soils of upper slopes and better on the deep soils of lower slopes and along stream bottoms.

Along mountain ranges, precipitation patterns are such that upper slopes receive more rain and snow than lower slopes. In the Rocky Mountains, it has been demonstrated that precipitation increases with increasing elevation. However, at highest elevations much of the precipitation that falls is blown into ravines and thus does not supply moisture to the vegetation on the summit (Figure 7.17a).

Finney, Holoway, and Heddleson (1962) found that soil microclimate and soil moisture are influenced by aspect and slope position and that in southeastern Ohio, soils on northeast slopes differed markedly from soils on southwest slopes. Northeast slope soils had deep litter, a mull humus, deep A_1 horizons, high base saturation, narrow C/N ratios, and less acid A and upper B horizons than soils on southwest slopes. Colluvial accumulations were greater on northeast slopes than on southwest slopes. Both physically and chemically, soils on northeast slopes were more favorable for tree growth than soils on southwest slopes.

Lee and Sypolt (1974) compared growth of hardwoods on north and south slopes in West Virginia. They suggest that slower growth of trees on south slopes is associated with: (1) higher air temperatures and net radiation, which would decrease net assimilation of photosynthate; (2) a lag in soil temperature, which would inhibit water uptake by roots and affect stomatal behavior; (3) higher air temperature, which would also result in greater transpiration demand (Table 9.2). They concluded that in the area studied, growth reduction on south

TABLE 9.2. Temperature, Radiation, and Vapor Pressure Difference Between North and South Facing Slopes (Lee and Sypolt, 1974)

	NORTH	SOUTH
Midday canopy air temperature	26.8°C	31.3°C
Midday soil temperature (15 cm)	18°C	15°C
Midday Net Radiation (28-day average)	0.85 ly	1.05 ly
Vapor Pressure Difference (30% R.H.)	24.7 mb	35.1 mb

slopes is associated with the microclimate, evapotranspiration, and air and soil temperature differences rather than to differences in physical or chemical soil factors.

Physiographic location modifies temperature and moisture conditions of a site; these in turn affect soil development. Physiographic location, latitude or longitude, or elevation are simply expedient ways of stating relative difference between sites and do not represent true cause and effect relationships for tree growth. However, until the temperature differences, precipitation differences, the interactions between temperature and moisture, and the corresponding growth response that occurs as a result of different physiographic situations have been determined, it will be necessary to use physiographic location to define site differences.

TOPOGRAPHIC INDEX: VARIATION IN SLOPE CONFIGURATION

Choate (1961) found it possible to estimate site quality for Douglas-fir from aerial photographs. He accomplished this by correlating site to a topographic index derived from profile and contour configuration patterns (Figure 9.1). The argument for this relationship is that the

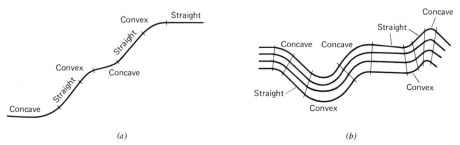

FIGURE 9.1. Topographic (a) profile and (b) contour classification (Choate, 1961).

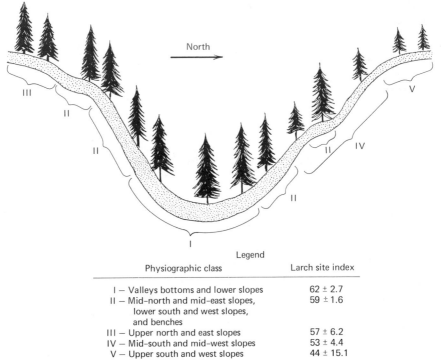

North

| Legend | |
Physiographic class	Larch site index
I — Valleys bottoms and lower slopes	62 ± 2.7
II — Mid–north and mid–east slopes, lower south and west slopes, and benches	59 ± 1.6
III — Upper north and east slopes	57 ± 6.2
IV — Mid–south and mid–west slopes	53 ± 4.4
V — Upper south and west slopes	44 ± 15.1

FIGURE 9.2. Physiographic site classes for western larch (Schmidt, Shearer, and Roe, 1976).

poorest sites are associated with convex land forms, the best sites with concave shapes, while straight slopes are intermediate. These arguments hold with respect to both profile and contour configuration.

PHYSIOGRAPHIC LOCATION

The physical location of a forest stand within the terrain of a particular province has been used effectively to describe site difference. Shallow slopes usually are better sites than steep slopes; north slopes better than south; lower slopes better than upper ones (Figure 9.2). In the case of western larch, five site classes can be recognized that combine various combinations of physiographic location. Other examples will be presented in Chapter 10.

Chapter 10

The evaluation of site quality, along with other stand factors, is essential to rational land-use planning. Land-use planning requires that some estimate of productivity be assigned to each locality. Forest land management in particular requires that a measure of timber production potential be assigned to each stand, especially where capital investments are proposed as part of a silvicultural program. Also, it is important to know the timber production potential of individual species where one species group is to be substituted for another on a site. When the transfer of a stand from timber production to some other use—recreation, wildlife, or watershed—is proposed, estimates of productivity for each use should be made. At the present time, timber production potential is the most readily measured of these.

Site Quality Evaluation

Jones (1969) recognized three general approaches to site evaluation: *site index, vegetation,* and *environmental*. He and others questioned the feasibilty of a single set of harmonized site index curves for a species (see Chapter 4). It appears that as a result of analyses of environmental factors that influence stand development, several site quality classes need to be recognized when defining growth of a number of tree species. For example, the height growth of trees varies with site; as a result, different growth curves are needed to represent these differences. The term used to designate these different height growth curves is *polymorphic*. Within a relatively well-defined climatic-soil province, however, one harmonized set of site index curves for a species can usually satisfy most management needs. Where regional needs require that climate and soil provinces be combined, then polymorphic site index curves will be required.

In Chapter 5, *site types* and their relationship to forest site classification were discussed. These represent the second of Jones' approaches—*vegetation*. The reader is encouraged to review the references in Jones, particularly the concept proposed by Bakuzis (1962).

The *environmental* approach to site quality evaluation, when it incorporates rigorous multivariate statistical analysis in conjunction with measurements of site index (or some type of height/age relationship), appears to meet the needs of evaluating site production on the basis of the total spectrum of factors that influence growth and development of forest stands. A number of empirical soil-site studies up to 1975 were summarized by Coile, (1952a), Ralston (1964), and Carmean (1975). Jones (1969) reviewed and compared the different site evaluation methods.

Nearly all site studies have dealt with pure even-aged stands. Stand density as a variable affecting growth was usually accounted for by selecting only well-stocked stands for study, thus tending to limit the effect of variation of stand density on growth. By controlling composition and age, and to some extent density, a researcher would then measure growth response of stands on a range of sites within limited climatic, soil, physiographic, and biotic conditions.

Empirical site studies, which will be reported, have considered stands that were relatively undisturbed by logging or silvicultural treatment, although in some instances stands that had been burned or "faced" for naval stores have been included. Presently, there is a need for more information on the effect of different silvicultural practices such as site preparation, introduction of genetically improved trees, and the use of fertilizer and drainage in relation to timber production.

The outline of the following discussion is: (1) tree growth on well-drained soils; (2) growth on soils with impeded drainage; (3) growth on organic soils (although this group might have been considered under the second, it is presented separately); (4) the effect of fire and gum naval stores.

WELL-DRAINED SOILS

Stoeckeler (1948) studied the growth of quaking aspen in the lake states and found it could be related to the texture of the soil, acidity of the subsoil, and fire history (Table 10.1).

Einspahr and McComb (1951) found that growth of oak in Iowa could be related to depth of soil to bedrock, steepness of the slope, and direction of slope. Examination of additional stands by other workers in other regions showed growth of oak increased as soil depth increased, as slope steepness diminished, and when stands occurred on north slopes as compared to those growing on south slopes. Gaiser (1950) was one of the first researchers to attempt to quantify the relationships of oak and site using multivariate regression analysis. This approach has been elaborated on by Trimble and Weitzman (1956).

TABLE 10.1. Site Index and Volume of Quaking Aspen in the Lake States (From Stoeckeler, 1948)

TEXTURE AND ACIDITY	UNBURNED STANDS AT 50 YEARS		BURNED STANDS
	Height (ft)	Volume (cu. ft./ac.)	Height (ft)
Sandy loam to silt loam (Ca substrate)	76	2900	64
Sandy loam to silt loam (acid substrate)	77	2400	65
Sandy loam (acid substrate)	68	1900	53
Fine sand and loamy sand (acid substrate)	57	900[a]	46
Coarse sand (acid substrate)	45	400[a]	

[a]Disease caused stands to deteriorate before reaching 50 years. Estimate of SI 57 was age 40; that for S.I. 45 age 35.

In order to be able to compare the relative growth of a number of stands it is first necessary to assign a value of growth to each stand; a stand height/age relationship is a convenient and easily adapted way of accomplishing this. The form equation most frequently used to establish the height/age relationship was developed by Schumacher (1939); it is logarithm height $= b_0 + b_1 \left(\dfrac{1}{\text{age}} \right)$. Deviations in height from the base curve for a stand are considered to be related to soil and physiographic and biotic factors as they have affected stand growth. If the height of a stand is higher than the base curve, then the quality of the site is better because of a difference in one or several factors affecting growth; if the height of a stand is lower than the base, the site is poorer because one or several factors limit growth, and so on. In the Trimble–Weitzman study it was possible to relate site index to several physiographic conditions as well as soil depth resulting in the expression:

$$
\begin{aligned}
\log \text{site index} = {} & 1.9702 - 0.0618 \,(\text{sine of the azimuth} \\
& \text{from southeast} + 1) \\
& + 0.0012 \,(\% \text{ distance from ridge}) \\
& - 0.0020 \,(\% \text{ slope}) \\
& - 0.1509 \left(\frac{1}{\text{total soil depth in feet}} \right)
\end{aligned}
$$

This expression using each of the site factors separately is shown in Figure 10.1.

In function (a) *aspect*, northern and northeastern aspects are better growth sites than southern ones. It can be shown from data collected as part of another study (Fritts, 1961) that south slopes can be as much as 6°F warmer than north slopes, which creates a vapor pressure difference equivalent to 9.13 mm Hg or about 20% increase in potential evapotranspiration. For function (b) *slope position*, a linear relationship is shown indicating that lower slope positions are better for growth probably because they are cooler and receive more water from subsurface flow. Function (c) *steepness of slope* indicates that steep slopes are not as productive as shallow ones perhaps because water run-off is more rapid. Function (d) *soil depth* shows an asymptotic shape where rather large increases in productivity occur with small changes in depth on shallow soils with lesser increases occur-

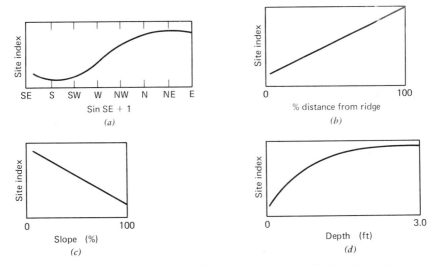

FIGURE 10.1. Graphic relationship between the individual site factors, and site index of oak. (a) Aspect. (b) Slope position. (c) Steepness of slope. (d) Soil depth.

ring as soil depth increases. However, deep soils (3.0 foot depth) are considerably more productive than shallow soils (1.0 ft).

Carmean (1965) and Hannah (1968) extended the studies to black and white oak and found a need to include more detailed measures of soil and topography, particularly in the definition of the configuration of various slope conditions. Their prediction equations add to the precision of site quality evaluation. They introduce a more complete definition of physiographic effect by using profile and contour topographic effects.

Coile (1948) found that growth of loblolly and shortleaf pine in the piedmont region of North Carolina and South Carolina could be related to surface soil depth and subsoil consistence. Stoehr (1946) was able to relate growth of loblolly pine and shortleaf pine to soil, latitude, and fire occurrence. The residual soils of the piedmont are characterized by a sandy loam surface soil (A horizons) with clay loam subsoil (B horizons). Past land-use caused an appreciable amount of surface-soil erosion resulting in considerable variation in surface-soil depth. Subsoils in the piedmont are derived from a variety of parent materials; their texture and consistence varies from friable sandy loams to plastic clays. These factors were used by Coile to explain growth differences of loblolly pine in the North Carolina and South Carolina piedmont region (Table 10.2).

The relationship between soil depth and tree growth in the loblolly pine study is similar to that of oak. Larger increases in site quality were associated with small increases in depth for shallow surface soil depth and less increase in growth on deep soils. The consistence of the subsoil shows the effects of different amounts and types of clay on root growth and availability of soil water for growth. Friable subsoils facilitate growth by providing more water than plastic subsoils where growth is inhibited because low oxygen supply reduces water uptake (Figure 10.2). The function for subsoil consistence is a negative linear relationship.

Carmean (1956) in his work with Douglas-fir found that height growth curves had different forms, depending upon the type of soil. Stands growing on soils of basalt, sandstone, and shale origin produced similar height curves; however, stands growing on gravels, sands, and on valley terraces and lacustrine soil (imperfectly drained) produced curves different from other parent materials and between themselves. This is in agreement with other studies dealing with species growth on a number of different soil types. In addition to differences in physical soil properties, Carmean (1954) found that site index of Douglas-fir could be related to precipitation and elevation of the site. These relationships are:

Basalts

$\log \text{S.I.} = 2.27634 - 0.000076881X_{16} + 0.000066894X_4 - 0.0030871X_3$

TABLE 10.2. Site Index of Loblolly Pine in the Southeastern Piedmont Region of the United States (from Coile, 1952*b*)

CONSISTENCE OF SUBSOIL	DEPTH OF SURFACE SOIL (IN.)			
	0–3	3–6	6–10	10 and Deeper
Very friable	60	80	80	90
Friable	50	70	80	80
Semi-plastic	50	70	70	80
Plastic	40	60	70	70
Very plastic	30	50	60	60

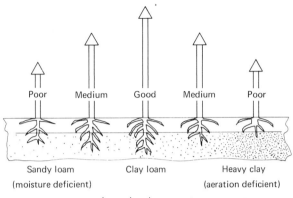

Poor   Medium   Good   Medium   Poor

Sandy loam   Clay loam   Heavy clay
(moisture deficient)   (aeration deficient)

→ → increasing clay content → →

FIGURE 10.2. Effect of subsoil texture on growth of loblolly pine in south Arkansas and north Louisiana (from Zahner, 1957).

Shales and sandstones
log S.I. $= 2.15080 + 0.0034946X_2$
Gravels
log S.I. $= 2.17521 - 0.0030871X_3 - 0.00033653X_{16} + 0.0014511\ X_{17}$
Sands
log S.I. $= 2.12633 - 0.0030871X_3 - 0.00033653X_{16} + 0.0034946X_2$
Lacustrine
log S.I. $= 2.20363 - 0.0030871X_3 - 0.000076881X_{16} + 0.073833X_{18}$
$+ 0.0014511X_{17}$

when X_2 = depth to C horizon in inches
X_3 = average percent gravel horizons above C
X_4 = average moisture equivalent horizon above C
X_{16} = elevation above sea level in feet
X_{17} = annual precipitation in inches
X_{18} = soil compaction, loose soil received value $+1$, slightly compact 0, and a compact soil a value of -1

Fralish (1968) found that in stands where silt plus clay of the soil substrate was less than 18% there was no sugar maple in the understory of mixed northern hardwood stands, while soils having over 65% silt plus clay had an abundance of sugar maple in the understory.

A number of studies have shown that as the thickness of the A_1 soil horizon increased, site quality improved. The A_1 horizon is the surface layer of mineral soil into which humus material is incorporated. Thickness of the A_1 horizon can be considered as an indirect relationship of site quality since the volume of organic matter a site produces depends upon the productivity of the site. A study in Vermont (Post and Curtis, 1970) showed that thickness of the $O_2 + A_1$ horizons was related to growth of sugar maple, yellow and paper birch, and white ash. This would also appear to be an indirect relationship, the conclusion being that white ash and sugar maple for best growth required a moist site, soils with a high percent silt and clay, and calcareous substrate, or some other causal factor. Mader and Owen (1961) show that growth of red pine can be related to the amount of nitrogen in the soil. Further examination might indicate that the proportion of clay, the type of clay, or the type of parent material that affects soil moisture was a more reliable measure of productivity. Microorganisms responsible for "fixing" soil nitrogen would find conditions for growth more favorable on moist sites, therefore, soil nitrogen could be expected to be more abundant on these sites. Much of the difference in tree growth between sites can be explained on the basis of deficiencies in the amount of water available during the growing season.

SOILS WITH IMPEDED DRAINAGE

The rate water is drained from the soil and the amount of water retained in the capillary pore space determines the rate of water up-take and root growth. Various classes of moisture drainage are recognized, depending upon the need for refinement. The terms for a catena of eight drainage categories are: (1) excessively well drained; (2) somewhat excessively well drained; (3) well drained; (4) moderately well drained; (5) somewhat poorly drained; (6) poorly drained; (7) very poorly drained; (8) permanently wet.* Fewer categories may be used by combining several classes where a broader classification is desired.

Rather than a seven drainage class catena (trees do not generally grow sufficiently well on drainage class 8 to permit the use of this class), a five class or a three class catena can be used depending on the need to recognize more or fewer drainage differences. Gaiser (1950) used a three class system for loblolly pine in the coastal plain of Virginia and North Carolina. As a result of his analysis, he found

---

*Definitions of each of the eight drainage categories are given at the end of this chapter. See also U.S.D.A. 1951 Soil Survey Manual, Agr. Handbook No. 18, pp. 169–172.

that data for well-drained and imperfectly drained soils could be pooled but it was necessary to separate poorly drained soils that had plastic subsoils from those with friable subsoils. These relationships are:

Well or imperfectly drained:
$$\text{log site index} = 1.692 + 0.110 \, (\text{log depth} + \text{log I.W.})$$
Poorly drained with plastic subsoil:
$$\text{log site index} = 1.715 + 0.100 \, (\text{log depth} + \text{log I.W.})$$
Poorly drained with friable subsoil:
$$\text{log site index} = 1.983 - \frac{0.772}{\text{depth}}$$

The term I.W. refers to the imbibitional water value of the subsoil. This measurement reflects the degree of plasticity (consistency) of the B horizons of the red-yellow residual soils (ultisols). An expression developed by Knudsen (1950a) for coastal plain soils is:

$$\text{I.W. of the B Horizon} = 0.28470 + 0.6936 \, (\text{silt in B})$$
$$+ 0.23610 \, (\text{clay in B}).$$

Husch and Lyford (1956) used a seven drainage class catena in relating site index of eastern white pine to various site factors. Their relationship was:

$$\text{log height of eastern white pine} = 1.817 - \frac{7.7552}{\text{age}}$$
$$+ 0.0058 \frac{(\text{basal area/acre})}{10}$$
$$+ 0.0064 \, (\text{drainage class})$$

Note that in this study stand density had to be taken into account. Ralston (1951) also included a measure of stand density in his analysis of longleaf pine site quality evaluation. This study will be discussed in another portion of this chapter.

Metz (1950) extended the study of loblolly pine into the coastal plain of South Carolina, Georgia, Florida, and Alabama. His analysis indicated a need to use a three drainage class catena different from that of Gaiser. The results of the analysis are:

log height of loblolly pine =

$$C + \frac{6.97}{age} + (0.000420) \text{ (I.W. of B)}$$
$$+ 0.000021 \text{ (\% clay in B)}$$
$$- 0.000077 \text{ (\% clay of B) (depth of A)}$$

The constants C are:

well drained   = 2.0605
imperfectly drained = 2.0729
poorly drained   = 2.0887

Zahner (1954) continued the study of loblolly pine into Louisiana and Arkansas. The relationships of this study are:

$$\text{log site index} = C + 0.00679 \text{ (I.W. of B)} - 0.000368 \text{ (I.W. of B)}^2$$
$$+ 0.00156 \text{ (depth + subsoil)}$$

Constants C are:

well drained   = 1.8872
poorly drained = 1.9124

Zahner's study completed the picture of the relationship between consistency of the B horizon and site index for loblolly pine, which Coile (1948) had shown to exist. In the original study by Coile there was a negative relationship between consistency of the B (I.W. of B) and in Zahner's study there is a positive relationship with friable subsoils that were not excessively drained, and site quality decreased as the plasticity of the B increased (Figure 10.2).

Post and Curtis (1970), in a study of variation of site index of northern hardwoods in Vermont, found that they could relate site index to a number of soil and physiographic factors. The most practical relationship that combined a reasonable precision of estimate without the need for detailed soil physical measurement was as follows:

composite site index, age 75 =

$$88.5 - 0.286 \left(\frac{\text{elevation}}{100} \right) (\text{latitude} -40)$$
$$- 133.6 \left(\frac{\text{soil group 1}}{\text{solum depth}} \right) + 8.32 \text{ (soil group 3)}$$

where soil group 1—shallow to bedrock soils: in Vermont these are
soils of the Lyman and Hollis series

soil group 2—deep well-drained soil: Berkshire, Marlow,
Charlton, Paxton

soil group 3—moderately well-drained: Woodbridge, Sutton

soil group 4—somewhat poorly drained: Ridgebury, Cabot

Note: For soil group 2 and 4 a zero (0) is substituted for each of the soil group variables to arrive at the site index. For soil groups 1 and 3, a one (1) is substituted for the appropriate soil group and a zero (0) for the other soil group, that is, where a soil is group 1 substitute as follows:

$$SI75 = 88.5 + 0.287\left(\frac{\text{elevation}}{100}\right)(\text{latitude} -40)$$

$$- 133.6\left(\frac{1}{\text{solum depth}}\right) + 8.32\,(0)$$

and for group 3

$$SI75 = 88.5 + 0.286\left(\frac{\text{elevation}}{100}\right)(\text{latitude} -40)$$

$$- 133.6\,(0) + 8.32\,(1)$$

This study brings out an important consideration dealing with mixed stands—the use of composite site index. This index facilitates measurements of growth potential, but requires that height and age be determined for a number of species in each of the stands—in this case sugar maple, white ash, and yellow and paper birch.

Physical and chemical measurements were made for various soil conditions in the northern hardwood study: N, P, K, Mg, and Ca amounts per acre for total solum and for the surface eight inches. These measurements did not show any correlation to site index differences as encountered in the study. Also texture and moisture holding capacity of the soil were not related to site index.

Among other equations derived to show relationships between environment and site index was the following:

composite site index 75 =

$$87.1 - 0.257\left(\frac{\text{elevation}}{100}\right)(\text{latitude} - 40)$$

$$- 8.75\,(\text{soil group 1}) + 9.02\,(\text{soil group 3}) + 4.19\,(\text{sine of azimuth})$$

This equation has an interesting similarity to those for oak, in that there is a limited need to include soils information. There is a primary need to include a physical description of the physiographic location of the stand. This last equation does not add much to the precision of estimating site index of northern hardwoods (the standard error of estimate is reduced from ±6.6 for the first equation to ±6.2 for the second); but, it does appear that aspect should not be overlooked as a site factor since it has been shown to be important to estimating site index of other species.

ORGANIC SOILS

Generally, on well-drained upland soils increase in organic matter in the surface horizon reflects increased production potential. In an early study Auten (1937) found a positive linear relationship between thickness of the A_1 horizon and tuliptree site index. On the other hand, on poorly drained soils an increase in organic matter indicates poor soil aeration and poor growing conditions.

Black spruce is a species generally associated with poorly drained peat soils with high organic content. On such sites, ten-year diameter growth of black spruce was measured at 0.4 in. and height growth at 3 ft; while on a muck soil, for the same period, diameter growth was 2.4 in. and height growth 13 ft.

Pond pine is another species that is found frequently on organic soils; however, best growth of the species occurs on mineral soils with deep surface soils (Table 10.3). The variable $A_1 \times$ 0.M. used in the table was developed from the product of the depth of the A_1 soil horizon and the percent organic matter in that horizon.

Rennie (1963), in summarizing site quality evaluation of heath and swampland sites, indicated that ground vegetation indicator types as developed by Cajander (1926) are the best indication of site productivity. Indicator plants (site types) are reliable guides to post drainage forest potential. Holman (1964) arrived at the same conclusion from his study of tree growth in swampland sites in Sweden.

CLIMATE

To account for climatic differences that result in differences in growth throughout the range of species, various approaches have been tried. Studies such as those by Gaiser (1950), Metz (1950), and Zahner

TABLE 10.3. Site Index (ft) of Pond Pine in the Southeastern Coastal Plain as Influenced by Soil Characteristics (From Coile, 1952a)

TEXTURE SUBSOIL	DEPTH TO MOTTLING (in.)	MINERAL SOILS		ORGANIC LOAMS $A_1 \times$ O.M. $= 400$	MUCK AND PEAT $A_1 \times$ O.M. $= 1000$
		$A_1 \times$ O.M. $= 20$	$A_1 \times$ O.M. $= 200$		
Sand	10	59	55	45	31
	40	61	58	48	34
Loamy sand	10	61	57	47	33
	40	63	60	50	37
Sandy loam	10	64	60	50	35
	40	67	64	53	38
Sandy	10	68	64	52	37
Clay loam	40	71	67	55	40
Sandy	10	71	66	54	39
Clay	40	74	70	57	42
Clay	10	74	69	56	40
	40	76	72	59	43

(1954) for loblolly pine divided the range of the species into three geographic zones in which there was considered to be a relatively homogenous climate.

Coile (1952b) found that when the data for the piedmont study of shortleaf pine were analyzed so that northern stands were identified ($N = +1$) from southern stands ($S = -1$), the relationship was

$$\text{log site index shortleaf pine} = 1.8878 - \frac{0.1580}{(\text{depth of A})} - \frac{0.0408}{\text{I.W. of B}}$$
$$+ 0.0053 \text{ (latitude)}$$

Shortleaf pine growing on the same soil type were two feet taller on northern than on southern piedmont sites. This difference has been a result of northern sites having lower temperatures or more precipitation during the growing season, or some other change in one or several site factors.

McClurkin (1953) studied the growth of longleaf pine in the gulf coastal plain and developed the following relationship:

log site index longleaf pine = 1.8697 + 0.002636 (rainfall Jan.–June) (depth to least permeable soil horizon) − 0.006734 (depth to least permeable horizon)

This study shows an interesting relationship between soil depth and precipitation. As rainfall increases, site quality increases for a given soil depth, but soil depth alone increases site quality independently of rainfall. Carmean (1956) found that growth of Douglas-fir on gravely and alluvial soils increased with a decrease in elevation and gravel content of the A and B horizons, and with increased annual precipitation.

Hofmann (1949) showed that growth of pond pine growing along the Atlantic coastal plain is related to the length of the growing season. The relationship is (shown in Table 10.3 but without the effect of growing season):

log site index pond pine =
$\quad$ 1.6775 + 0.00347 (moisture equivalent of subsoil)
$\quad\quad$ − 0.000276 (depth of A_1) (organic matter A)
$\quad\quad$ + 0.00071 (depth of mottling horizon)
$\quad\quad$ + 0.000312 (length of growing season)

EFFECT OF FIRE AND GUM NAVAL STORES

Stoeckeler (1948) showed that fire reduced growth of trembling aspen. On a site with 55% silt and clay the site index for unburned sites is 82 ft, light to moderate burning, 75 ft, and severe burning, 70 ft. Stoeckeler concluded that the pathological rotation should vary with site and that for Site Class I, 60 years should be the time for harvest; for Site Class II, 55 years; Site Class III, 50 years; Site Class IV, 40 years; and Site Class V, 35 years. Slower growth and disease activity made it impossible to grow aspen for long periods of Site Class V.

In the southeastern United States Coile (1952b) found that on burned areas the loblolly pine site index was reduced an average of 25 ft and on burned areas shortleaf pine as much as 20 ft.

Ralston (1951), in his study of longleaf pine, separated the data into Carolinas (+1) and Georgia and Florida (-1). In addition, it was necessary to designate whether trees had been faced for gum, which resulted in an average loss of 4 ft in site index for trees that had been faced. The results: for imperfectly and poorly drained soils

$$\text{log total height} = 1.886 - \frac{11.20}{(\text{age})} + \frac{136.0}{(\text{age})^2} + 0.00244 \ (\text{MEq of B}) -$$

0.00191 (depth of mottling) + 0.000384 (stand density) - 0.0072 (latitude)

for well-drained soils

$$\text{log height} = 1.915 - \frac{11.11}{(\text{age})} + \frac{136.0}{(\text{age})^2} + 0.00118 \ (\text{MEq of B}) + 0.000374$$

(stand density) - 0.014 (round vs. faced) - 0.22 (north vs. south) - 0.008 (faced north, round south vs. round north, faced south)

Knudsen (1950b) showed a reduction of about three feet in site index of slash pine when trees had been faced for naval stores.

ESTIMATING TREE GROWTH FROM EVAPOTRANSPIRATION AND SOIL MOISTURE

Bassett (1964) proposed a method of estimating tree volume growth using soil-moisture availability indices that were estimated by determining the number of growth days and the number of no-growth days for a particular year. A growth day unit was computed as: growth day = 1 − (potential evapotranspiration (in.) × moisture tension (atm.) in the surface foot of soil). If on a particular day potential evapotranspiration is 0.24 in. and the moisture tension in the soil is equivalent to 0.33 atm., the growth day index = 1 − (0.24 × 0.33) = 0.92, a good growth day. When soil moisture tension exceeded 3–4 atm, diameter growth stopped. The number of growth days and no-growth days can be related to basal area or volume growth of a stand.

$$\text{cubic foot growth} = -30.9 + 1.26 \text{ (growth days)}$$
$$\text{basal area growth} = -1.07 + 0.045 \text{ (growth days)}$$

For no-growth days the relationship was:
$$\text{cubic foot growth} = 119.7 - 1.20 \text{ (no-growth days)}$$
$$\text{basal area growth} = 4.3 - 0.043 \text{ (no-growth days)}$$

Supplied with data for tree growth, soil moisture, and evapotranspiration, it should be possible to estimate annual growth of a forest with reasonable precision. Moisture tension is a function of soil texture and bulk density, and it might be possible to evaluate growth potential on the basis of the amount of precipitation a site received and the amount of water the soil could supply.

SUMMARY

The relationship between tree growth and environment is relatively complex; however, by using empirical data, with a multiple variate regression analysis it is possible to approach the problem objectively and with precision. Each formula derived to predict site index on the basis of environment alone has a considerable error of estimate. In some cases this error may include almost two site index classes (100 ± 10 ft); site index for a particular set of conditions could only be predicted as occurring between 90 and 110 ft. Part of this error is associated with use of temporary plots whose growth periods may have had different weather histories. Also, genetic races of a species can be pooled as if no differences existed. It is often difficult to know exactly the form of relationship that exists between a site factor and tree growth. It is also difficult to measure the increment of change for factors in a realistic manner so that the increments reflect change in tree growth. Empirical site quality studies do have considerable value in that they show whether relationships exist between climate, soil, physiography, and biota. The task remains to expand these studies of growth and environment so that a more complete understanding of the forest tree phenotypes may be known.

DRAINAGE CLASS DESCRIPTIONS

1. *Excessively drained soils.* These soils are commonly considered to be very droughty; in other words, there are periods during most growing seasons when the amount of moisture in the soil is not sufficient for optimum plant growth. Soil moisture is especially dependent on the frequency and amount of rainfall because of very low water storage capacity within the root zone.

   Soils with coarse, sandy, or gravelly textures throughout, and with no mottling, and soils with sandy or gravelly textures over gravel or bedrock are examples of those considered to be excessively drained.

2. *Somewhat excessively drained soils.* These soils are droughty but not so droughty as the excessively drained soils. There are periods in many, rather than most, growing seasons when the amount of moisture in the soils is deficient for optimum plant growth. These soils have a low water storage capacity, but it is sufficient to allow growth of plants except in the longer droughts.

   In this class are coarse textured soils over coarse textured unstratified deposits, coarse textured soils over fine textured materials, moderately coarse textured soils, and soils with a relatively high content of coarse fragments.

3. *Well-drained soils.* Periods when plants suffer either lack of adequate moisture or too much moisture are rare in these soils. Excess moisture from heavy rains is removed rapidly, yet enough is retained because of fineness of texture, depth to a moisture restricting horizon or other soil characteristics so that there is relatively high water storage capacity within the root zone. This provides the plants with adequate moisture except during the most severe droughts. Mottling, in these soils, does not occur within the soil.

4. *Moderately well-drained soils.* These soils have adequate moisture for plant growth during most of the growing season, but are water logged at depths of 14 to 18 inches for short periods following heavy rains or for longer periods in the spring and fall. The waterlogging causes local reduction of the iron compounds in the soils and the soil is mottled with gray and strong brown colors in the lower B horizons. The major effect of the fluctuating water table on the soil is to cause mottling, but, in addition, the surface of plowed soils and the subsoil may be slightly darker

than that of well-drained soils. Presence of mottling within the solum is used as the principal criterion for assigning a soil to the moderately well-drained class.

5. *Somewhat poorly drained soils.* Saturation of soils below the upper eight to ten inches occurs for a rather long period during the spring and fall and may be more or less permanent during an unusually wet season. During the drier part of the growing season, the water table may not occur within four or five feet and when this is the case, the soils are generally ideal for plant growth. The water fluctuates, however, rather widely and for this reason root growth is generally restricted in depth.

Depth to temporarily saturated horizons is judged by depth to mottling. In addition, the surface soil is darker and the B horizon in sandy soils may be such a color as to mask mottling. Presence of mottling in the subsoil is the principal criterion for assigning a soil to this drainage class, but dark brown color of surface soil is also taken into account.

6. *Poorly drained soils.* Poorly drained soils are waterlogged to the surface for a larger part of the year. Only in mid-summer or during periods of extreme drought is the water table more than eight to ten inches below the surface. For this reason, root growth of many plants is restricted. These soils are mottled in both surface and subsoil horizons. Colors below the surface tend to be dull, but in some sandy soils unusually bright colors and even a cemented bright subsoil may occur below an almost dead-gray surface of iron segregation under these poorly drained conditions. Gray surface soils or mottling within the upper few inches are the criteria used for assigning a soil to this drainage class.

7. *Very poorly drained soils.* These soils are waterlogged to the surface a good deal of the time. Water actually stands on them for short intervals in wet seasons, but during the growing season the water table is generally just under the surface so that the surface is wet but not saturated.

These soils generally have peat or muck surfaces ranging from two to ten inches in thickness with black or very dark brown colors in contrast with the gray surface soil of poorly drained soils.

In most forested areas, these soils are characterized by frequent wind-throw mounds and the soil on the mounds is generally better drained than the soil between them. Trees become established on the mounds and grow in spite of the saturated soil condition in the areas between them.

8. *Permanently wet soils.* Water stands on the surface of these soils during a large part of the year, but in periods of dry weather the surface may be wet but not saturated. In general, these soils support only water-loving vegetation such as weeds sedges.

Part Four

Forest Biotic Populations and Influences

Chapter 11

COMPETITION AMONG PLANTS

Competition between plants occurs when two plants make demands upon a site factor in excess of the ability of that factor to satisfy the needs of both plants. If two plants of the same species and age occupy adjacent positions in a stand, they compete for water, light, and soil minerals. They have evolved the same niche requirements and as they continue to grow and increase in size they will crowd the growing space each occupies and eventually crowd each other. If two plants of different species occupy adjacent positions in a stand, they may compete for light and moisture, but their mineral requirements might be slightly different, or they may complement their mineral and other needs, and little or no competition may occur since their niche re-

Biotic Communities

quirements evolved in different ways. Competition then occurs among individuals of the same species (intra-species) or among individuals or groups of species (inter-species).

The concept of niche assumes that species, and races within species, evolve needs that are met in a set of environmental situations that differ from the needs of other species and races. There are common needs such as light, water, and minerals; each plant group is equipped to obtain these in different ways. The quantity of these needs required for growth and reproduction varies among and within the groups. Some plant species accumulate nitrogen in greater amounts than others; consequently, the litter under them is enriched in this element. One species may be more tolerant than another, some individuals being able to produce sufficient carbohydrates in reduced light to sustain themselves. If a soil is deep, one plant species because of a deep rooting habit could extend its roots deeper in the soil, thereby permitting it to obtain water from a larger volume of soil than species that had a more spreading or shallow root system.

The evolutionary development of plants has caused them to assume physiological and morphological characteristics that permit survival and growth under a variety of conditions. They must be able to compete within the environment that favors their growth and reproduction, otherwise they will become extinct. Therefore, the continued presence of a species on a site is evidence that the conditions favor its growth and reproduction more than those of another site where the species is absent. Most tree species have a range of environments within which they can grow, although each species and race has an optimum environment in which growth and reproduction is best. The term absolute tolerance implies a species ability to grow and reproduce—a niche. To determine the competitive ability of a particular species, individuals of the species must be observed in a variety of sites and their performance determined.

From general observation it has been noted that intolerant species require open space in which to reproduce. Such species reproduce best in areas that are clear-cut, burnt, or in abandoned fields, and they can continue to compete only if they are able to maintain a height growth advantage over other species in a stand. There is a tendency for intolerant species to form stands of lower density than do the more tolerant species (Smith, 1962). On the other hand, tolerant species tend to require some protection to reproduce, and do well under the canopy of an already existing stand. In order to have maximum growth, individual trees of both tolerant and intolerant species must be released from overhead competition. However, stands of tolerant species at maturity may reach 300 square feet of basal area per acre, compared to 150 square feet basal area or less for intolerant species.

In each generation, a forest stand experiences a type of niche selection. At the start of stand development many tree seedlings begin growth, but few survive to sawtimber size. During the period needed to bring a forest stand to maturity, a severe reduction in the number of trees takes place. In some cases, because of an abundance of seed and satisfactory seedbed, the number of seedlings at the beginning of stand development may number in the hundreds of thousand of stems per acre; within a short time (20–30 years), as a result of competition, the number of stems is reduced to several hundred or to not more than several thousand, and at the time of physical maturity 100 to 150 trees per acre, or even fewer large trees may be present in the stand. In attempting to reproduce a stand, foresters hope to be able to establish between 600–1000 seedlings of the desired species with the expectation that when the stand is harvested 80–100 trees may remain (depending on rotation). Trees can thus be grown to a larger size in less time than it takes for natural stands. If the niche requirements of the trees cannot be adequately met to establish a desired species, however, then some other tree species or groups of trees and other vegetation will invade the site. Less vigorous trees are eliminated during the period of stand development. However, during this period the total number of other organisms growing on the site may have actually increased because of environmental changes that occur as the stand canopy becomes stratified. These organisms produce competition that can reduce tree growth. There appears to be no way in a natural setting to produce tree crops without encountering a degree of competition. In fact, to produce a desired bole form requires that trees be crowded so that the development of knot producing branches, which affect wood quality, is reduced.

In addition to the reduction in the number of stems per acre as a stand gets older, there is the need to consider mixed stands, and also uneven-aged ones. In uneven-aged and in some mixed stands, the forest develops a layered structure with each layer occupying a different niche in vertical as well as horizontal space.

By having evolved different life forms, plants that grow in forests are able to capture energy at different levels in the storied forest thus permitting a number of species to occupy different niches within the stand. Seasonal difference in growth and reproduction enable more species to occupy the same growing area. Thus, there is a tendency for some species to complement the needs of others, thus providing a mutually beneficial environment.

Competition is somewhat less complex where there are pure, even-aged stands to consider; the ecosystem is less complex since there are fewer species present. However, genetic diversity among individual trees and variation in site quality produce growth differences that make general rules of procedure difficult to formulate. There is concern that continued production of single species stands (monoculture) for a number of rotations on the same site can cause site deterioration because there is no complementing of site requirements in pure stands, as can occur in mixed ones. However, where species are grown within their natural range in the United States, there have been no outstanding examples where site deterioration has occurred when pure stands have been grown for successive generations. With artificially pure stands it is important that a number of genotypes be included in the population to assure that some unforeseen insect attack or disease outbreak does not eliminate the entire stand as can be the case when too few genotypes are used.

Sjolte–Jorgensen (1967) concluded that where plantation spacing is concerned individual trees attain different size according to the amount of space they have available to them. He observed that:

1. Mean height of stand increased with increased spacing
2. Mean diameter also increased with increased spacing
3. Form factor is slightly decreased with increased spacing

Balmer, Owens, and Jorgensen (1975) reported that 15 year old loblolly pine in the piedmont of South Carolina had more total cubic foot volume when they had been planted at spacings of 6 × 6 and 8 × 8 ft than at 10 × 10 and 12 × 12 ft. At the wider spacings, however, the mean stand diameters were larger, indicating it was possible to grow trees to a larger size in less time when planting at wider spacing. In general, this experience with loblolly pine is similar to that of a number of other species.

Herbaceous and woody weeds often prevent the establishment of a desired tree species. Research on regeneration of a number of tree species has demonstrated that establishment and initial growth are generally best when there is little or no competition to trees from grass

---

or low shrub growth. During germination a number of species require some degree of protective shade, but once seedlings are established and have formed secondary needles or leaves, they grow best where they are free of root and overhead competition. Delayed initiation of height growth of pine and spruce planted in old fields has been attributed to root competition from grass; and, in addition, some grasses produce a substance or substances that inhibit growth of a number of tree species (allelopathy).

TREE DISEASES AND FOREST INSECTS

In the production of timber crops for commercial use, parasitic organisms reduce tree growth and destroy some trees and stands. Insects and diseases are a primary reason the number of trees in a stand decrease with time. A tree may be killed by drought or by competition for growing space, but it will not disappear from the stand until attacked by fungi, insects, and other animals and its organic remains reduced and recycled. If all seedlings that are produced in a year could be brought to maturity, there would be little room on this earth for any other forms of life; so in this sense parasites and saprophytes are helpful. However, there are times when the "overabundance" of a particular fungus or insect does severely limit, if not prevent, the growth and reproduction of a desired species.

A good example of a fungus that has almost completely eliminated natural stands of an entire species in the eastern United States is the one causing chestnut blight. In less than 50 years this virulent pathogen nearly destroyed the American chestnut in the eastern United States. It is not often that a tree species comes in contact with a pathogen for which some resistance is not natural. Chestnut sprouts are still found in the forest, but few reach the size where they are of commercial value because the stems become infected with blight and soon die.

White pine blister rust and the southern pine fusiform rust are diseases to which some trees, or races of the affected species, are more resistant than others. Blister rust affects all five-needled pines while fusiform rust attacks the southern pines, primarily slash and loblolly pine. Both blister rust and fusiform rust reduce the number of trees growing in a stand, but generally not all trees are killed.

Wood rot fungi do not constitute a hazard to the survival of any particular species because they do not affect growth until after the trees are able to reproduce. Diseases with the ability to kill trees before they reach reproductive age have the capacity to eliminate a species. A drastic reduction in the number of breeding individuals of a species endangers its chance of continued survival.

Nonparasitic fungi can benefit tree growth. Saprophytic fungi are instrumental in breaking down raw litter, producing humus, and releasing minerals permitting them to be recirculated in the soil. In fact, some leaves and needles are attacked by fungi or infested by insects before they fall from the trees. Saprophytic fungi attack dead limb stubs, causing them to weaken and drop from the tree. The degree of self-pruning a species demonstrates depends upon the rate of decay by fungi. The faster a limb is detached, the sooner the limb stub is overgrown. Some fungi act as both saprophytes and parasites. The beneficial effects of mycorrhizal fungi have been mentioned.

Insects can cause damage to trees, or they may be of benefit by attacking other insects, promoting the processes of litter breakdown, or enhancing tree reproduction. An example is the honey bee, which is important in pollinating flowers of tuliptree, locust, and basswood.

Bark beetles attack a number of tree species causing a considerable loss in timber volume. If there is a range in age and vigor among trees in a stand, mortality from bark beetle attacks is not always complete since some of the more vigorous trees in the stand may survive. Vigor guides for ponderosa and lodgepole pine have been developed to provide a method of identifying a tree's relative resistent to bark beetle attack (Figure 4.7). On the other hand, the eastern white pine weevil attacks the more vigorous trees in a stand apparently because the micro-climate or anatomy of the terminal leader of dominant trees is more favorable for egg laying and larval development. The gypsy moth and the Douglas-fir tussock moth occur in stands where there are large numbers of the host species and the insect populations continue to increase as there are host trees to feed upon, unless checked by natural factors.

Baker (1972) indicated that insect populations can be limited by several abiotic and biotic factors. Temperature is an important factor for when temperatures exceed 120°F or drop to 0°F or below, larval death can occur. Late spring frost may destroy tender foliage needed by insect larvae as food, or frost may delay emergence from hibernation prolonging exposure of larvae to parasites, predators, and disease. Some insects require dry conditions for best growth and development while others require high moisture. If the number of host or decadent trees is greatly reduced, larval losses can be high. Parasites, predators, and pathogenic microorganisms act to limit the spread of insect attack. Operating singly or in combination, biotic factors can limit the duration and magnitude of an out-break, prolong the interval between outbreaks, or prevent outbreaks entirely.

Growth of red and white spruce may be damaged more by the spruce budworm when grown in mixture with balsam fir. Mature balsam fir are attacked with greater severity than spruce. Batzer (1969) identified three forest stand characteristics that help predict mortality of balsam fir from spruce budworm attack. These characteristics are: (a) percent basal area of spruce, (b) percent basal area of nonhost species, and (c) total basal area balsam fir. These were combined into a predicting equation which states: M(percent mortality balsam fir) basal area (arc sine transformation) = 70.144 + 0.529 (percent basal area spruce) − 0.636 (percent basal area nonhost species) − 0.272 (total balsam fir basal area, f^2/ac.). As the basal area in stands increased, more balsam fir died because spruce rather than fir is preferred for egg laying; an increase in nonhost species reduced budworm attack; survival of large larvae increases with the number of large fir and high density stands result in small crowned trees.

The fact that past agricultural practices in New England created environments favorable for the development of many nearly even-aged pure white pine stands has no doubt contributed to the continued presence of both the white pine cone beetle and the white pine weevil. A large number of stands of nearly the same age permits a pest to increase its numbers quite rapidly and to maintain a large base population. The white pine cone beetle has been so prevalent that only infrequent regional crops of white pine seed mature in New England. Cone insects of this genus are responsible for cone crop failures of the western five-needled pines. Damage from seed and cone insects is one of the most important pest problems in forestry, particularly in seed orchard management.

The white pine weevil has in a similar manner been able to maintain a large population in New England because of the large number of young white pine stands. In the origianl New England forest it was estimated that eastern white pine represented less than 10% of the number of trees present; today in some localities the proportion of white pine trees is much higher, providing the weevil with ample material on which to sustain a large population. Where there are large acreages of nearly pure even-aged stands of a species, it can be expected that insects and disease will present problems to forest management.

To reduce the effect of insect damage is not simple. Growing only vigorous stands or producing mixed stands does not provide a sure cure, nor is dependence upon chemical and biological control the sole method of controlling insect damage. A combination of several approaches—utilizing silvicultural, chemical, and biological controls—is needed. Even though most insects have natural parasites—either other insects or diseases that tend to restrict population increase—an epidemic should not be permitted to go unchecked by man until natural forces limit the infestation. There does not appear to be enough land available to simply allow an insect epidemic to run its course with the expectation that other resistant tree species and escape trees of the host, or natural parasites of the insect, will so limit the effect of the epidemic as to assure a sufficient supply of wood for national needs.

It is most important that trees not be planted outside their natural range, nor should they be planted "off-site." There are instances where attempts have been made to extend the range of species, or to plant trees on sites to which they are not adapted. In these instances the trees do not grow well and are subject to attack by insects and disease. For example, attempts were made to extend the range of red pine to a zone south of its natural range and to grow it on soils too fine textured for its best growth (Stone, 1954). These beyond the range and off-site plantations often declined in vigor and suffered root damage from fungi and severe attack of shoots by European pine shoot moth (*Rhyacionia budinana*.) Attempts have been made to extend the range of slash pine. These have not always been successful especially where trees have been planted on deep sandy soils to which the species is not well adapted. Plantations on these sites may grow well for several years but soon decline in vigor and become infested with fusiform rust (*Cronartium fusiforme*) (Kellison, Heeren, and Jones, 1976). In short, it is important to know the site requirements (niche) of tree species so that trees are not grown on sites or in stand structures to which they are not adapted, else they may be subject to attack by pests that damage them, perhaps severely so the trees are eventually destroyed.

FOREST INSECT POPULATIONS

The relative abundance of a particular species of insect depends upon (a) inherent factors, (b) host species, (c) parasites and predators, and (d) climate.

The inherent factors that are important are the size of brood which can be produced, the number of generations produced in a year, and the length of each life cycle. A species that has large broods, a short larval period, and a long life span is capable of producing large numbers of offspring each year. In addition, some insects are capable of spontaneous reproduction (parthenogenesis), while others produce eggs with multiple embryos (polyembryonic).

There are insects that are quite host-specific, so a pure stand of trees would encourage the development of a large population of such a species. For example, the white pine cone beetle (*Conophthorus coniperda*) attacks only the cones of eastern white pine. On the other hand, most insect species are not restricted to single tree species; rather they are capable of finding food on several tree species. For example, the white pine weevil (*Pissodes strobi*) is known to attack eastern white pine, Norway spruce, jack pine, Scotch pine, pitch pine, red pine, Sitka spruce and other conifers. Generally, thrifty uninjured trees are most susceptible to attack by this insect. Mattson and Addy (1975) state that leaf feeding insects usually do not impair plant production unless plants are stressed or senescent. Pine bark beetles and other phloem feeders attack trees weakened or stressed by some type of disturbance, such as a lightning strike, fire, a wind storm, an ice storm or a freeze, and drought (Lorio and Hodges, 1977).

The abundance of insect parasites and predators determines the size of an insect population. Common insect parasites are bacteria, fungi, and protozoa. Predators include birds, such as flycatchers, warblers, thrushs, sparrows, cardinals, and woodpeckers. The most important predators among insects are the beetles; however, fly maggots, aphid lions, ants, and wasps only feed on other insects. Nematodes and mites will also feed on insect adults and eggs.

As it true with other biological populations, insects are sensitive to extremes in temperature since they are exothermal, and unlike mammals, are incapable of regulating their body temperature. Therefore, they are quite sensitive to seasonal and diurnal extremes of temperature. Some insect species are phototrophic (light sensitive) and most larvae require a moisture environment.

Types of Insects

Distinctions among the different insects can be made by indicating the different types of injury they cause to trees: defoliators, sap suckers, bud, twig, seedlings, and root feeders, and cone and seed insects. These approaches can lead to some confusion since some insects as larvae feed on one part of the tree and as adults on another part. Also, within orders there are families that feed on different tree parts than others within the order. Briefly, the orders* according to their importance to forest trees are:

Coleoptera—beetles

Lepidoptera—moths and butterflies

Hymenoptera—ants, bees, wasps, sawflies, horntails

Hemiptera—bugs: lacewings, cicadas, tree hoppers, leaf hoppers; spittle bugs

Homoptera—aphids, scales, mealy bugs

Isoptera—termites

Orthoptera—grasshoppers, walking sticks

Diptera—flies

Among the *Coleoptera* are the bark beetles (family *Scolytidae*). This family accounts for much of the insect-induced mortality of forest trees. The southern pine beetle (*Dendroctonus frontalis*) is the most damaging insect in the southeastern United States. It is responsible for killing large numbers of southern pines over wide areas. Also in the genus are the western pine beetle (*D. brevicomis*) and the mountain pine beetle (*D. ponderosae*) that cause extensive damage to the western pines—ponderosa, Jeffrey, sugar, western white, and lodgepole. The spruce beetle (*D. rufipennis*) is prevalent throughout Alaska, Western Canada and the United States where it has inflicted severe damage to Englemann spruce, killing extensive stands of this species.

The other major groups within *Coleoptera* are borers, the round-headed wood bores (family, *Cerambycidae*), and the flat-headed wood bores (family *Buprestidae*). Some of these insects commonly attack

---

*The suffix *ptera* means winged. Insects belong to the Phylum Arthropoda (meaning jointed legs), class hexapoda (six legs).

living trees, although others live during part of their life cycle in fresh slash or in dying trees. Damage to trees from bark beetles and wood bores occurs during the larva stage. Young larvae mine the inner bark and the wood, creating galleries that girdle infested trees, eventually killing them.

The two species that are the principal carriers of the dutch elm disease (*Ceratocystis ulmi*) are included in Coleoptera: the small elm engraver (*Scolytus multistriatus*) and the elm bark beetle (*Hylurgopinus rufipes*). They carry spores of the elm disease on their bodies and inoculate unaffected elm trees as they bore into the bark.

Two groups of Coleoptera that do considerable injury to young trees are the weevils: the white pine weevil (*Pissodes strodi*) and the reproduction weevils (*Cylindrocopturus spp.*), particularly the pine reproduction weevil (*Cylindrocopturus eatoni*). The seedling bark feeders *Hylobius spp.*, especially the pales weevil (*Hylobius pales*), does considerable injury to young pine reproduction in newly cut areas. The difference between the injury caused by these groups is that the white pine weevil causes major damage in the larvae stage, while the reproduction weevil adults girdle young trees while feeding.

Within Coleoptera are the nut weevils (*Curculio spp.*) that attack the fruits of oak, hickory, walnut, butternut, and hazelnut. The cone beetle larvae (*Conophthorus spp.*) are especially damaging insects since they attack one-year conelets of the five-needled pine species.

Larvae of moths and butterflies (order *Lepidoptera*) eat the foliage of both conifers and deciduous species; however, greatest damage occurs from those that attack conifers. These are the spruce budworm (*Choristoneura fumiferana*), which infests large areas in the northeastern United States and eastern Canada, and the Douglas-fir tussock moth (*Orgyia pseudotsugata*), which occurs in the northern Rocky Mountains. Also included are the gypsy moth (*Porthetria dispar*), which defoliates oaks primarily, but larvae also feed on aspen, basswood, paper birch, and willow. This pest is prevalent in the eastern United States. The tent caterpillars (*Malacosoma spp.*) are included in this order, and larvae of this group feed on deciduous species. They live in tentlike silken webs when they are not feeding. Trees that are infested by tent caterpillars are not generally killed as a result of defoliation by these insects. Several genera within Lepidoptera infest cones and nuts, and after the eggs hatch the caterpillars eat the cone bracts, scales, and seeds, or the acorns or nuts. Most damaging are the species within the genera *Dioryctria,* which attack the cones of pine, fir, and spruce, and *Eucosma,* which attack pine cones.

Sawflies are in the order *Hymenoptera,* and the most important species in the order is the larch sawfly (*Pristiphora erichsonii*). Larvae of this species so weaken all species of larch by repeated annual defoliation that the trees become easy victims to other insects if not killed outright. Reproduction is parthenogenetic.

The true bugs (order *Hemiptera*) consist mainly of insects that are terrestrial in habit. The lace bugs (family *Tingidae*) live and feed on the under-surfaces of leaves of deciduous species. Assassin bugs (family *Reduviidae*), damsel bugs (family *Nabidae*), and flower bugs (family *Anthacoridae*) are predaecous on other insects, mainly aphids, lepidopterous larvae, some borers, and soft-bodied plant feeding insects.

The sap suckers (order *Homoptera*) include aphids (family *Aphididae*) and scales (family *Coccuidia*). The important beech scale (*Cryptococcus fagi*) is a vector for a beech bark disease (*Nectria coccinea*).

The other orders (*Isoptera, Orthoptera* and *Diptera*) include insects that in various ways affect tree growth, but these do not include as many serious pests as those listed above.

ECOLOGICAL EFFECTS OF INSECTS

According to Graham (1963), the main ecological effects of insects are:

(a) Clear away dead wood and litter (termites, wood borers)
      Initiate breakdown of wood residue
(b) Affect stand composition and structure
      Hasten succession from pioneer to climax by removing pioneer species (mountain pine beetle killing lodgepole pine; Gypsy moth)
      Obstructing succession by perpetuating pioneer species, causing large opening in stands and creating conditions for wide-spread fire (various bark beetles, spruce budworm, Douglas-fir tussock moth)
      Change balance of competing tree species (spruce budworm attacking balsam fir first and then spruce)
(c) Set in motion other destructive agents
      Provoke beetle attack (larch sawfly exposing trees to beetle attack)
      Increase fire hazard
(d) Cause increased wind-throw
(e) Increase rate of snow melt and run off

TREE DISEASE POPULATIONS

Noninfectious diseases are those caused by atmospheric agents (see Chapter 7). Infectious diseases are caused primarily by heterotrophic pathogens that live and reproduce on various tree species. Also included as pathogens are green plants which may depend upon trees for physical support or some other means of life support. Pathogens are plant diseases that belong in part to the plant subdivision *Eumycetes*—the fungi. In addition, bacteria (subdivision, *Schizomycetes*), slime molds (subdivision, *Myxomycetes*), viruses, and in some instance nematodes (order *Nematoda*) are included among disease causing organisms and are classed as pathogens. Nematodes are members of the animal kingdom but are grouped by some forest pathologists among pathogenic agents because they are damaging to trees directly or they may be predators on other animals and some plants. Viruses occupy a position between the abiotic and biotic although they are composed primarily of protein; they do have the capacity to duplicate. Viruses have been identified as causing some tree diseases; however, their main effect may be attributed to their pathogenicity of bacteria, fungi, insects, and organisms other than trees.

Forest Tree Fungi

Fungi are grouped into four classes: (1) *Phycomycetes*—stains and some of the damping-off diseases; (2) *Ascomycetes*—the mildews, stains, and cankers; (3) *Basidiomycetes*—the rusts, rot fungi, and most of the mycorrhizal fungi; (4) *Deuteromycetes* (fungi *Imperfecti*)—a group whose taxonomy is in doubt because they do not have a definite spore stage; they include wilts, damping-off, and mycorrihizae.

Among the Phycomycetes is the order *Mucorales*, which contains a black stain fungus (*Rhizopus* spp.). More important are the *Phythium* spp. and *Phytophtora* spp. in the order *Peronosporales* since each of these includes species of damping-off fungi. *Phytopthora cinnamomi*, which causes root rot, is thought to be a cause of the little leaf-disease in shortleaf pine. This disease causes serious damage to at least 44 species, many of them important timber species.

Ascomycetes includes in the series *Plectemycetes* (powdery mildews) the genera *Aspergillus* containing the brown stains, *Penicillium* containing the blue-green stains, *Ceratocystis* including the Dutch elm disease (*Ceratocystis ulmi*), and the oak wilt (*C. fagacearum*), which affects both red and white oaks, and chestnut, tanoak, and chinkapin. Dutch elm disease was introduced to this country about 1930 and has eliminated American elms in a number of locations in the United States and Canada. In the series *Pyrenomycetes* (leaf and stem cankers) are the *Gnomonia* spp., which includes anthracnose that attack the leaves of sycamore (*Gnomonia veneta*) and also to some extent is found on hickory and oak leaves. When defoliation occurs several years in succession crown dieback may occur. The *Nectria* spp. ("target" canker) cause stem cankers on red and sugar maple, sweet and yellow birch, and some of the other birches, and basswood and American elm. Stems seriously affected may eventually break off entirely or may be reduced in value as well as vigor. Hypoxylon canker (*Hypoxylon mammatum*) is considered a most serious disease of aspen in the Great Lakes region. The disease may become so severe on poor sites that rotations must be kept to less than 30 years. It also occurs throughout the eastern United States and in the Rocky Mountains. *Endothia parasitica,* the chestnut blight, also under Pyrenomycetes, is parasitic to American chestnut, chinkapin, and post oak. It is, however, saprophytic to the hickories and most oaks, and because of this it is able to maintain itself in the forest ecosystem. The series *Discomycetes* (needle blights) includes a number of diseases of both conifers and deciduous species. *Rhytisma* spp. causes tar spot disease on maple, yellow poplar, and willows. *Rhabdocline pseudotsugae,* a needle disease, attacks young Douglas-fir trees. This disease limits the use of Douglas-fir as a Christmas tree in the eastern United States. *Scirrhia acicola,* the brown spot needle disease, attacks longleaf pine seedlings wherever they occur, and although not so damaging, is also found on shortleaf, slash, spruce, Austrian, pitch, pond, eastern white, scotch, loblolly, and Virginia pine. Control of brown spot on longleaf pine can be accomplished by using prescribed fire. Included in *Discomycetes* is the canker (*Strumella coryneoidea*), which is a major disease of oak in the Great Lakes region and parts of the northeast and central United States. When occurring on stems of trees, Strumella cankers can weaken them so that they may break, or limit the use of trees for timber.

The Basidiomycetes includes the order *Uredinales,* the stem and leaf rusts having the two important genuses *Cronartium* and *Gymnosporangium,* the stem and leaf rusts of conifers. The white pine blister rust (*Cronartium ribicola*) is the most reported perhaps of any tree disease. It is particularly damaging to both the eastern white pine and western white pines as well as to sugar pine, and may eliminate the possibility of growth of these species on some sites. The southern fusiform rust (*C. fusiforme*) attacks slash and loblolly pine, but short-leaf and longleaf pine are very resistant. The alternate host for the blister rust are *Ribes* spp., while the alternate hosts for the fusiform rust are the oaks, especially the black oaks. The *Gymnosporangium* species includes leaf rusts of serviceberry, apple, and redcedar. The best known species is the cedar-apple rust *Gymnosporangium juniperivirginianae,* which attacks redcedar, the western junipers, and the alternate host apple.

The Basidiomycetes include also the decay and rot fungi of the trunk and root, which are grouped in five families: Agaricaceae, Clavariaceae, Hydnaceae, Polyporaceae, and Thelephoraceae. Among each of these are a number of fungi that attack a variety of tree species. Listed below are the five families and the principle genera that are included among them (Boyce, 1961; Hepting, 1971; Smith, 1970; Talbot, 1970):

Agaricaceae (gill fungi)
    Armillaria—1sp.
    Clitocybe—6spp.
    Collybia—1sp.
    Lentinus—4spp.
    Pholiota—6spp.
    Pleurotus—4spp.
    Schizophyllum—1sp.

Clavariaceae (club fungi)
    Sparassis—1sp.

Hydnaceae (hedge hog fungi)
    Echinodontium—3spp.
    Hydnum—5spp.

Thelephoraceae (leather fungi)
    Corticum—13spp.
    Coniophora—6spp.
    Hymenochaete—2spp.
    Peniophora—6spp.
    Stereum—15spp.

Polyporaceae (pore fungi)
    Daedalea—6spp.
    Fistulina—1sp.
    Fomes—40spp.
    Ganoderma—4spp.
    Lenzites—4spp.
    Merulius—4spp.
    Polyporous—71spp.
    Poria—39spp.
    Trametes—9spp.

Wood decay fungi are limited in their ability to attack through living tissue; as a result, spores of this group can only successfully grow on exposed heartwood. Exposed heartwood may be the result of logging damage, fire injury, or damage resulting from a number of weather related events such as wind, ice, and snow breakage. Insects may produce entry courts for spores of wood decay fungi. Once successfully established, growth of wood decay fungi may be quite rapid, and within several years characteristic fruiting bodies that contain spores are visible on the exterior of infected trees. When fruiting bodies become evident, they indicate an advanced stage of decay. It is only possible to outline the effects of a few important species of Basidiomycete wood decay fungi.

Armillaria root rot (*Armillaria mellea*) is a widespread root rot disease of forest and orchard trees. At first infection, trees may show only slight symptoms of growth decline; however, when a profusion of fruiting bodies are evident at the ground line on the lower trunk, the nature of the malady is quite apparent, and infected trees may exhibit more rapid growth loss. Hydnum (*Hydnum erinaceus*) is a heartwood fungus occurring on various species of oak that have been damaged by fire or in some other way. The disease affects primarily the lower trunk of a tree. Stereum (*Stereum sanguinolentum*) enters the roots of white spruce after injury, sometimes after attack by Pales weevil, but injury can occur following flooding, root cankers, or simply as normal mortality as a result of a growth decline after a period of severe environmental stress. As indicated, Fomes is quite prevalent, especially red heart (*Fomes pini*). This fungus is prevalent among but not limited to old growth soft and hard pines as well as other conifer species in the western United States, especially northern California and the northern Rocky Mountains. It also is found on eastern pines and hemlock. *Fomes annosus* infects stumps of newly cut trees and then may progress through root grafts to nearby stems. It is a disease that increases after pine stands have been thinned. Trunk rot of a number of western conifers may be attributed to several Polyporaceae, especially *Polyporus schweinitzii* and *P. sulphureus*. These produce brown root rot in subalpine fir and western larch as well as a number of the western conifers. Each of the species mentioned may cause serious damage to trees; however, progress of infection will vary as a result of a number of factors.

Deuteromycetes contains several orders. The orders *Sphaeropsidales* includes species of *Cytospora*, one of the agents that is thought to contribute to decline in maple, and causes cankers in black walnut, aspen, pear, willow, poplar, and sassafras. The order *Monoiliates* includes the species of the genus *Phyllosticta*, which contains 55 species that cause leaf spots on deciduous species. More important are the species of *Fusarium* and *Verticillium*. *Fusarium* includes many species that cause damping-off and root rot among the conifers. *Verticillium* species cause rot in some conifer seedlings, but more important are the wilts that affect maple, yellow-poplar, osage orange, magnolia, apple, and elm. The order *Myceluimsterilia* contains the genus *Rhizotonia,* which includes a number of damping-off species.

In addition to the fungi, there are parasitic higher plants that damage trees, mainly the mistletoes. The true mistletoes grow primarily on deciduous species, and include the genera *Loranthus, Phoradendron,* and *Viscium.* The dwarf mistletoes, genus *Arceuthobium,* do extensive damage to conifers. The true mistletoes are apparently able to carry out photosynthesis and mainly derive physical support and water for growth from host trees. They are somewhat epiphytic in nature, although they are not epiphytes, but more like a parasite. The dwarf mistletoes do not have sufficient photosynthetic potential for their needs and must depend upon their hosts for a large part of their food and water. They are considered truly parasitic.

Lichens are a rudimentary plant made up of a fungus body with a green or blue-green algae enclosed in the hyphae. Three general forms of lichens are recognized: crustose, foliose, and fruticose. The crustose form a crust on the branches and boles of trees. The foliose are leaf-like also growing on boles and branches. The fruticose are bush-like and grow mainly on branches, and are found especially on conifers growing on poor sites. Lichens do not appear to cause damage to trees using them mainly for support, deriving their food from their own metabolic processes. Trees on which the lichens grow do exhibit poor vigor, not apparently because of lichen growth, however.

Ecology of Tree Diseases
Fungi respond mainly to differences in temperature, moisture, substrata pH, and minerals. Since they do not require light they can grow under the soil surface and in the inner structure of the tree. Fungi are able to grow on substrate that is lower in pH than can bacteria and viruses. For this reason fungi are common in the litter layers of the forest soils, especially on the acid soils in the northern coniferous forest. Competition is important in the inoculation of a host. Success in inoculation depends upon: (1) rapid spore germination; (2) good enzyme production; (3) production of substances toxic to other organisms; (4) tolerance of antibiotic substances produced by the host and other organisms. Survival depends upon the ability to continue growing, to undergo dormancy, and to reproduce before the population declines to too low a level. Fungi are parasitized by other fungi, viruses, bacteria, lichens, nematodes, ants, and mites, and by some higher

animals. Fungi in turn may be parasitic to the host, but this need not be the case. For example, the mycorrhizal fungi do not damage tree roots, and there are other fungi that live in the tree root zone subsisting on secreted substances, dead parts, and from substances in the soil mass. Plant leaves also secrete substances that can support fungal spores which do little or no damage to the leaves.

There appears to be a type of succession among fungi similar to that which occurs in higher plants. Succession can be separated on two levels: substratic succession and seral succession. Condition of the substrate and the physical environment determines which fungi species will succeed in an initial inoculation. Wounds are the primary inoculation surface. These can occur as a result of wind, ice, snow, or other atmospheric agents and from fire damage. Such physical injury provides entry for Basidiomycetes wood rots (*Fomes, Polyporus, Poria*). *Daldina, Ustulina, Nectria, Fusarium,* and rusts can enter pruned branch stubs and leaf scars. Elm bark, ambrosia and beech bark beetles, and ants provides entry courts for a number of decay and stain fungi. Wounds infected with Ascomycetes or bacteria, or both, provide entry later for Nectria and for Phytophthora, which in effect is a seral succession. Logging and fire scars are inoculum surfaces for a number of fungi, primarily the Basidiomycetes.

Entry may be made also through natural openings such as stomates and lenticels. Damping-off fungi enter through a continuous surface since the tissues of young seedlings are soft, while hard woody tissue prevents hypha from penetrating the epidermal layer of older seedlings. Other species apparently have hypha that are fine enough to grow through the cuticular surface and outer epidermal wall.

Fungus spores are disseminated primarily by wind, although insects and other animals may also carry spores. Moving plants are another means of transport. Dutch elm disease was introduced into the United States on infected logs and white pine blister rust on white pine seedlings. Mycelial growth is a primary means of transport in the soil and on parts of trees.

The ecological effects of disease organisms are in general similar to those of insects, and as has been indicated the two groups may function in concert to cause minor or major disruptions to the forest. Keep in mind that in a natural setting fungi are essential to the recycling of energy through the ecosystem, and it is only when they compete with man for wood that they are considered destructive.

ANIMALS OTHER THAN HUMANS

To a forester it may be quite frustrating to have a young stand of aspen or paper birch eaten by deer.* The loss of a pole stand as a result of beaver flooding can cause concern. The increase in the size of a porcupine family in a hemlock stand may require action to reduce or eliminate the pests. Rabbits nip the tops of newly planted seedlings, and gray squirrels can collect enough acorns to prevent the establishment of a new oak stand.

The deermouse and the white-footed mouse have been labeled million dollar seed eaters. Mice and voles consume large quantities of conifer seed; particularly palatable are seeds of the more valuable Douglas-fir, pine, and spruce (Hooven, 1956; Graber, 1969). In addition to rodents, migrating flights of meadow larks and blackbirds can eat so much longleaf pine seed that few seeds are left and only a partial stand develops (Derr and Mann, 1959). White-throated sparrows, juncos, grosbeaks, and mourning doves are birds that also eat large quantities of seed.

Red and gray squirrels and ruffed grouse eat tree buds in the late winter and early spring. Flower buds of species that begin to mature early—such as elm, aspen, and willow—are fed upon with greater frequency, causing a reduction in seed produced. It has been noted, too, that ruffed grouse drumming logs are located near large aspen trees. Apparently the male grouse establishes his drumming log near large trees because females feed on buds in these trees (Gullion, 1972).

Burton, Anderson, and Riley (1969) reported that yellow birch showed excessive deer browse causing a failure in this species establishment. Seedlings selected for browse were the most vigorous in the stand. They estimated that where yellow birch failed, there were 10–15 deer per square mile, about 15 times the number normally expected. This number of deer occurred because the area had been closed to deer hunting. In other areas where the deer population was almost one deer per square mile, browse damage decreased to an acceptable level and regeneration was successful. Marquis (1973) noted, as have others, that excessive deer browsing on Allegheny hardwoods, particular black cherry, since the 1920s has delayed and often prevented successful regeneration following cutting.

On the other hand, there are instances where animals have been

*See Appendix II for scientific names of wildlife species referred to in text.

used to reduce the number of hardwood stems in a stand so that the conifer element was favored. At one time cows and sheep were permitted to graze in woodlots since they would eat the hardwoods and leave the conifers. Studies have since indicated that this may be an unsatisfactory way to control composition, particularly when the value of the individual cow or sheep may be quite high and the forage value of tree sprouts quite low. In conifer stands where there is a grass understory or grass glades interspersed throughout the forest, it is possible to graze cattle and sheep for part of the year. In pine stands in the southeastern states, spring grazing of cattle can be done where prescribed burning has been carried out to control undesired woody vegetation to establish native grasses. Halls and Shuster (1965) noted that as density of tree cover increased, the pounds of grass for cattle decreased in pine-hardwood stands of Texas. Pinehill bluestem (*Andropogon divergens*, Hack) was most abundant where trees were sparse, but gave way to longleaf uniola (*Uniola sessiliflora* Poir.) as tree cover increased. More important, perhaps, was the increase of two-eyed berry (*Mitchella repens* L.) in closed stands. This species is an important deer food. It would appear that open stands are better for supplying cattle forage while closed stands favor forage for deer. In the western mountain states, summer grazing of cattle and sheep on high elevation pastures is a standard practice. In both cases, however, forage production may be so limited that animals need to be moved to other forest stands or on to cultivated pasture when the capacity of a native pasture is reached. In general, it is not advisable to permit cattle to use hardwood stands for pasture.

Watershed values also can be affected by animals occupying forest stands. Johnson (1952) found in western North Carolina that grazing affected total porosity of the soil. In the most heavily grazed type—cove hardwood—total porosity of the surface two inches of soil was reduced 43%; in the lightly grazed oak-hickory type, 15%; and in the pine-oak ridge type, 6%. These changes occurred over an eight-year period during which time the cattle were in the forest only during the summer months.

ANIMALS THAT INHABIT THE FOREST

The concept of niche applies to animals and insects as well as plants. Through time herbivorous animals and insects have evolved niches as a result of their need to utilize different types of plants. Animals and insects are consumers rather than producers and, therefore, depend upon green plants for food. As a result their niche structure is perhaps more recognizable than it is for plants. Herbivores have evolved a rather complex niche arrangement. The omnivores and carnivores are, in turn, dependent upon the number of herbivores that occupy a forest stand.

Pease (1926) stated that the animals that inhabit the forest differ from those that range the tundra, desert, and grassland. Further, forests can be separated into coniferous and deciduous forest. In the coniferous forest he indicated the identifying animals to be:

Chickadee	Woodpeckers
Jays	Porcupine
Cross bills	Mites
Red squirrel	Beetles
Pine martin	Ants

In the deciduous forest the identifying animals are:

Warblers	Woodpeckers
Flycatchers	Spiders
Vireos	Assassin bugs
Gray and fox squirrels	Ants
Raccoons	

Trippensee (1948) lists as forest wildlife the following:

Black bear	Wild turkey
White-tailed deer	American woodcock
Ruffed grouse	Prairie chickens, sharptails,
Varying hare	and sage grouse

Of these species deer and bear are common to both the deciduous and coniferous forests. The ruffed grouse and varying hare are the more typical of the coniferous forest, while turkey and woodcock are confined to the deciduous forest. However, some woodcock migrate and may be found in conifer stands during the breeding season. Prairie chickens, sharptail, and sage grouse are border inhabitants, ranging between forest and grassland. Trippensee lists squirrels and bobwhite quail as part of farm wildlife; however, these species inhabit fringe areas between forest and open lands.

Trippensee confines the American elk to an area he designates as "Wilderness." An animal that inhabits the wooded areas of the western mountains, the elk does use the coniferous forest for shelter, and in mature stands with open glades it can find food. The western counterpart to the white-tailed deer is the mule deer. The moose is found both in the eastern and western United States and inhabits areas bordering the coniferous forest.

Kendeigh (1961) developed a more inclusive list of birds and mammals for the deciduous, boreal, and montane forests (Table 11.1).

ANIMALS WITHIN THE FOREST ECOSYSTEM

Animals should be considered an integral part of the forest ecosystem, and it is important to an understanding of the system that their place be identified and their function understood. The concept of biological

TABLE 11.1. Lists of Animals Characteristic of Deciduous, Boreal, and Montane Forest (Kendeigh, 1964)

DECIDUOUS	BOREAL	MONTANE
Interior Species	Interior Species	Interior and Edge Species
Mammals	*Mammals*	*Mammals*
Mountain lion	Gray wolf	Grizzly bear
Bobcat	Black bear	Mountain lion
Gray fox	Red squirrel	Bobcat
Black bear	Northern flying	Wolverine
Raccoon	squirrel	Western marten
Opossum	Water shew	Shrews
Shorttail shrew	Porcupine	Elk
Gray squirrel	Snowshoe rabbit	Mule deer
Southern flying squirrel	Deer mouse	Beaver
Eastern chipmunk		Yellow bellied marmot
White-footed mouse		Golden-mantled ground
Eastern mole		squirrel
		Douglas squirrel
		Bushy-tailed woodrat
		Western chipmunks
		Red-backed mice
		Heather vole
		Long-tailed vole
		Western jumping mice

TABLE 11.1 Lists of Animals Characteristic of Deciduous, Boreal, and Montane Forest (Kendeigh, 1964)

DECIDUOUS	BOREAL	MONTANE
Interior Species	Interior Species	Interior and Edge Species
Birds	*Birds*	*Birds*
Tufted titmouse	Goshawk	Golden eagle
Downy woodpecker	Pigeon hawk	Flammulated owl
Black-capped chickadee	Great-horned owl	Pigmy owl
White breasted nuthatch	Saw-whet owl	Blue grouse
Hairy woodpecker	Ruffed grouse	Williamson's sapsucker
Red-bellied woodpecker	Yellow bellied sapsucker	White-headed woodpecker
Ruffed grouse	Hairy woodpecker	Hammond's flycatcher
Barred owl	Black-backed three-toed woodpecker	Western flycatcher
Great-horned owl	Traillo's flycatcher	Western wood pewee
Pileated woodpecker	Olivesided flycatcher	Stellar's jay
Ovenbird	Grey jay	Clark's nutcracker
Red-eyed vireo	Common raven	Mountain chickadee
Redstart	Red-breasted nuthatch	Pigmy nuthatch
Wood thrush	Brown creeper	Varied thrush
Eastern wood pewee	Winter wren	Mountain bluebird
Cerulean warbler	Hermit thrush	Townsend's solitaire
Scarlet tanager	Swainson's thrush	Audubon's warbler
Acadian flycatcher	Golden-crowned kinglet	Townsend's warbler
Yellow-throated vireo	Ruby-crowned kinglet	Hermit warbler
Whip-poor-will	Solitary vireo	Western tanager
	Nashville warbler	Evening grosbeak
	Wilson's warbler	Cassin's finch
	Purple finch	Oregon junco
	Pine grosbeak	Grey-headed junco
	Pine siskin	
	Red crossbill	
	Lincoln's sparrow	

pyramids has been utilized to explain the relationship between the animals and the forest, and among animals themselves. Up to this point the forest has been discussed as a closed system, which in effect it is because it could continue to carry out most of its functions without some parts of the animal segment. One might argue that animals are important to certain stages of forest development only, but are unnecessary for continued forest production. To animals, the forest is an essential link in their life cycle because it provides the essentials of food, water, shelter, and natal sites—the habitat. Within this habitat animals occupy different niches that can be explained as levels of the food chain (sometimes referred to as trophic levels). The forest is autotrophic, a producer of food; the animals are heterotrophic depending upon plants and other animals in the forest for food (Table 11.2). There are three animal groups in the food chain: the herbivores, which use plants directly as food; the omnivores, which use both plants and other animals as food; and the carnivores, which use other animals as a primary food (Figure 11.1). In moving from a lower level in the chain to a higher level (e.g., from plant to herbivore), energy is lost because of animal respiration. Ultimately, total plant biomass will be consumed and cycled through the ecosystem. In a perfectly balanced system plants would fix each year an amount of energy required for autotrophic respiration, and an amount needed to heterotrophic organisms. If a greater or less amount of energy is fixed by plants, the system can be considered out-of-balance. An economically productive forest is one that is biologically out-of-balance since the forest must produce more energy in the form of wood fiber each year than is con-

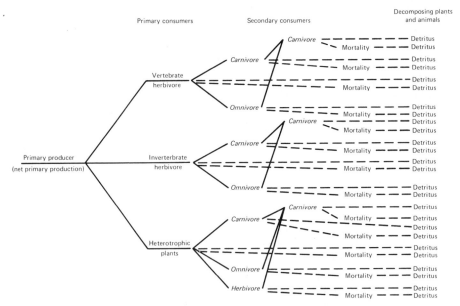

FIGURE 11.1. Simplified outline of an energy transfer network within a forest ecosystem.

TABLE 11.2. Animal Food Chain in a Forest Ecosystem

	VERTEBRATES				INVERTEBRATES
	MAMMALS	BIRDS	REPTILES	AMPHIBIANS	
Carnivores—	Bobcat Lynx Fox Martin Weasel	Hawks Owls			
Secondary Omnivores—	Bear Opossum Raccoon Skunk Tree and ground squirrels				
Primary Omnivores—	Moles Mice Shrews	Birds—leaf, stem ground feeders	Snakes Turtles	Frogs Toads	Centipedes Spiders Preying Insects
Herbivores—	Deer Hare Porcupine	Birds— seed feeders			Earthworms Millepedes Plant Eating Insects—leaves twigs, fruits, etc.
Plants and plant parts—	Seeds Nuts Fruits	Fungi Fruiting Bodies	Grasses Herbs Ferns	Twigs Bark Buds and stems Leaves	

Herbivores feed mainly on plants and plant parts.
Vertebrate primary omnivores feed on herbivores and primary invertebrate omnivores. Invertebrate omnivores feed on invertebrate herbivores and on plants and plant parts.
Secondary omnivores—These are all mammals that will feed on plants, herbivores, and primary omnivores.
Carnivores feed primarily on herbivores and omnivores and on other carnivores, but may at times eat plants or plant parts.

sumed. This need in itself encompassses a major objective of forest management.

The primary omnivorous birds are a complex group because of their nesting and feeding habits. Some birds spend most of their time on or near ground level where they feed on various insects, worms, millepedes, and plant parts, and they nest at or near ground level. Other birds occupy shrubs and small trees while others occupy higher layers in the lower and upper tree canopy (Table 11.3).

TABLE 11.3. Typical Nesting Habits of Selected Nongame Forest-dwelling Birds[a]

SPECIES	USUAL NEST HEIGHT (ft)	CONCEALMENT	REMARKS
Louisiana waterthrush	0	Roots, logs, banks	Near water
Ovenbird	0	Dead leaves	Usually in dry soil
Whip-poor-will	0	Dead leaves, brush	Deep woods, ravines
Brown thrasher	0– 7	Thickets	Prefers thorny vines
Carolina wren	0–10	Cavities, thickets	Often in a building
Rufous-sided towhee	0– 3	Grass, forbs	Brushy openings or deep woods
Song sparrow	0– 6	Grass, thickets	Edges of woods
Yellowthroat	0– 2	Grass, vines	Moist locations
American goldfinch	5–15	—	Forks of shrubs, saplings, vines
American redstart	5–15	—	Forks of shrubs, saplings, vines
Cardinal	3–15	Thickets	Prefers vine tangles
Gray catbird	2–15	Thickets	Prefers vine tangles
Chipping sparrow	1– 4	Thickets	Often near building
Indigo bunting	1– 3	Thickets	Brushy areas in, near woods
Yellow-breasted chat	3– 10	Thickets	Often in thorny vines
Yellow warbler	2–10	—	In shrubs, saplings
Red-eyed vireo	2–15	—	Suspended from forks
Robin	2–15	—	Often on a building
Wood thrush	4–15	Thickets	In forks or on a limb

TABLE 11.3 Typical Nesting Habits of Selected Nongame Forest-dwelling Birds[a]

SPECIES	USUAL NEST HEIGHT (ft)	CONCEALMENT	REMARKS
Black-capped chickadee	8 +	Cavities	Often in old woodpecker hole, bird box
Downy woodpecker	6–30	Cavities	Dead tree or dead part of live tree
Screech owl	6–30	Cavities	Woodpecker hole, tree cavity, building, bird box
Tufted titmouse	8 +	Cavities	Often in old woodpecker hole, bird box
White-breasted nuthatch	2–60	Cavities	Stump, snag, old woodpecker hole, bird box
Northern (Baltimore) oriole	20 +	None	Prefers broad-crowned trees
Blue jay	10–15	—	May prefer conifers
Broad-winged hawk	20–80	None	Builds in a large crotch
Crow	20–80	None	Usually in a large crotch
Eastern wood pewee	20–60	None	Often on edge of clearing
Great crested flycatcher	6–15	Cavities	Tree or stump, woodpecker hole, bird box
Scarlet tanager	16–55	None	Usually in mature woods

[a]From Gill, Degraaf, and Thomas, 1974.

Following a major disturbance in a closed forest where large openings occur, the bird population can change because upper canopy birds have no feeding or nesting sites and must therefore seek food and shelter elsewhere. The number of ground and shrub layer birds, however, may increase while other birds appear indifferent to change (Table 11.4).

TABLE 11.4. Response of Most Abundant Species of Songbirds to a Commercial Clearcut of Mature Northern Hardwood Forest on the Archer and Anna Huntington Wildlife Forest, Newcomb, N.Y. Initial Population Response Indicated by: X Some Change; XX = Substantial Change. Trends Cover an 11-Year Period Following Logging. Statistically Significant Change at the 5% Level of Probability = *; at the 1% level = ** (from Webb, 1973).

SPECIES	INITIAL RESPONSE TO LOGGING			TREND OF POPULATION AFTER INITIAL RESPONSE
	INCREASE	NO CHANGE	DECREASE	
Ovenbird			XX**	Increase**
Black-throated Green warbler			XX*	Do*
Red-eyed vireo		0		Do**
Black-throated Blue warbler		0		Do
Olive-backed thrush		0		Do
Woodpecker species		0		Do
Scarlet tanager		0		Do
Redstart	XX*			Decrease
Chestnut-sided warbler	XX*			Do
Chimney swift	XX**			Do
White-throated sparrow	XX*			Do
Veery thrush	X			Increase

ENERGY TRANSFER WITHIN THE FOREST ECOSYSTEM

Energy stored by plants within the ecosystem is cycled through the system at various intervals. As pointed out in an earlier chapter, residence time of plant material may be less than one year. For example, new leaf or branch material can be consumed by a herbivores (primary consumer). The organism will utilize a part of the energy for growth and reproduction, a part for respiration, and a part as detritus material. The amount of energy distributed in these three ways varies with the type of organism (Table 11.5).

Using the information contained in Table 11.5, if a warbler were to feed upon caterpillars—feeding on the newly developed leaves of

TABLE 11.5. Biomass, Food Source, and Energy Transfer of Animals in a Northern Deciduous Ecosystem (Adapted from Gosz, Holmes, Likens, Borman, 1978)

SPECIES	BIOMASS (g/m²)	FOOD SOURCE			TOTAL CONSUMPTION (kcal/m²/yr)	TRANSFER (% CONSUMPTION)		
			PLANT				ASSIMILATED	
		ANIMAL	AUTO.	HETERO		EGESTION	RESPIRATION	STORAGE
Caterpillars	160.4	—	100%	—	576.00	86	9	5
Chipmunks	0.21	5	85	10	31.00	18	80	2
Salamanders	0.20	100	—	—	1.06	19	32	51
Birds	0.12	81	19	—	7.40	30	69	1
Mice	0.08	14	74	12	11.10	17	81	2
Shrews	0.02	100	—	—	7.10	10	89	1
Deer[a]	—	—	100	—	4.5	—	—	—
Snowshoe hare[a]	—	—	100	—	71.0	—	—	—

[a]Transfer energy equivalents were not given for these two animals.

young birch—there could be an energy exchange where the caterpillars contained within their bodies 5% of the energy available in the leaf material. The warblers would retain only 1% of the energy available to them from the caterpillars. If we can assume that one gram of leaf material contained about 4600 calories, a warbler would only be able to store (0.05)(0.01)(4600) = 2.3 calories of the original energy. It becomes apparent that the ability of organisms to retain original energy depends upon their metabolic rate and the quality of the food material upon which they feed. The more digestible a food, the more will be assimilated and less will be egested. The lower the respiration rate the less energy will be consumed to maintain body heat.

Gosz et al. (1978) point out that an adult shorttail shrew (a carnivore) weighing 15 grams required about 15.2 kilocalories (kcal) of food energy per day. This food consisted of spiders, beetle larvae, and other invertebrates that contained only 0.11 kcal per individual. A shrew would have to kill each day about 138 insects, and of this total only about 1% of the energy would be used for body growth and reproduction. On the other hand, a salamander—a cold-blooded vertebrate carnivore—would be able to store 51% of available energy. It would appear that a habitat having a capacity to support both salamanders and shrews would be able to support more salamanders than shrews.

Returning to the original example of the caterpillar and the warbler, note that caterpillars were only able to assimilate 14% of the energy contained in the leaf material, and 86% of the energy being egested, the part the caterpillars could not digest. This detritus material would fall to the forest floor where it could be used by other organisms for their metabolic needs. Here a new food chain would begin because the bacteria, molds, and mildews would be able to utilize the energy contained in the undigested fecal material. These microorganisms would in turn be used by larger animals for food, perhaps reaching the shrews and salamanders later down the food chain.

It is not too difficult to trace the transfer of energy through an ecosystem when dealing with the beginning stages of the chain. In the Gosz et al. (1978) study there were only about 34 acres (13.23 hectares) contained in the study area. An understanding of the third and fourth levels of a food chain requires larger areas and an ability to monitor the habits of larger omnivores and carnivores. However, we might question the need to go very far in this direction once we recognize that the amount of energy available to predator species on such small areas is limited, with much of the energy used in the early stages. Large animals require a much greater area from which to obtain food.

It is necessary now to go back and consider Figure 11.1. In order to use the diagram, it is necessary in some instances to enter a species at more than one point. For example, a chipmunk will feed at one time on plant material as a primary consumer, another time on fruiting bodies of fungi as a secondary consumer, and another time on invertebrates (Table 11.5). In the same manner, the warbler, used as an example before, could feed on plant material as a primary consumer and at another time as a secondary consumer when feeding upon caterpillars. Obviously it is difficult to devise a simple scheme to illustrate all the combinations of primary and secondary consumers and their interactions. It will require much work before a complete understanding of energy transfer within a forest ecosystem will be completely brought about.

CARRYING CAPACITY

The carrying capacity of a forest is a way in which its productivity for animals can be viewed. Carrying capacity depends upon how well habitat needs of each animal group can be satisfied, and, therefore, how many animals of a particular group can be contained in a particular forest condition. One measure of carrying capacity is expressed in density (number of animals per unit area). In the present case, number of animals per square mile provides one way to view the entire range of different animals (Table 11.6). In other cases number of animals per acre, or per square meter, can be used. Since the range in animal size is rather large, number of animals per square mile does not seem too awkward a measure. Home range is usually a more restricted area that represents the need to support a single family of animals or an animal colony; in this instance an acre unit of measurement appears appropriate. It must be recognized that home range is confined to units of a particular habitat, although in some instances the habitat may vary since some animals range over a wide area feeding at different times in different habitats. This is particularly noticeable with bears. On the other hand, some birds confine their activities to a relatively small area. It would appear that if a maximum number of animal species is to be attracted to a forest situation, it is essential that there be a variety of habitats available, that is, open areas interspersed among seedling, sapling, pole, and sawtimber stands. Also a

TABLE 11.6. Density, Home Range, and Habitat of Some Forest Wildlife (USFS, 1971)[a]

	AVERAGE DENSITY	HOME RANGE	HABITAT
Mammals			
Squirrel, fox and gray	320/mi²	1½–8 ac.	Mixed hardwood, Hardwood
Cottontails	1280/mi²	15–20 ac.	Openings
Deer	26/mi²	300 ac.	Hardwood, conifer, mixed wood, openings
Bear	1/mi²	50,000 ac.	Hardwood, conifer, open
Raccoon	10/mi²	1000 ac.	—
Fox	10/mi²	1200 ac.	—
Birds			
Quail, bobwhite	16/mi²	40 ac.	woodland, brush open, swamp
Grouse, ruffed	21/mi²	40–50 ac.	Hardwood, conifer, upland brush
Wood duck	0.3/mi²	30 mi feeding	Small ponds
Dove, mourning	21/mi²	—	Openings
Woodcock	—	2 mi feeding	Thickets
Turkey	10/mi²	1 mi feeding	Mixed hardwood, hardwood, conifer, open understory
		RESIDENTS	
Songbirds	1800/mi² sapling	5–15 ac. warblers	—
	1200/mi² pole	10–200 ac. woodpecker	—
	3200/mi² sawtimber	10/mi² raven	—

[a]These data are approximate, and are presented as indicative only of the numbers that might occur in these two categories.

variety of food species of plants and animals is needed. A single age class of trees with a single stratum—like pure stands—would provide habitat suitable for only a limited number of animals.

Gullion (1972) showed the need of ruffed grouse for seedling hardwood stands during the brood period, pole-sized stands for breeding cover, and older, mixed stands for winter cover. Nesting cover occurred over a range of stand conditions (Figure 11.2). Conner and Adkisson (1976) found that downy woodpecker nests were most abundant in stands averaging 91 years old, which had 17 m²/ha basal area with 401 stems/ha, and having both living and dead trees infected with heart rot. Pileated woodpecker nests were found in stands averaging 143 years old, with 27 m²/ha basal area and 475 stems/ha, and with 20–60 cm dbh dead trees that retained most of their bark. Flicker nests occurred most often in stands averaging 93 years old with 2 m²/ha basal area and 49 stems/ha. Stands where open ground was readily available for foraging for insects were favored by flickers.

White-tailed deer will find some food in forest stands that are at different stages of development, but seedling-sapling stands supply more winter browse than do older stands (Barnes, 1976). Winter browse is particularly important to deer survival in northern regions

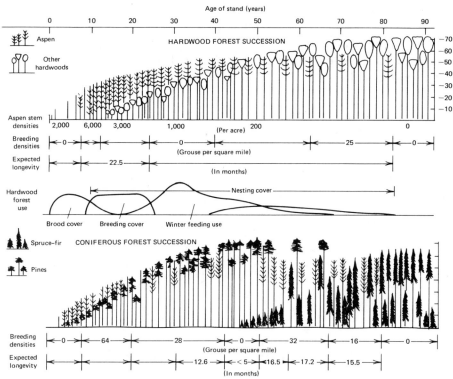

FIGURE 11.2. Forest structure, density, and survival of ruffed grouse (from Gullion, 1972).

because the animals are hampered in their movement by deep snow and browse must be abundant and easily obtained in winter months to assure maximum survival of yearling deer. Barnes found that seedling-sapling stands in all forest types contained 19.7 lb/acre of browse, while pole and sawtimber stands contained only 9.7 and 9.6 lb, respectively. Barnes detected only slight differences among the U.S. Forest Service designated timber-types, which were white-jack pine, spruce-fir, pitch pine, oak-pine, oak-hickory, elm-ash-red maple, maple-beech-birch, and aspen-birch (see Table 6.1).

Oak sawtimber stands provide hard mast in the form of acorns which deer and other animals feed upon in fall and spring, and in open winters. Acorn production varies with tree size and species (Table 11.7). For example, a mixed oak stand with 80 ft² basal area per acre, a mean stand dbh of 12 in., with half the stand northern red oak and the other half scarlet oak (using the data from Table 10.4), could produce, on the average, (2.8)(40) + (4.9)(40) = 308 lb/acre of oak mast. On the other hand, such a stand would provide only 55,000 twigs of browse per acre (Figure 11.3). Using a conversion rate of 1500 twigs per pound of browse (Shaw and Ripley, 1965), the stand would provide 36 lb of browse. If a mature deer required 7 lb of browse or acorns per day, the oak stand would provide only 5 deer-days of

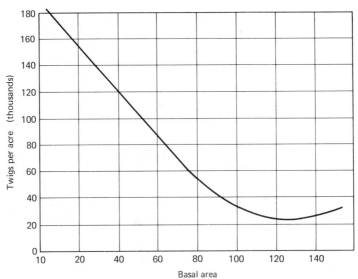

FIGURE 11.3. Number of twigs per acre as related to basal area (USFS, 1971).

TABLE 11.7. Expected Acorn Yields Pounds (Air-dried) per Square Foot of Basal Area at Indicated Diameter (USFS, 1971)

dbh	(ba)	CHESTNUT OAK	WHITE OAK	POST OAK	NORTHERN RED OAK	SOUTHERN RED OAK	SCARLET OAK	BLACK OAK	WATER OAK	BLACKJACK OAK	SANDJACK OAK
4	(0.09)		1.2								6.1
6	(0.20)			2.9							6.5
8	(0.35)			3.0							5.9
10	(0.55)	1.8	1.3	2.8	0.7	0.6	4.5	2.0	0.8	0.9	5.1
12	(0.79)	3.7	1.9	2.5	2.8	1.0	4.9	2.2	2.6	2.3	
14	(1.10)	4.5	2.5	2.3	5.0	1.4	5.1	2.1	3.4	2.9	
16	(1.40)	4.5	3.1	2.1	7.1	2.0	5.7	2.0	5.1	3.0	
18	(1.80)	4.5	4.8	1.9	8.0	2.7	6.7	1.9	4.0	3.3	
20	(2.19)	4.0	4.8	1.8	7.2	3.6	6.8	1.8	4.0	2.7	
22	(2.64)	3.7	4.8	1.7	6.5	4.6	6.6	1.7	3.9	2.7	
24	(3.14)	3.2	4.3		4.9	5.8	5.7	1.7	3.8	2.6	
26	(3.69)	2.8	4.0		3.7	6.5	5.0	1.6			
28	(4.28)	2.8	3.6		2.9		4.3	1.5			
30	(4.91)	2.5	3.0		2.0		3.7	1.4			

browse compared with 44 deer-days of mast. If a stand had less than 50 ft² of basal area, the browse potential would be greater but acorn production would decrease. Deer and other herbivores require forest conditions that provide both browse and mast; these can be provided in seedling-sapling stands and sawtimber stands, respectively. Stands in the large sapling to medium pole stages of development provide only minimum amounts of food for herbivores (Figure 11.4).

The nutritional value of plant food varies in different parts of the same plant and between different plant species (Martin, Zim and Nelson, 1961). Leaves are particularly digestible and are high in carbohydrates, but are low in nutritional value with respect to vitamins, proteins and fats. Buds and flowers contain mostly carbohydrates; while roots, bulbs and corms contain both carbohydrates and proteins. Inner bark has about the same carbohydrate content as leaves. Seeds and nuts have the greatest concentration of food containing not only carbohydrates and vitamins but proteins, minerals and vitamins. Fats

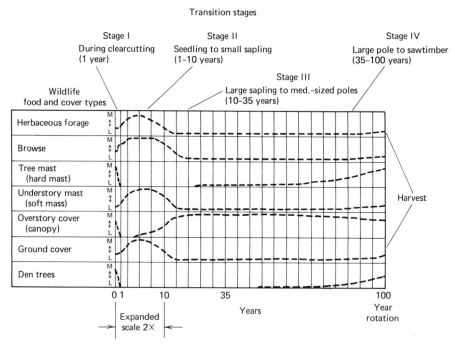

FIGURE 11.4. Relative abundance (M = more, L = less) of major wildlife food and cover types in a clearcut during and after clearcutting (Hassinger, Liscinsky, and Shaw, 1975).

represent a moderate proportion of the carbohydrate reserve in seeds and nuts. In the example with deer, it is not totally realistic to compare browse food value to acorn, because it will require less weight of acorns to supply the same carbohydrate amount of woody browse, and in addition, the acorns would provide additional food value in protein, vitamins and minerals.

The animal food chain—fragile in appearance although in truth quite strong—is sometimes referred to as a web of life. If a food chain gets out of balance where too many animals of one type attempt to occupy a habitat insufficient for their numbers, some animals weaken and die, as with trees in the forest. This has occurred when a deer herd exceeded the carrying capacity of a habitat. When humans interrupt the food chain, crowding among herbivores may occur. Overhunting of the mountain lion, the gray wolf, the lynx, and the bobcat has contributed in some regions to overpopulation of both large and small herbivores. Their increased numbers may then exceed the carrying capacity of habitat. Human activities have been responsible for eliminating habitats for some species—such as the ivory billed woodpecker and the Bachman's warbler—while some species respond positively to land use by man—such as the robin and cardinal. Endangered species are usually those with very specific habitat requirements; the nonendangered species are able to adjust to a variety of habitats.

Interspecies competition among animals can occur when two species attempt to utilize a common segment of a habitat. This may occur among herbivores, omnivores, or carnivores. Korschgen (1967) indicated that turkeys (the eastern and southern races) depend upon food produced in forest habitat during most of the year and they compete with livestock. Livestock consume large quantities of turkey foods and may eliminate food producing plants by over-grazing. White-footed mice, and fox, gray, red, and flying squirrels consume large quantities of mast, fruit, and seeds used by turkeys. During the summer, opossum, raccoons, and foxes consume quantities of preferred turkey food as do grouse, quail, crows, jays, and a number of song birds. Deer in particular compete with turkeys for mast in the autumn and winter seasons. Feeding by competing animals can cause serious problems for any or all species when food is scarce.

Cliff (1939) found that where the Olympic elk and mule deer occupied the same winter range in Oregon, the mule deer population reached a high of 19,500 animals but collapsed during a severe winter and was reduced to 8600. The elk population did not suffer because they were able to find food on trees and shrubs that was out of reach to the deer. Murie (1951) stated that Rocky Mountain elk not only compete with deer but also with livestock, mountain sheep, and antelope, for both summer and winter food.

The ultimate disposition of any forest animal in a natural setting is to become a part of the total system following death when its organic remains are used as energy by decomposer animals and plants, and the mineral elements recycled through the system. If too many animals remove too much plant material and the animals themselves are later removed from the forest, then it can be expected that there will be a diminution of the system's mineral reserves. Such a situation would not appear to occur with forest wildlife, but could be more likely with domestic grazing animals.

TERRITORY

Animals have developed a means of controlling population density. The term used for the concept is *territory*. It is defined as an *isolated area defended by one individual of a species or by a breeding pair against intruders of the same species and in which the owner of the territory makes itself conspicuous* (Lack and Lack, 1933). The functions of territory are the following:

Isolate pair or group to provide a place for courtship and mating without interference

Assure an even distribution of animals over available habitats so carrying capacity is not exceeded

Facilitate feeding

Provide protection from predators

Isolation among birds is used by mating pairs where the sex ratio is unbalanced and there are more males than females, as with ducks (Wallace, 1963). With herd animals, such as deer, a dominant male maintains the breeding group of females. Woodcock males locate a singing spot to attract females and ruffed grouse males use a drumming log. An even distribution of animals is important because a particular habitat may not be suitable for a large number of breeding groups at one time. Where there is an abundance of food, families may be able to live close together. Kendeigh (1947) found an unusually large concentration of Tennessee, Cape May, and Bay-breasted warblers during a spruce budworm outbreak, where some territories were less than one-half acre. In uniform terrain a territory may be oval. Where animals occupy habitat near lakes and streams, the margin of the water area will tend to define a territory pattern. During the feeding periods, if the young require food to be carried, as with birds, it is important that the territory be sufficiently large to meet food needs of adults and young. For mammals, the territory must supply the needs of a nursing female and one or more adult males if the family remains intact during the rearing period.

A territory is defended against intruders of the same or other species if they appear to represent a threat. Also, a familiar territory is one where retreat from a superior predator is possible. Squirrels soon learn the quickest way to travel from one area to another within a territory when pursued by another squirrel or a predator.

It is difficult to generalize on the length of time a breeding group will occupy a particular territory. Post-natal care varies with each group and among species and within a group of the same species. Some animals care for their young for one year after birth, or until the next breeding season, but they do not necessarily occupy the same territory all this time. Most young animals have the inherited ability to develop skills required for survival even when separated from adults, so it is difficult to say how long infants require post-natal care once weaned or fledged, or how soon they would be forced to vacate the natal territory.

Not all animals are territorial. Some animals live in colonies with several breeding groups inhabiting a common territory. After weaning, animals may use the same territory they were confined in during the natal period, or they may be driven away by the adults. Winter conditions may change habitat and territory needs particularly where animals migrate over a long distance or where winter range is widely separated from summer range.

HUMANS

Unlike their ancestors, modern humans have found that they need not adapt their life style to conform to the demands of the environment. They can modify an environment to meet their physical and biological needs. Humans no longer must suffer hunger, heat or cold, or disease since they have the capacity to control the environment to ward off most of these dangers. Since they have been successful in meeting the biological dangers of their environment, humans have been able to pursue the needs of their "intellectual" self (Consciousness II and III*). They seek to satisfy these needs in some instances through accumulation of modern "gadgets" that satisfy the need for "status." Consciousness II needs do not meet the physical–biological needs of the individual, but perhaps these are social–psychological needs. Evolution of humans has been, therefore, from that of primarily a biologically dependent species to a more independent one, but cannot entirely escape a genetic heritage nor ignore dependence upon the environment. Man's impact on the environment is that of rapid exploitation of the natural resource base and a sacrificing of the environment that surrounds him.

In addition, human populations no longer subjected to ravages of periodic plagues have increased at tremendous rates and apparently this increase is limited only by the amount of food available. Since the amount of food increases each year, humans have not reached a maximum population the resources of Earth can sustain. However, without some change in the rate of population increase exploitation of the timber resources, pollution of the environment, and increase in the amount of land people can live on will continue unabated. The demand for space on which to establish cities, build roads, and construct dams has removed land from crop and timber production. Salo *et al.* (1977) found that between 1967–1975 there was a net reduction of forest land in the United States from 444.6 to 375.4 million acres (not all of the reduction was land capable of commercial timber production). There was also a 30 million acre reduction in cropland. Urban and water acreage increased by 24 million acres; range and pasture increased 64 million acres.

Current public policy for forest land use in the United States is quite different from that of previous generations. In the early settlement period there was an initial need for open land to grow food crops and raise animals. This resulted in a massive effort in clearing trees to meet those needs. In the eastern United States following the early settlement era, as much as 75% of the land was cleared of trees for agriculture (Figure 5.2). Since then modern agriculture has made it possible for fewer people to produce, on fewer acres, the food needs of

*Charles A. Reich, *The Greening of America*. Random House. New York, 1969.

the United States. As a result, farms in many sections of the country have reverted to trees and the volume of wood produced today probably exceeds the volume produced in pre-settlement time since many of these stands are young and at the peak of their growth capacity. However, the pattern of ownership of land has changed, and many woodland owners do not utilize the wood produced on their land so less land is available for timber harvest than in the past. The demands that Consciousness II and III humans make on the forest remove much land from timber exploitation; it is held for aesthetic and other nontimber use.

Large urban population centers require large forest areas be managed primarily for water yield and recreation, and only secondarily for timber production. Modern humans have more leisure time, can travel long distances in a single day, and enjoy getting away from dense urban centers where they work. Frequently trips are made to forested areas where they can hike, hunt, and "hide out." Much of the use of the forest by urban dwellers takes the form of "looking at it." These people become upset when they see what appears to be a man-made disturbance—forest fire damage and recent logging. Many people in the Unites States no longer consider the forest as primilarly a source of raw material for fuel and fiber, but regard it as an area of scenic beauty to be enjoyed in a undisturbed natural state. Also, the demand for water for domestic and industrial use continues to increase making it necessary for people to concern themselves with ways of increasing available water supplies. Setting aside large areas of forest for watershed appears to be a logical solution.

Demand for wood continues to increase in the United States, and the amount of land available for growing it decreases each year. In an attempt to plan for increased wood demand upon a shrinking land area, the wood-using industries have begun to intensify cultural practices upon their own lands and to encourage better timber management by nonindustrial landowners. The development of the *Tree Farm* program by the wood-using industries is an effort to demonstrate that timber is a crop and that the forest industry can emulate agriculture in farming the land by using scientific management to meet the wood requirements of future generations. This program appears to have had only limited success because many individuals and conservation groups continue to voice concern over the management of both public and private forest land, and many nonindustrial landowners can take only a short-term approach to land management. There is, therefore, a danger that man someday may demand more of our forests in wood, water, and recreation than they are capable of supplying.

Present-day use of forests by humans takes many forms and makes many demands upon them. The multiple-use concept has not had a chance to reach its full potential because of the rapid changes in human needs. It appears certain that there will be conflicting demands made upon forests in the future, as they are made today, and it is quite important that foresters understand all the biological implications that current silvicultural practices have upon the forest, and to attempt to adjust these practices to meet multiple-use goals of the future.

FOREST FIRES AND HUMANS

Man's interest in fire spans many thousands of years. From the cooking fires of primitive people to fires to open the forest for hunting and travel, through the period of fires for land clearing in the early period of settlement in the United States, to the beginning of control of wildlife in this country, man and his use and abuse of fire have had their effect upon forests. Forest fires are capable of accomplishing a number of things both good and bad. It has only been since 1950 that foresters have begun seriously to consider fire as a silvicultural tool as well as an enemy (Riebold, 1971). It is now recognized that natural fires have been responsible for establishing and maintaining several important forest types in North America and other continents, and that the use of fire by native populations was in some sense an enlightened form of land management.

Cooper (1960) and Brown and Davis (1973) indicated that fire is responsible for the continued existence of stands of such species as longleaf pine, jack and lodgepole pine, ponderosa pine, Douglas-fir, western white pine, eastern white pine, red pine, and aspen. Records of past fire frequency in the Great Lakes region show fires occurring every 6–8 years. Some of the resistance to the use of controlled or prescribed fire stems from the knowledge that at infrequent intervals there have been large and disastrous fires in several forest regions. Brown and Davis (1973), summarizing the work of others, showed that since 1900 there have been five large well-publicized forest fires. The last fire occurred in Maine in 1947. So it has been difficult for people to accept the use of fire as a management tool in the forest. It has been necessary for researchers to study and to explain the effect of different types of fire on trees, wildlife, and streamflow.

There are three kinds of fire: surface, crown, and ground. A *surface fire* burns the surface litter and small plants of the low shrub and herb layer. Many seedling trees are killed by fires of this type. A *crown fire* moves through the canopy of a stand burning from tree crown to tree crown. These fires usually occur in dense conifer stands or in dense sapling-pole stands of deciduous species. They burn during periods of high wind and in hilly terrain when a fire burns up-slope. Crown fires also have the capacity to create air updrafts contributing to their spread. A *ground fire* burns deep into the organic layers of the soil. Usually these soils are too wet to burn; however, fires will occur in them after prolonged drought. Ground fires may burn for long periods—some for as long as several months. Although not extremely active during the entire time, hot spots smolder in tree stumps, roots, and deep peat layers, and are extinguished only when the water table rises, or after a prolonged period of precipitation, although some survive the winter and break-out into surface fires the following spring. Of the three types, the most destructive is the crown fire because it is capable of burning over large areas in a short time. Ground fires are confined to smaller areas but cause severe damage when they do occur since they destroy valuable soil organic matter. Surface fires often cause minimal damage but may be extremely damaging when they occur in stands not resistant to fire.

ADAPTATION OF TREES TO FIRE

Some trees are known as "fire species" because they have developed a resistance to fire, or are able to take advantage of the site conditions created by it. A number of species have thick bark that insulates the sensitive meristematic tissue from the heat of fire. Two of the best examples of fire resistant conifers are redwood and western larch; others are the southern pines (longleaf, loblolly, slash, shortleaf), pitch and pond pine, ponderosa pine, and Douglas-fir. All of these species, except the longleaf, are nonresistant in the seedling-sapling stages of development. Among the deciduous species that show moderate bark resistance is chestnut oak; other oaks and deciduous species are not so well adapted to withstand the heat of fire. Some species are able to recover rapidly following fire because of their vigorous sprouting ability. These are bear oak, black-jack oak, and dwarf chinkapin oak (Little, 1974). These sprouting species along with pitch pine form the famous pine-scrub oak stands of New Jersey. Species that produce serotinous cones are well equipped to reproduce after fire even though the trees on which the cones are born are killed by it. Some examples are pond, pitch, sand, jack, and lodgepole pines, and black spruce. Then there are species such as aspen that find a burned site quite suitable for reproduction. Black cherry and Atlantic white cedar, aspen, gray and paper birch, yellow-poplar, and sweetgum often reproduce vigorously after a surface fire that opens a stand so that these species can become established. When a surface fire does not burn too deeply into the surface organic soil horizons, Atlantic white cedar seed stored in the lower organic horizon may be able to germinate (Little, 1974).

Trees that are not resistant to fire, that do not sprout readily, or do not produce small easily disseminated seed are at a disadvantage. Although not always killed by fire, these nonresistant species may be injured as a result of bark scorch, root injury, or crown scorch. Bark scorch occurs on trees that are not completely girdled by fire, but where one or more faces of the lower bole are burned severely. A "cat" face thus develops. This lower-bole injury will expose the inner wood to attack by insects and disease. Root injury is of a similar form where surface roots are damaged and exposed to insects and disease. Crown scorch may be complete where all of the foliage is killed, or it may be only partly destroyed. If scorch occurs in the dormant season, trees may survive, but if it is in the growing season most trees will be killed. Spruce, fir, and hemlock are particularly susceptible to crown scorch when young because their crowns are deep, reaching nearly to the ground, and as a result are largely consumed by even the lightest surface fire. Mature hemlock stands can be quite resistant to surface fires, however. In addition, there are thin barked species that are particularly susceptible to fire. When a fire occurs it does not generally burn evenly; as a result there will be hot spots where even the more resistant species are killed, and in other areas local flareups of short durations may do damage to only the more sensitive species.

EFFECT OF FIRE ON SOIL

Fire will affect both the physical and chemical properties of soil. Briefly, the physical effects most important to tree growth are the reduction in the thickness of the organic surface layer and the exposure of the surface soil to direct solar radiation as a result of the removal of the overstory. In effect, fire exposes the surface soil to more direct solar radiation which may or may not aid in reproduction and growth of trees. Shearer (1974) reported that western larch and Douglas-fir reproduced well on a seedbed prepared by prescribed fire after clearcutting; it was best where either living or dead trees shaded the burned seedbed. Regeneration was sparse on poorly burned seedbeds where there was still a residual layer of organic litter, and on hot, open slopes that primarily faced south. Surface soil temperature, as high as 79°C, occurred on unshaded slopes facing either east, south, or west from mid-June to early August. Maximum surface soil temperature on north-facing slopes was 55°C. Many seedlings were lost to low surface soil moisture and some were killed later by frost heaving.

Ralston and Hatchell (1971) note that soil surface temperature may go as high as 500–1000°C under hottest surface fire conditions, and that these high temperatures may affect the crystal structure of soil particles. They also found that some affected soils have reduced percolation rates caused by water repellency and resistance to re-wetting. Organic matter not consumed resists re-wetting after fire. Unconsumed organic matter indicates that surface temperatures were no greater than 350°F, and that there were no changes in pore space and water infiltrating capacity. Viro (1974) noted that young stands on burned sites in Finland did not suffer from lack of water if the humus layer had not been totally destroyed. Where humus was destroyed, water penetration was impeded and higher evaporation occurred, which was particularly damaging to trees on coarse textured soils.

Grier (1975) reported large soil nutrient losses can occur during wildfires as a result of oxidation. As wood burns, organic N, Ca, Mg, K, and Na oxidize into the atmosphere and were carried away in the ash by strong winds. Grier estimated the loss to be on the order of 809 lb/ac. of N; 67 lb of Ca; 29 lb of Mg; 275 lb of K; 623 lb of Na. He noted further that following fire there was no change in stream water chemistry that indicated increased nutrient losses. If losses did occur, they were probably diluted by increased run-off from the watershed and therefore were not measured by techniques used for monitoring water chemistry. Replenishing lost minerals would occur primarily by the slow weathering of subsoil material and nitrogen fixation by ceanothus and other plants with nodules, and from precipitation.

Repeated prescribed burning reduces surface organic matter; but organic matter may increase in the mineral horizon tending to offset surface reduction (Figure 11.5). Stone (1971), using the work of Wells (1971) and others, showed that although the annual loss of nitrogen from prescribed burning exceeded 400 lb/acre over a 20-year period, there was, however, an increase in N on burned-over unburned plots (Figure 11.6). Stone surmised this increase to be the result of partial replacement by rainfall, with the major amounts coming from non-symbiotic nitrogen fixation, primarily by blue-green algae and azoto-bacter (study does not support this supposition). Long-term effects of burning were to increase somewhat the amounts of P and Ca in the surface layers. Although the long-term effect on pH did not indicate

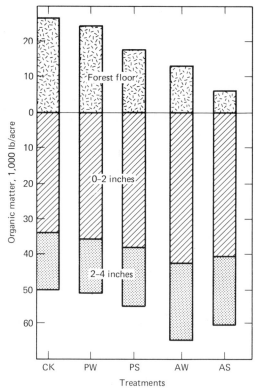

FIGURE 11.5. Organic matter in the forest floor, 0–2, and 2–4 inches of mineral soil for check (CK), periodic winter (PW), periodic summer (PS), annual winter (AW), and annual summer (AS) treatments after 20 years (from Wells, 1971).

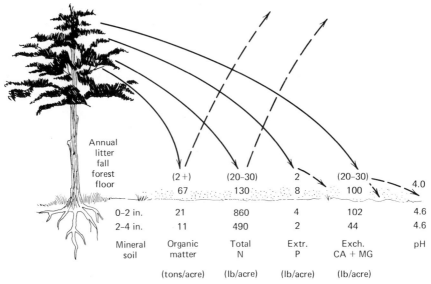

Mineral soil	Organic matter	Total N	Extr. P	Exch. CA + MG	pH
Annual litter fall forest floor	(2+) 67	(20–30) 130	2 8	(20–30) 100	4.0
0–2 in.	21	860	4	102	4.6
2–4 in.	11	490	2	44	4.6
	(tons/acre)	(lb/acre)	(lb/acre)	(lb/acre)	

FIGURE 11.6. Partial nutrient cycle in 50- to 60-year-old stands of loblolly pine after 20 annual winter fires. Solid arrows and parenthetical values indicate litter decomposition. Dashed arrows indicate loss or liberation by annual fire; the amounts roughly approximate litter additions. Liberation of Ca + Mg also raises the pH of the forest floor surface. Ca, Mg, and P in the litter and forest floor are total amounts (from Stone, 1971).

much change, the immediate effect of fire was to raise the surface soil pH to 8, which, combined with changes of P and Ca, resulted in an environment suitable to blue-green algae and azotobacter—two nonsymbiotic nitrogen fixing groups. Other studies showed that these two groups were present in large numbers in forest soils and Stone suspected they were responsible for the increase in N on burned sites.

EFFECT OF FIRE ON ANIMALS

It is not clear whether many animals are killed during a fire. Some reports indicate that many are able to flee or to find safe shelter during fire, while others claim that relatively large numbers of animals die, primarily from suffocation (Bendell, 1971). In both cases some mortality was reported and it can be expected that fires can kill, particularly large wildfires. Seasonal differences in mortality can be expected. Spring fires are more lethal because birds are in brood and new-born mammals are not able to move quickly.

Fire has a profound effect upon all factors of the habitat. As a result of burning, an area is subjected to greater extremes in temperature, higher wind speeds, more snow, and drier summer conditions. A reduction of nesting sites for birds and squirrels occurs along with an increase in low food for deer, elk, moose, bear, and ground feeding birds and rodents. Stoddard (1931) makes a strong case for use of fire to maintain and improve habitat for bobwhite quail. Periodic fire improves conditions for turkey to forage for food. The Kirkland warbler depends on young jack pine stands for nesting areas. These stands can be produced either by clearcutting and planting, or by natural seeding of pine following fire. Bendell (1971) showed that there was a net increase of 7% in the number of bird species and a net increase of 2% in the number of mammal species occurring after fire. In effect, it would seem that fires are not responsible for greatly increasing the number of animal species that utilize a burned forest area, but are mainly responsible for shaping species composition of animals.

GENERAL REFERENCES

Ainsworth, G.C. and A.S. Sussman (eds.). 1968. *The Fungi and Advanced Treastise,* Vol. III. Academic, New York.

Anderson, R.F. 1960. *Forest and Shade Tree Entomology.* Wiley, New York.

Baker, W.L. 1972. *Eastern Forest Insects.* USDA Forest Service Misc. Publ. No. 1175. U.S. Govt. Printing Office, Washington, D.C.

Boyce, J.S. 1961. *Forest Pathology,* 3rd ed. McGraw-Hill, New York.

Brown, H.A. and K.P. Davis. 1973. *Forest Fire Control and Use.* McGraw-Hill, New York.

Furniss, R.L. and V.M. Carolin. 1977. *Western Forest Insects.* USDA Forest Service Misc. Publ. No. 1339. U.S. Govt. Printing Office, Washington, D.C.

Graham, K. 1963. *Concepts of Forest Entomology.* Reinhold, New York.

Golley, F.B., D. Petrusewicz, and L. Ryszkowski. 1975. *Small Mammals: Their Productivity and Population Dynamics.* I.B.P.S. Cambridge Univ. Press, Cambridge, London, New York.

Hepting, G.H. 1971. *Diseases of Forest and Shade Trees of the U.S.* U.S.D.A. Forest Service Agr. Hndbk. No. 386.

Kozlowski, T.T. and C.E. Ahlgren. 1974. *Fire and Ecosystem.* Academic, New York.

Report of the Forest Study Team. 1975. "Pest Control: An Assessment of Present and Alternative Technologies," Vol. IV. Forest Pest Control. Natl. Research Council, Natl. Acad. of Sci., Washington, D,C.
Ricklefs, R.E. 1973. *Ecology.* Chiron Press, Newton, MA.
Smith, W.H. 1970. *Tree Pathology: A Short Introduction.* Academic, New York.
U.S.D.A. Forest Service. 1971. *Prescribed Burning Symposium.* Asheville, N.C.: Southeastern Forest Expt. Station.
Wallace, G.J. 1963. *An Introduction to Ornithology,* 2nd ed. Macmillan, New York.
Wheeler, H. 1975. *Plant Pathogenesis.* Springer-Verlag, Berlin.

Chapter 12

Trees affect the physical and biological environment above and within a forest stand, the environment of adjacent areas, and other areas where they are used as windbreak and sound barriers. The effect extends from above the crown canopy deep into the soil and to a distance from the stand edge. Temperature, moisture, mineral elements, and the living organisms in the soil, in the stand, and in the canopy are affected by the presence of trees. Tree windbreaks are used to protect field crops and to act as sound barriers in urban areas. They also affect soil water relations and water runoff from forested watersheds. In some locations in the United States water values from forests exceed and take priority over the value of the timber these areas can produce.

Water, Air Movement, and Sound

PRECIPITATION

In the past the argument was made that forests were responsible for increasing precipitation by their presence, some authors claimed that forests could increase precipitation by as much as 25% and that extensive logging or land clearing was responsible for reducing precipitation received in a locality. A number of studies (see Kittredge, pages 94–98) have considered the problem and the current concensus is that forests may be responsible for minor increases in precipitation, but they do not increase it to the extent that was originally proposed.

The main influence that forests appear to have on precipitation is not upon the large air masses associated with frontal movement, but through minor disturbance to air movement, mainly friction. This results in more precipitation (about one percent) occurring over forests than over open terrain, particularly in forest openings. In the case of fog drip, the forest may be instrumental in causing greater increases in total annual precipitation.

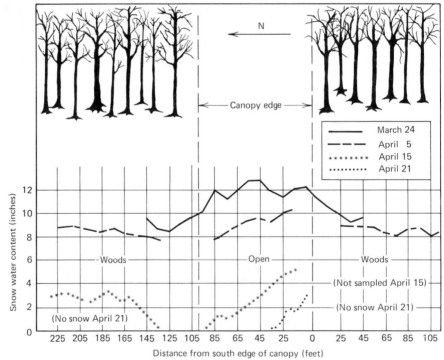

FIGURE 12.1. Pattern of snow melt in a northern hardwood stand (Sartz and Trimble, 1956).

If the forest is not able to catch more precipitation, then what influences does it have upon precipitation? The answer is found in the effect trees have upon interception of precipitation, on the rate of infiltration of water into the soil, and upon evaporation of water and snow. In general, because of interception, less precipitation reaches the forest floor under the canopy of a forest stand than in open areas. Some of the intercepted precipitation reaches the ground as stemflow; the remainder evaporates into the air. It is important to watershed management to be able to determine the amounts of precipitation which actually reach the ground.

From the standpoint of accumulating snow, a closed stand is not as efficient as a lightly stocked one, or as a stand with openings (Table 12.1). Presence of a forest canopy does slow snow melt because water from the snow pack is released over a longer period than in openings. If forest openings are not too large, the adjacent stand may influence the melting snow within the opening. As shown in the Figure 12.1, snow melt in the opening is reduced along the south edge because of the shade of the trees.

TABLE 12.1. Snow Accumulation and Melt in Openings and on Forested Slopes in Oregon[a]

| FOREST CONDITONS | SNOW WATER | | TOTAL MELT |
	APRIL 22 (in.)	JUNE 1 (in.)	JUNE 1 (%)
In forest openings	63.2	26.8	58
In forest adjacent to openings	56.0	25.1	55
In dense forest (80–100% crown cover)	48.1	19.1	59
In moderately stocked forest (50–80% cover)	50.3	25.0	50
In lightly stocked forest (20–50% cover)	57.7	25.0	57

[a]From Progress Rept. 1958–59. Cooperative Snow Management Research. H.W. Anderson and Lucille G. Richards. U.S.D.A. Forest Service Pacific Southwest Forest and Range Exp. Sta.

The most effective method of causing snow pack accumulation and prolonging the melt period appears to be by strip cutting rather than by removal of individual trees throughout the entire stand. Weitzman and Roy (1959) report that cutting strips in an east–west direction was the most effective method of promoting snow accumulation and prolonging melt since relatively large amounts of snow accumulated on the strips and the snow pack was protected from direct solar radiation (Table 12.2)

TABLE 12.2. Accumulation and Duration of Snow Melt in Northern Minnesota (Weitzman and Roy, 1959)

DATE	CLEAR-CUT[a] STRIP	SINGLE TREES CUT	UNCUT	LIGHTLY STOCKED	OPEN PATCH
		(Water equivalence in inches)			
February 14	2.9	2.0	1.6	2.2	2.9
March 7	4.0	2.6	2.5	3.2	4.5
March 29	4.3	3.2	2.7	3.5	4.0
April 3	4.2	2.8	2.7	2.9	3.6
April 9	3.6	3.2	2.7	2.8	2.4
April 16	3.2	2.9	2.5	1.8	1.5

[a]Oriented in east–west direction.

Not all of the snow water in a snow pack infiltrates the soil for some water is lost from the snow pack through evaporation. West (1959) reported that evaporation from a closed stand was 0.40 in. of water in one year and 0.95 in. the following year. In an opening in the forest evaporation was 1.07 in. the first year and 1.68 in. the following year. It would appear that openings in the forest are more subject to evaporation loss, and measurement of snow water in the snow pack does not indicate the amount of water expected to reach soil surface and appear as runoff. This is particularly true on exposed ridges and large openings where the annual loss of snow water through evaporation can be as much as four to five inches on ridges and two to three inches in large forest openings.

In those sections of country where a large part of the precipitation is in the form of rain, the relative amounts of interception, stemflow, and throughfall need to be considered. In snow country, the depth of snow pack reflects the net throughfall of precipitation; snow water content of the pack reflects net precipitation. Even here interception loss is an important factor. Leonard (1961), working in a northern hardwood stand in New Hampshire, reported the following relationships between gross rainfall and throughfall.

Leaf period, throughfall   = 0.8984 (gross rainfall)* − 0.030
Leafless period, throughfall   = 0.9424 (gross rainfall − 0.029
                                        or snowfall)
Stemflow, leaf period   = 0.0563 (gross rainfall) − 0.0024
Net-precipitation for one year = 0.9547 (gross rainfall) − 0.324

Several authors studied interception for different stand conditions: Boggess (1956) reported on a shortleaf pine stand in Illinois; Hoover (1953), on a young loblolly pine stand in South Carolina; Stuart and Sopper (1968) in Pennsylvania, mixed oak; Patric (1966), on western hemlock and Sitka spruce in southeast Alaska. The relationships for each of these for total annual accumulation (in.) follows:

Illinois
Throughfall = 0.8957 (rainfall in open) − 0.0562
Stemflow   = 0.0982 (rainfall in open) − 0.0045
Net rainfall = 0.9939 (rainfall in open) − 0.0607

South Carolina
Throughfall = 0.732 (rainfall in open) − 0.016
Stemflow   = 0.222 (rainfall in open) − 0.018
Net rainfall = 0.954 (rainfall in open) − 0.034

Pennsylvania
Net rainfall = 0.940 (rainfall in open) − 0.033

Alaska
Throughfall = 0.77   (rainfall in open) − 0.086
Stemflow   = 0.0084 (rainfall in open) − 0.0004
Net rainfall = 0.78   (rainfall in open) − 0.086

Proportions of net rainfall occurring as throughfall and stemflow do not differ much between regions; in fact, the New Hampshire and South Carolina net rainfall relationships show a striking similarity considering the difference in species and the difference in the patterns of precipitation. Reynolds and Henderson (1967), in comparing throughfall and stemflow with rainfall above the crown canopy, showed that leaf-fall often produced surprisingly little effect on inter-

*Gross rainfall was measured as inches of depth in an opening in the forest near the study area.

ception loss in deciduous species. In the studies presented above, interception loss appears to be greatest in Alaska where conifers dominated the stand and only about 78% of the precipitation reached the forest floor and 22% was intercepted. In contrast, in Illinois were deciduous species dominated only 1% net rainfall was intercepted.

FOG DRIP AND DEW

Fog can be a regular contributor to the precipitation of a forest stand. In regions of frequent fog, high elevations in mountains, and coastal mountains, trees are a barrier to air movement, and their stems and leaves provide a surface upon which water droplets can condense. When the water droplets coalesce, they fall from the branches and leaves and produce a steady drip. The fog belt of California, where redwood and sequoia forests are found, are excellent examples of where fog contributes a considerable amount of water to forest precipitation.

Kittridge (1948) cites the data of Hoge, who recorded the amount of precipitation caught in rain gauges in the open and under several vegetation conditions (Table 12.3).

TABLE 12.3.   Fog Drip under Vegetation at 5850 feet Altitude on Mount Wilson, California (See Kittridge, 1948, p. 118)

LOCATION AND COVER	GAGE CATCH (in.)	FOG DRIP (in.)
October–May, 1916–1917		
Open	22.81	0.0
Under dense ceanothus, 8 ft high	22.67	0.0
Under dense canyon oak, 45 ft high	47.74	24.93
Under big cone spruce,[a] 40 ft high	48.05	25.24
Under ponderosa pine, 80 ft high	60.53	37.72
January–May, 1918		
Open	27.23	0.0
Under ponderosa pine, windward, S.	52.51	25.28
Under ponderosa pine, windward, E.	56.62	29.39
Under ponderosa pina, leeward, N.	85.38	58.15
Under ponderosa pine, leeward, W.	87.50	60.27

[a]Douglas-fir (big cone)

From these data it is evident that the amount of precipitation accumulated depends upon the height of the vegetation. Vegetation such as grass and low brush is not effective in intercepting the moisture-laden air, while trees, particularly the taller species, are more efficient in intercepting moisture. It is interesting to note the difference in the amount of accumulation under the ponderosa pine on different aspects, leeward slopes receiving greater amounts of precipitation from fog than windward slopes.

Oberlander (1956) measured fog drip under selected trees exposed to winds from the Pacific Ocean. The first line of trees to seaward receive the greatest amount of moisture and the trees to the east exhibit a gradual decline in fog precipitation (Table 12.4). This appears to be in contradiction to Hoge's data.

TABLE 12.4. Precipitation under Fog Exposed Trees (Oberlander, 1956)

SPECIES	APPROX. HEIGHT (ft)	EXPOSURE TO FOG	INCHES OF PRECIPITATION
Redwood	200	Forest tree; no front line exposure	1.8
Tan oak	20	Direct exposure	58.8
Douglas fir	125	Partly protected by ridge line	7.2
Douglas fir	125	Little protected by ridge line	8.9
Douglas fir	125	Direct exposure	17.1

Vogelmann and co-workers (1968) compared fog drip accumulation at 1800, 2800, and 3600 ft elevations in the Green Mountains of Vermont. At 1800 ft there was less precipitation in the screened rain gauge than in the open gauge; at 2800 ft there was a 5% increase in the screened gauge; At 3600 ft there was a 66.8% increase in precipitation in the screened gauge as compared with the open gauge. Screens were used on the rain gauges to form a condensation surface.

Although plants are not able to take in large amounts of water through their leaves, there is very good evidence that some trees may absorb sufficient water through their leaves from dew by which they are able to sustain themselves during prolonged drought (Stone, 1957). Very little data are available to show how much precipitation can accumulate from dew and frost in forest stands; however, some data that indicate the amounts of precipitation which can accumulate on grass is available. One set was obtained from lysimeter studies in Ohio (Table 12.5). In Ohio the critical drought periods of July and August are not times when large amounts of dew occur. However, during September and October, months when vegetation can suffer from lack of water, there appears to be a significant increase in dew precipitation. These two months are part of the fall forest fire season; frequent dew maintains fuel moisture, resulting in reduced fire danger. A West Virginia study shows that dew occurs frequently but in lesser amounts than was measured in Ohio (Table 12.6).

TABLE 12.5. Precipitation Accumulated from Dew on Grass Cover (in.)

Jan.	Feb.	Mar.	Apr.	May	June
0.85	0.85	0.49	0.35	0.12	0.14
Aug.	Sept.	Oct.	Nov.	Dec.	Total[a]
0.13	0.34	0.55	0.53	0.65	4.92

[a]No data were collected during July.

TABLE 12.6. Dew Accumulation in West Virginia (Hornbeck, 1964)

1962	ACCUMULATION FOR PERIOD (in.)	OCCURRENCE OF DEW DEPOSITION (no. of days)	AVERAGE MEASURABLE DEPOSIT (in.)	MAXIMUM DEW DEPOSITION DURING PERIOD (in.)
May	0.030	19	0.0016	0.004
June	0.051	19	0.0027	0.005
July	0.044	19	0.0023	0.005
August	0.055	23	0.0024	0.005
September	0.032	13	0.0025	0.005
October	0.030	10[a]	0.0030	0.007
November[b]	0.017	8	0.0021	0.004
May–November	0.259	111	0.0024	0.007

[a]10 days missing because of instrument difficulty.
[b]November values include frost deposition.

There is no indication from these studies that taller vegetation such as shrubs and trees would be more efficient in increasing dew accumulation, but there is reason to believe that they can and that the effects would be similar to those causing fog drip (Lloyd, 1961).

AIR TEMPERATURE

The crown canopy of a forest stand intercepts solar radiaton, and, as has been shown, some energy is reflected, some is used in transpiration, and some of the remaining solar energy causes air movement in the crown canopy. Energy accumulation will be less beneath the crown canopy and air temperatures are lower during the summer within a forest stand than in openings in the forest. As an example, Schomaker (1968) measured solar radiation in May and June under both conifers and hardwoods and found that compared to an open field conifers intercepted 92% of the solar beam and hardwoods 85% when in leaf and 41% before flushing.

Greatest differences in air temperature occur during early afternoon when maximum solar intensities are reached. Minimum air temperature differences between forest stands and openings are often difficult to detect and may not exist since they occur during the early morning hours before sunrise when they have become equalized during the night over large areas, except where there are differences due to local radiation.

Physiographic location of a site, other factors being equal, will determine the degree of difference in maximum air temperature between a closed stand and on open areas. Soil moisture supply can also affect maximum air temperature differences, as considerable heat (590 cal/g) is required to evaporate water. Fritts (1961) found he could relate maximum air temperature in closed stands to several site and

TABLE 12.7. Maximum Air Temperature in Forest Stands at Several Physiographic Locations (Fritts, 1961)

LOCATION OF STAND	WHEN THE MAXIMUM AIR TEMPERATURE IN THE CLEARING WAS:		
	80°F	90°F	100°F
Upper south slope	78°F	87°F	97°F
Lower south slope	74	84	93
Bottom or ravine	75	82	90
Middle north slope	75	83	91

atmospheric conditions. He measured maximum air temperature differences in two stands occupying a ravine in Illinois. Data included wind velocity (seven miles from study area), maximum air temperature (in an opening near the area), solar radiation (in an opening), and the product of vapor pressure deficit times soil moisture (VPD × SM%). Maximum air temperature of the stands and the openings were compared and are shown in Table 12.7.

As would be expected, upper south-facing slopes were relatively warmer than other sites and differences here between maximum air temperatures under trees and in the open were less than the other three sites. When an analysis of temperature differences was computed using wind velocity, maximum air temperature in openings, solar radiation, and VDP × soil moisture, significant relationships were computed (Table 12.8).

TABLE 12.8. Correlation Between Meteorlogical and Soils Data and Stand Air Temperature (Fritts, 1961)

VARIABLES	UPPER SLOPE	SOUTH LOWER SLOPE	BOTTOM	NORTH MIDDLE SLOPE
Wind	Rel[a]	Rel	Not Rel[b]	Not Rel
Clearing max. temp.	Rel	Rel	Rel	Rel
Solar radiation	Rel	Rel	Not Rel	Not Rel
VDP × soil moisture%	Rel	Rel	Rel	Rel

[a]Rel indicates a statistical relationship to maximum air temperature.
[b]Not Rel indicates no statistical relationship to maximum air temperature.

From these results it would appear that air temperature and soil moisture are related to stand maximum temperature. Maximum air temperature and vapor pressure deficit (VDP) are related since the vapor pressure deficit will approximately double (Table 7.6) with an increase of 18°F, so it might be questioned whether the VDP × soil moisture relationship is a true cause–effect relationship. However, in the study it does appear that VDP × soil moisture as a separate variable contributes to a relationship beyond that of air temperature alone. Cochran (1969) calculated that the thermal properties of a moist soil produces a more stable temperature pattern (less diurnal change) than of a dry soil. For example, he found that if one calorie of heat is removed or supplied to 1 cc of A_1 horizon material at a moisture content of 30% by volume, the resulting temperature change was 2.2°C; however, if the soil horizon was dry, the temperature change was 6.7°C. This is based on the following relationship:

$$\text{temperature change} = \frac{\text{heat quantity}}{\text{volume} \times \text{volumetric heat capacity}}$$

McHattie and McCormack (1961) report that evaporation of water from Piche evaporators was from one-half to one and one-half times greater on cleared sites as on forested sites. Apparently open sites cause a temperature "built-up" greater than that which occurs on the forested sites around the openings. They report that the forested sites averaged from 1–7°F lower in air temperature than cleared sites. McHattie and McCormack also observed that flowering on a south-facing slope occurred seven days earlier than on a north-facing one. A conclusion that can be drawn is that south-facing slopes have a longer growing season and thus are compensated somewhat for their generally more severe moisture conditions. Therefore, the difference in growth between southern and northern aspects may be confounded by a growing season difference. However, at more southern latitudes south aspects may not show a beneficial effect as has been demonstrated in the site studies of oak. A number of studies have shown that growth is slower on southern aspects than on northern ones.

SOIL TEMPERATURE

A mantle of litter or snow over a soil surface acts as a barrier to radiational heat loss from the soil and as insulation to an exterior heat source. A snow mantle reduces soil heat loss, thereby buffering the effects of air temperature fluctuations. Hart and Lull (1963) state that under a snow pack there was no soil freezing at minimum air temperature 4°F if there was 18 inches of snow. Under 12 inches of snow, soil freezing did not occur when air temperature minimums were 5°F and under 6 inches of snow freezing did not occur with air temperature minimums of 25°F. In addition to snow, surface organic litter provides a mantle, reducing radiational heat loss. Soil temperature increases with increase in depth below the surface. When snow and litter act as a buffer to heat loss, the soil maintains a relatively uniform temperature compared to the air above it. Presence of litter on a forested site also limits frost penetration to a depth of four to six inches in late autumn and early spring while on grass sites frost penetrates to the same or greater depths.

Soil texture is important in governing the type of freezing that takes place in a soil. Fine textured soils, high in silt and clay, having fewer noncapillary pores and the capacity to retain greater amounts of water than coarse textured soils, are more inclined to exhibit concrete frost. This is a condition where the soil particles are bound together by ice into a hard mass resembling concrete in density. Coarse textured soils, sands, and loams, exhibit a frost pattern that is granular in nature. Trimble, Sartz, and Pierce (1958) demonstrated that although concrete frost is impermeable to water, its occurrence in forest stands is "traversed in places by large open holes that allow water to enter the soil" (Table 12.9). It would appear also that soil organic matter in the hardwood stand on a loam soil and in the white pine stand on a sandy loam soil caused a sufficient change in soil structure so that soil frost was granular, producing a loose textured condition that permitted more rapid infiltration of water than where no granular frost occurred. Concrete frost in all conditions did not permit water to enter the soil. This difference in soil frost condition between forested and nonforested is important to watershed management.

TABLE 12.9. Mean Infiltration Rates in Frozen Soil (Inches of Water per Hour) (Trimble, Sartz, and Pierce, 1958)

| | | | TYPE OF FROST | |
COVER	SOIL TEXTURE	NO FROST INFILTRATION	GRANULAR FROST	CONCRETE FROST
Abandoned pasture	sandy loam	9	—	0
Lawn	sandy loam	11	—	0
Red Spruce	sandy loam	14	—	0
Hardwood	sandy loam	30	—	0
Hardwood	loam	14	66	0
White Pine	sandy loam	50	204	0

It can be concluded that forested soils do not freeze to depths as great as nonforested ones. Structure of forest soils is such that they tend to be granular in texture when frozen, not solid ice as in the open. Hence, forest soils can take in large volumes of water even when they are frozen. Water from heavy rains and snow melt infiltrates forest soils because of their more open texture, while water will in many instances flow over the surface of frozen grass-covered or bare soils.

The tree canopy and organic surface soil layer act as agents to modify temperature and moisture regimen in the upper soil layers. In an experiment that simulated a forest system, Bilan (1960) found that shading and mulching of two-year loblolly pine seedlings caused an increase in root mass near the soil surface, whereas seedlings grown in an open unmulched condition had their root mass deeper in the soil. Experience in applying fertilizers to forest soils has shown that the "feeder" roots of trees are concentrated in the H and A soil horizons, although there are roots which extend throughout the solum. Apparently, the growing conditions for tree roots are favored by the soil temperature and texture of the surface soil layers under forest stands.

MEASURING WATER RUNOFF FROM LAND AREAS

Several approaches have been proposed for estimating water runoff from land areas. One approach was by measuring streamflow directly and subtracting that amount from the precipitation that fell on the watershed. The difference is consumptive use and represents water lost to various factors. One can measure water vapor flow over the vegetation growing on the watershed and estimate water loss in that manner. The concern is to assign to each component of input and loss a specific value (inches of water equivalent) so that estimates of change in water yield can be made should the vegetation on a watershed be changed in some manner.

Water Balance Method

Using this method the water entering the soil system as precipitation is budgeted to the different pathways of use. The following equation is also used in watershed management studies:

$$\Delta W = P - (Q + E + U)$$

where ΔW is soil moisture storage during period of measurement, P is precipitation, Q is runoff, U is deep storage, and E is evaporation and transpiration (E/T).

Metz and Douglas (1959) and Zahner (1956) used the water balance approach in their studies of soil water depletion introduced here in Chapter 8.

Empirical Formulas
With this method the E/T loss from vegetation is correlated to the loss of water from a free water surface (usually an evaporation pan). Measurements of air temperature, humidity, and wind speed are needed to derive the correlation of evaporation, and it is assumed that loss of water from E/T is a proportion of the loss from free water. In the case of forest vegetation the proportion is about 0.7 (Penman, 1948; Kohler, Nordenseon, and Fox, 1955; Van Bavel, 1966). Kitching (1967) measured water use in a number of conifer plantations in England and found that ratio of evaporation of pines to open water averaged 0.70 ± 0.25; the highest ratio was 0.81 ± 0.23; the lowest was 0.49 ± 0.18.

Energy Balance
The loss of radiant energy between that which is received and that which is used by a plant community is budgeted. The equation to be balanced is that of Rose (1966), which is (see also Heat Transfer, Chapter VII)

$$Rs(1-\alpha) = R_L + G + H + LE$$

where
Rs = flux density of total short wave radiation
α = reflection coefficient of plant and soil surface (albedo)
R_L = net flux density long wave radiation—the difference between that emitted and absorbed
G = heat energy flux density of soil
H = sensible heat energy flux density into atmosphere
LE = latent heat energy of vaporization of water

The difference between incoming radiation (Rs) and outgoing radiation (R_L) is.

$$R_n = Rs(1-\alpha) - R_L \qquad R_n = \text{net radiation}$$

Substitution for the above, energy exchange can be equated to $R_n = G + H + LE$, now $R_n + G + H + LE = 0$. G is the amount of energy dissipated within the soil, H is the amount of energy utilized in increasing temperature of plant above ground biomass, and LE is the amount of energy used in evapotranspiration.

The amount of energy used in E/T can then be estimated by monitoring incoming and outgoing radiation and integrating this exchange with soil energy potential, and plant-temperature energy increase. By knowing the energy used in E/T, it is possible to determine the volume of water lost. There is an error in this equation because it does not account for the energy used in photosynthesis and released in respiration (aA). These represent a small amount of energy; however, the equation should include them; therefore,

$$R_n = G + H + LE + aA$$

Vapor Flow
In this method the increase in water vapor in the air is measured above the crop (stand). A humidity gradient can be determined at different levels in the stand.

In whichever way the E/T loss is measured, the total amount of water used in the process depends upon the amount stored in the soil, the difference in vapor pressure of the leaf and the air, and the wind velocity. Concern over the E/T potential originates from either of two sources: (1) that the loss of water to the atmosphere represents on a national basis, about 70% of the water budget; (2) moisture stress in a plant created by losses of water for transpiration can represent loss of growth. It is therefore important, both from the standpoint of watershed management and from a timber production standpoint, to be able to control the E/T loss of forest stands.

Streamflow and Runoff
In the humid climates of the eastern United States average annual runoff of streams equals about 48% of the annual precipitation. Lull (1965) provides the following equation for estimating runoff from 43 watersheds in New England (period 1940–1957) as:

$$\text{Annual Runoff} = 1.13 + 0.0431(\% \text{ forest}) + 0.4839$$
$$(\text{annual precipitation})$$

Amount of runoff for a particular month varies depending upon the state of precipitation—snow or rain—and upon the evapotranspiration demands of vegetation and other factors. A study of 137 basins from Maine to Maryland using 17-year streamflow records showed annual water yields to range from 19 to 24 inches with a mean of 22 inches; the region has an average annual precipitation between 35 to 50 inches. Seasonal distribution of runoff was: winter (December–February), *six* inches; spring (March–May), *ten* inches; summer (June–August), *three* inches; autumn (September–November), *three* inches. Differences did occur between and within the regions (1963 Annual Rept. Northeastern For. Exp. Sta. 1964. pp. 35–36).

The equation used to equate streamflow to the difference between precipitation and depletion loss, using the water balance equation of Kramer (1969), is

$$\Delta W = P - (Q + E + U)$$

where ΔW is the change in soil moisture, P is the precipitation, Q, the runoff, U, the deep drainage to ground water, and E, the evaporation for soil and plants. Rearranging to be able to equate runoff to the other factors:

$$Q = P - (E + U + \Delta W)$$

E includes transpiration (t), soil evaporation (e), and interception loss (i). Substituting for E,

$$Q = P - (t + e + i + U + \Delta W)$$

It should be noted that the three factors in the equation representing depletion loss of water, before it can show up as streamflow runoff, are transpiration, evaporation, and interception. Although vegetation can reduce evaporation loss from the soil, it more than makes up for this reduction by transpiring large volumes of water back into the air. A single tree may transpire as much as 35 gallons of water per day. An evapotranspiration demand of 0.20 inches per day by a forest stand amounts to a loss of 726 cubic feet (or 5430 gallons) of water per day. If 8% of 40 inches annual precipitation is intercepted, this is equivalent to a loss of about 3.2 inches of water, or 11,761 cubic feet of water loss per acre per year. To make a watershed yield a maximum volume of water, it is important to be able to reduce losses from transpiration, evaporation, and interception to a minimum without destroying the infiltration capacity of the soil.

Kittredge (1948) presents several numerical examples of the streamflow equation:

$$Q = P - (t + e + i + U + \Delta W)$$

Swiss experiment[*]:

> Forested = 37.1 in. = 62.6 in. − (11.8 in. + 4.6 in. + 9.1 in. + 0 + 0)
> 2/3 grass = 40.4 = 65.1   − (5.3 + 11.7 + 7.7 + 0 + 0)

Wagon Wheel Gap, Colorado[*]:

> Forested = 6.2 = 21.2 − (5.0 + 7.0 + 3.0 + 0 + 0)
> Deforested = 7.3 = 20.8 − (4.0 + 9.0 + 0.5 + 0 + 0)

Coweeta, North Carolina[*]:

> Hardwood intact = 19.3 = 62.4 − (19.0 + 15.1 + 6.4 + 0 + 2.6)
> Hardwood cut = 36.9 = 62.4 − (0.0 + 18.2 + 3.9 + 0 + 3.9)

It is relatively easy to obtain measures of runoff and precipitation compared to estimates of transpiration, evaporation, interception, deep seepage, and soil moisture recharge. Often, the amounts of water allocated to these sources of depletion are made on a proportional basis. From precipitation and streamflow records a gross depletion is estimated. For example, in the Coweeta study rainfall in the forest situation was measured rather accurately at 62.4 inches and the runoff measured as 19.3 inches. The difference (62.4 − 19.3 = 43.1) was then allocated to the various sources of depletion. Change in the water levels of deep wells provides a measure of whether deep seepage is occurring, but it is very difficult to measure the volume of water that is retained in voids below rooting depth simply by observing the rise and fall of the water level in a well, for there is no precise way of measuring the amount of voids into which water can flow. Interception studies provide an estimate of the amount of water lost from this source. Transpiration and evaporation studies are rather cumbersome

[*]One acre inch of water = 3630 cubic feet; all results are given in inches of depth.

to accomplish on an entire watershed and estimates are subject to considerable error, as are estimates of the soil moisture recharge. However, in temperate climates, soil moisture reaches a maximum during the winter dormant season and generally it can be assumed that soil moisture, like deep seepage recharge, is zero during a year's time. During the growing season there may be periods when soil moisture is reduced below field capacity, so soil moisture supply is needed to balance the equation. If the ground water reserve on a watershed is used for domestic or other purposes, ground water recharge may not balance out over the year.

Generally, changes in streamflow from uninhabited forested watersheds during the season are influenced mainly by interception, transpiration, and evaporation. By changing stand density or by changing to some other type of vegetation, water loss to i, t, and e can be reduced. For example, it was found that in southern California when grass cover was substituted for mature chapparal, annual interception loss was reduced from 12.8% to 7.9%. Translated to water yield increase, if watersheds were converted from chapparal to grass it could be expected that 1.3 inches of gross interception loss would be saved each year (Corbett and Crouse, 1968).

In the southern Appalachians a vigorous stand of grass consumed nearly as much water as did a hardwood stand. As the grass lost vigor, runoff increased over forested watersheds and by the fifth year runoff from the grass area was 5.8 inches greater than forested areas (Douglas and Swank, 1975). These same authors reported that on a watershed converted to white pine, runoff was less than when the watershed had been in hardwood; a reduction of about 6.3 inches of annual runoff. Losses occurred in the winter when pines intercepted more water than hardwoods and in early spring when the pines transpired water earlier than hardwoods. In Rhode Island, white pine did reduce runoff more than oak early in the growing season, but evapotranspiration in oak surpassed that of pine later in the growing season with the result that total E/T in oak was 19.32 inches (daily E/T 0.153 inches) and in white pine 17.59 (daily E/T 0.130 inches).

Current practice in watershed research is to use a number of watersheds, with various treatments being carried out on each and with at least one watershed being maintained as a control against which runoff from all treated watersheds can be compared. If there are only a few watersheds that can be used for study, maintaining a control watershed may be expensive. Changes can also occur on the control watershed (fire, disease, insects, etc.) that alter its character and thus result in a different runoff pattern. Reigner (1964) points out the difficulty of trying to use climatic data to calibrate watersheds and proposes a single watershed approach. The single watershed approach would be more useful if the statistical manipulation of data could be made more accurate. However, there is a serious question whether such studies can be as useful as those where both treatment and control watersheds are used because of long-term trends which can occur and which could not be detected in a single watershed study.

Runoff from a particular watershed depends on how much water the vegetation intercepts and transpires and how much infiltrates the soil. Surface runoff should be kept to a minimum because water from surface flow is usually of poorer quality (high turbidity) and causes

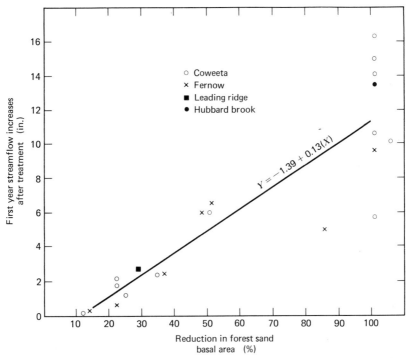

FIGURE 12.2. Increased streamflow in the first year after cutting is more or less proportional to the percentage of cutting (expressed by basal area). Removing less than 15% has no detectable effect. Combined results from four experimental areas in the eastern U.S. (Douglass and Swank, 1975).

excessive peak flow and rapid rise and fall of water level in streams and rivers. On watersheds where detention storage capacity is high, surface flow should not constitute more than three percent of runoff. In general, total runoff can be directly related to the basal area of the stand that is growing on the watershed (Figure 12.2).

WATER QUALITY
Water quality varies with the intended use, but will depend upon three general criteria: turbidity, mineral content, and temperature.

Turbidity depends upon the amount of organic and inorganic solids that enters and is suspended in water. Normal geological erosion removes solids from all land; however, different cover types afford more or less protection against excessive erosion (Figure 12.3). Forest stands in good condition afford better protection against excess erosion than those in poor condition, or cultivated or abandoned fields.

When water contains excessive amounts of silt and clay-size material, either organic or inorganic, the biological balance can be displaced. Fine soil material in water can settle out and smother fish eggs, but may not harm adult fish. On the other hand, if excessive amounts of organic material are deposited in slow moving water systems, increased activity of microorganisms that break down the material may reduce the oxygen level in the water to a point where adult fish and other animals which require oxygen may be affected. This can be so particularly when there is an accompanying increase in nitrogen and phosphorus in the water.

Mineral content of water depends upon the amount of organic matter deposited directly into the water, amount of surface runoff during snowmelt and high rainfall periods, subsoil exchange, and biological and physical activity within the water system itself. Each of these sources is difficult to assess individually. It is easiest to assess the total amount of minerals in streams that are influenced by different land management practices, and at the same time it is necessary to recognize that precipitation can deposit minerals on the soil and in water. Generally, undisturbed forest stands are not responsible for introducing large amount of N and P into water. Cropland and urban systems will in some cases be responsible in introducing greater amounts of N and P into water bodies than would forested lands (Figure 12.4). However, when forest stands are logged, the soil surface is exposed to increased solar radiation and moisture with the result that microorganisms in the soil surface increase the decomposition of the organic residues. This can hold the potential for introducing excess

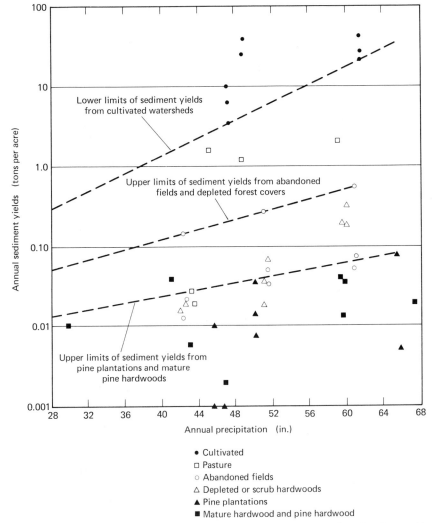

FIGURE 12.3. Annual soil loss rates from these individual small watersheds (mostly in northern Mississippi) depend on type of vegetation or treatment, and secondarily on amount of precipitation. Note that loss rates are on a logarithmic scale. The pine plantations were established on abandoned fields and pastures, and so demonstrate the possibility of rapid rehabilitation (from Ursic and Dendy 1965, added to by Ralston and Hatchell, 1971).

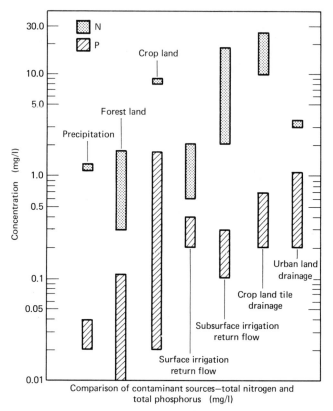

FIGURE 12.4. Nitrogen and phosphorus concentrations in outflowing water are affected by land use. The range of concentrations in forest land streams is normally somewhat more or less than in precipitation. Note that scale is logarithmic. Values from U.S. literature (after Loehr, 1972).

amounts of nitrogen into stream water. The Hubbard Brook example was cited earlier, and it was stated that the experiment was unusual in nature since the vegetation on the watershed was sprayed with a herbicide for three years after the clearcutting. There was insufficient vegetation on the watershed to take up the minerals that were mineralized in the decomposition process and thus some of the excess ended up in the stream water. Studies on other watersheds near the one in question showed that cutting alone did increase stream water nitrate amounts, but the increase was not enough to warrant serious concern for public safety. On one watershed nitrate concentration following cutting was 15 mg/ℓ the first year and 26 mg/ℓ the second year, while on the Hubbard Brook study maximum concentration was 60–80 mg/ℓ (Pierce et al., 1972). Also downstream from the Hubbard Brook watershed nitrates were diluted by waters from other untreated and uncut watersheds, reducing their concentration to a safe level.

RIPARIAN VEGETATION

Extremely large volumes of water are used by riparian vegetation (greasewood, *Sarcobatus vermiculatus*; salt cedar, *Tamarix gallica*; cottonwood, *Populus* spp.; baccharis, *Baccharis glutinosa*; willows, *Salix* spp.; mesquite, *Prosopis juliflora*), vegetation that grows along streams and rivers. Riparian vegetation is made up of plants, called *phreatophytes*, that transpire water in luxury amounts. Estimates

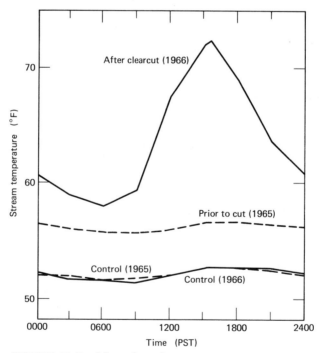

FIGURE 12.5. Mean hourly temperatures on an unlogged (control) watershed and a clearcut watershed in Oregon's Coast Range before and after clear-cutting for the period August 1–15 (G.W. Brown and J.T. Krygier, 1967).

made of the depletion rates by vegetation along stream channels and during one period from April to May, show 13 inches of water were lost through transpiration. Over a longer period, May–October, the rate increased to 54 inches, and over a two-year period an estimated annual loss of 50 inches of water occurred (Qashu and Evans, 1967). Although removal of stream bank vegetation will reduce water loss, water temperature may be adversely affected. Stream bank vegetation thus regulates water temperature, which is important to fish habitat, and excessive exposure of water to high light intensities may cause a buildup of algae and other water plants. Brown and Krygier (1967) measured maximum water temperature in a mountain stream in the Pacific Northwest both before and after the adjacent stand was logged. During this study stream temperature, although below the lethal range for fish, approached the critical 75–77°F range for salmonoids acclimated to a 69°F temperature (Figure 12.5).

It was found that clearcutting or maintaining land in field crop production in the southern Appalachians increased stream water tem-

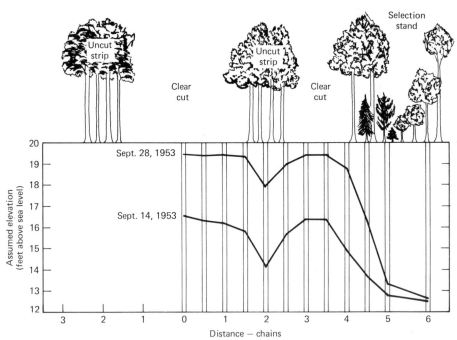

FIGURE 12.6. Depth to water table under uncut stand and clearcut stand (Trousdell and Hoover, 1955).

perature in summer from 6–11°F (Swift and Messer, 1971). Stream water temperatures rose under these conditions to 72–77°F—temperatures unsuitable for trout, but not detrimental to warm-water fish.

MAINTAINING GROUND-WATER TABLES

Following cutting of a forest stand on poorly drained sites, the water table can rise covering the soil surface over the entire area.

Wilde et al. (1953) noted that a 40-year-old oak-jack pine stand on a coarse sand depressed the ground-water table during the growing season nine inches more than a table under a grass covered field (Table 12.10). Difference in water table depth diminished later in the growing season. In the same report it was stated that clearcutting aspen stands on poorly drained sites caused the water to rise to the surface, changing the drainage class of the site. Maki (1968) noted that trees on wet sites can lower the water table during the growing season by as much as 0.19 ft/day.

It is difficult for trees to become established on wet sites; however, establishment can be enhanced through the technique of "bedding" (Mann and McGilvray, 1974), and subsequent growth improved by drainage ditches (Miller and Maki, 1957; Maki, 1968). When trees are to be planted on wet land, their roots must have a suitable place to grow. Once established, trees may lower the water table, keeping it below the soil surface; however, best growth is obtained where the water table is maintained at some depth below the soil surface by a system drainage ditches.

Results of a study by Trousdell and Hoover (1955) indicate that during the growing season a pine stand that had been cut, under the selection method, drew the water table down to seven feet from the surface as compared to six feet under an uncut strip, and only four feet under a clearcut strip. Following a heavy (2.81 inch) rain in late September, the watertable did not change appreciably under the selection stand, but came to within one foot of the surface under the clearcutting (Figure 12.6). Ground-water tables can on some sites provide additional amounts of water for tree growth, as can be noted in studies of site quality evaluation where growth was better on somewhat poorly drained soils as compared with well-drained soils (see Chapter 10). Trees grow poorly on permanently wet land. Where there is a large amount of such land, it is often necessary to drain the water before timber management can be made an economical activity.

TABLE 12.10. Difference in Depth to Water Table Between Jack Pine-oak and Grass[a] (Wilde, 1953)

DATE	INCHES
May 25	+ 1.0
June 18	+ 4.0
July 1	+ 7.5
July 9	+ 9.0
August 26	+ 4.0

[a] + denotes greater depth to water table under pine-oak than under grass.

Forested regions, unlike grassland, occupy areas of relatively high precipitation and the forests, as a result, are important in regulating streamflow. To achieve maximum runoff may require the use of clearcut strips in snowpack areas and large clearcuttings or stand conversion in areas of high rainfall. The main watershed problem in either case is to obtain a maximum throughfall of precipitation and to assure a maximum of infiltration of water into the soil. Trees, because of their greater rooting depth compared to grasses, are more satisfactory for watershed cover where flood control is a major concern since grass does not control surface flow as well as forest soil conditions. Undisturbed forest stands reduce surface flow of water over the soil, but they do not always provide the best condition for maximizing streamflow because of high evapotranspiration and interception water loss.

WIND AND SHELTERBELT

The forest is greatly affected by wind and in turn can influence the effect of wind. Overmature trees, trees with unstable root systems, trees in exposed stands, and trees with stem or butt rot are susceptible to wind throw or wind breakage. Wind speeds of 40 miles per hour or more occur regularly in most forested areas. Hurricane and tornado-force winds are confined to relatively smaller-regions; however, no forest region escapes an occasional hurricane-force (75 mph) wind.

Trees reduce wind velocity, even in the leafless state, by as much as 40% and have been used as windbreaks in regions where it is important to reduce wind velocity to protect growing crops, houses, barns, and livestock. Woodruff and Zingg (1953) found that ten rows of trees proved the most effective shelterbelt when compared with five and seven row design, although seven row design was nearly as efficient and might be less expensive to install (Table 12.11). However, it was necessary to arrange the ten rows so that the slope of the belt increased gradually with the prevailing wind and then decreased rather abruptly (Figure 12.7).

Prevailing wind

FIGURE 12.7. Arrangement of plant rows in a ten row shelterbelt planting.

TABLE 12.11. Effective Distance of Shelterbelts of Different Number of Plant Rows (Woodruff and Zingg, 1953)

| NO. OF ROWS | DISTANCE IN TREE HEIGHTS OF THE TALLEST TREE | | |
	75% REDUCTION	50% REDUCTION	25% REDUCTION
10	11.5 times	16.1 times	29.0 times
5	9.1	14.5	26.4
7	10.3	13.9	24.8

Hogg (1965) found that wind speed was reduced 61% at 3.5 tree heights and 40% at 7 tree heights to the leeward of a seven row, nine meter wide shelterbelt of pine, larch, and birch. Read (1965) reports that crop yield increases occur generally at distances between 2 and 15 tree heights from windbreaks.

Wind speed within a closed forest stand is considerably reduced in comparison to wind speed in the open. Cooper (1965) found that young slash and loblolly pine stands reduced wind speed to less than 1/10 the wind speed in the open. He showed a relationship between wind speed in the open and wind speed in closed stands as:

$$Wf = 0.0642 - 0.0097(Wo) + 6.2742 \frac{1}{(BA)} + 0.0074(Ht)(Wo)$$

where Wf = wind speed in forest, 4 foot level

Wo = wind speed in open, 20 foot level

BA = stand basal area, in square feet per acre

Ht = average height of dominant trees (ft)

Leikola (1967) found that in a Scots pine stand wind speed was less at 2 meters than at 9 meters. Total stand height was 11 meters; height to green crown was 6.5 meters.

Trees serve not only as windbreaks. Because of reduced wind speed in the interior of stands there is a reduction of water loss and of daytime maximum temperatures. A reduction in wind velocity and a more moderate air temperature improves the environment for growth of those plants and other organisms that occupy the forest understory.

SOUND ABATEMENT

Tree borders along highways and streets serve to alleviate sound from cars, trucks, and buses. Cook and Van Haverbeke (1971) report on the

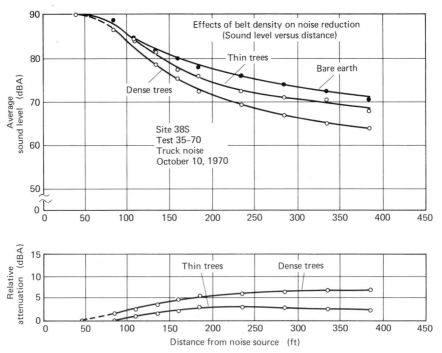

FIGURE 12.8. Effects of belt density on noise reduction (Cook and Van Haverbeke, 1971).

results of a three-year study into the effectiveness of trees as sound "barriers." It is interesting to note that the study was carried out in Nebraska, where trees had been planted in the 1930s and early 1940s as shelterbelts from wind. The results of the Cook–Van Haverbeke study showed greenbelts along streets and highways can reduce sound levels in the order of 8 decibels (approximately one-half of that at the source (Figure 12.8). As with wind speed, sound levels were reduced most by tree barriers that are tall and several rows deep; however, soft ground surfaces, such as plowed soil or grass, absorbed sound more readily than hard surfaces, such as paving, gravel, or stone layers. Evergreen species were more effective in year-round noise abatement than deciduous trees. In general, Cook and Van Haverbeke recommend that to reduce high-speed car and traffic noise, the green belt

TABLE 12.12

Sound Levels and Human Response

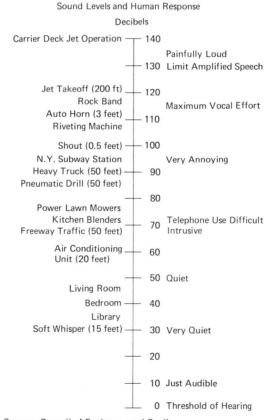

Decibels	
Carrier Deck Jet Operation	140
	Painfully Loud
	130 Limit Amplified Speech
Jet Takeoff (200 ft)	120
Rock Band	Maximum Vocal Effort
Auto Horn (3 feet)	110
Riveting Machine	
Shout (0.5 feet)	100
N.Y. Subway Station	Very Annoying
Heavy Truck (50 feet)	90
Pneumatic Drill (50 feet)	
	80
Power Lawn Mowers	
Kitchen Blenders	70 Telephone Use Difficult
Freeway Traffic (50 feet)	Intrusive
Air Conditioning	60
Unit (20 feet)	
	50 Quiet
Living Room	
Bedroom	40
Library	
Soft Whisper (15 feet)	30 Very Quiet
	20
	10 Just Audible
	0 Threshold of Hearing

Source: Council of Environmental Quality

in rural areas should be 65 to 100 feet wide and the edge should be within 50–80 feet of the center of the nearest traffic lane. In urban areas, the green belt should be 20–50 feet wide with the edge 20–50 feet from the center of the nearest traffic lane. To achieve a quiet sound level of 50 dBA (Table 12.12) requires not only a dense tree barrier and bare earth surface, but also a considerable setback from the sound source.

GENERAL REFERENCES

Kittridge, J. 1948. *Forest Influences.* McGraw-Hill, New York.

Kramer, P.J. 1969. *Plant and Soil Water Relationships: A Modern Synthesis.* McGraw-Hill, New York.

Montheith, J.L. (ed.). 1975. *Vegetation and the Atmosphere.* Academic, New York.

BIBLIOGRAPHY

Ainsworth, G.C. and A.S. Sussman (ed.). 1968. *The Fungi: An Advanced Treatise,* vol. III. Academic, New York.

Alexander, R.R. 1972. Partial cutting practices in old growth lodgepole pine. A field guide. *USDA For. Serv. Res. Pap.* RM-92A.

Allen, P.H. 1965. White pine understory in New Hampshire. Ph.D. thesis, Duke University, Durham, N.C.

Allen, R.M. 1953. Release and fertilization stimulate longleaf cone corp. *J. For.* 51: 827.

Allen, R.M. and N.M. Scarbrough. 1969. Development of a year's height growth in longleaf pine saplings. *USDA For. Serv. Res. Pap.* SO-45, New Orleans.

Anderson, R.F. 1960. *Forest and Shade Tree Entomology.* Wiley, New York.

Ashe, W.W. 1915. Loblolly or North Carolina pine. *N.C. Geol. and Econ. Survey Bull.* 24.

Auten, J.T. 1937. A method of site evaluation for yellow poplar, based on depth of the undisturbed A horizon. *USDA For. Serv. Centl. Stat. For. Expt. Sta.* Note 33.

Baker, F.S. 1950. *Principles of Silviculture.* McGraw-Hill, New York.

Baker, W.L. 1972. *Eastern Forest Insects.* USDA Forest Service Misc. Publ. No. 1175. U.S. Govt. Printing Office, Washington, D.C.

Bakuzis, E.V. 1962. Synecological coordinates and investigation of forest ecosystems, in *13th Congress Internatl. Union of Forestry Research Organiz. Proc.,* vol. I, Part 2, Sec. 21/1-2. Vienna, Austria.

Balmer, W.E., E.G. Owens, and J.R. Jorgensen. 1975. Effects of various spacings on loblolly pine growth 15 years after planting. *USDA Forest Service Res. Note* SE-211.

Barber, J.C. 1966. Variation among half-sib families from three loblolly pine stands in Georgia. *Res. Pap. Ga. For. Res. Coun.* No. 37, p. 5.

Barnes, R.B. 1976. A quantitative evaluation of winter deer browse in southern New Hampshire forests. MS thesis, University of New Hampshire.

Barnett, J.P. 1976. Delayed germination of southern pine seeds related to seed-coat constraint. *Can. J. For. Res.* 6: 504–570.

Baskerville, G.L. 1965. Dry matter production in immature balsam-fir stands. *For. Sci. Monogr.* 9.

Bassett, J.R. 1964. Tree growth as affected by soil moisture availability. *Soil Sci. Soc. Am. Proc.* 28(3): 436–438.

Bates, C.J. and J. Roeser. 1928. Light intensities required for growth of coniferous seedlings. *Am. J. Bot.* 15: 184–195.

Batzer, H.O. 1969. Forest character and vulnerability of balsam fir to spruce budworm in Minnesota. *For. Sci.* 15: 17–25.

Baumgartner, A. 1956. Investigations on the heat and water economy of a young forest. *Ber. Deut. Wetter dienst* 5: 4–53.

Beaufait, W.R. 1970. Some effects of high temperature on the cones and seeds of jackpine. *For. Sci.* 6: 196–199.

Becking, R.W. 1957. The Zurich–Montpellier school of phytosociology. *The Botanical Rev.* 23(7): 411–488.

Bella, I.E. 1967. Crown width/diameter relationship of open-grown jackpine on four site types in Manitoba. Bi-monthly Res. Notes, *Can. Dept. For. and Rural Develop.* 23(1): 5–6.

Bendell, J.F. 1974. Effect of fire on birds and mammals, in Kozlowski, T.T. and C.E. Ahlgren. *Fire and Ecosystems.* Academic, New York.

Berg, A.B. and A. Doerksen. 1975. Natural fertilization of a heavily thinned Douglas-fir stand by understory red alder. Oregon State U. For. Res. Lab. Res. Note 56.

Bethune, J.E. 1970. Distribution of slash pine as related to certain climatic factors. *For. Sci.* 6(1): 11–17.

Bilan, M.V. 1960. Root development of loblolly pine in modified environments. Dept. of Forestry Bull. No. 4. Stephen F. Austin State College, Nacogdoches, Texas.

——— 1960. Stimulation of cone and seed production in pole-size loblolly pine. *For. Sci.* 6: 207–220.

———. 1966. Low temperature as a limiting factor of root growth of loblolly pine seedlings. *Bull. Ecol. Soc. Am.* 47(3): 102.

Bilan, M.V. and Shan Wu Jan. 1968. Needle moisture content as indicator of cessation of root elongation in loblolly pine seedlings. *Bull. Ecol. Soc. Am.*, p. 108.

Bingham, R.T., R.H. Hoff, and R.J. Steinhoff. 1972. Genetics of Western White Pine. *U.S.D.A. For. Serv. Res. Pap.* WO-12.

Bochinger, L.S. and W.W. Heck. 1969. An ozone-sulphur dioxide synergism produces symptoms of chlorotic dwarf of eastern white pine. *Phytopathol.* 59: 399.

Boggess, W.R. 1956. Amount of throughfall and stemflow in a shortleaf pine plantation as related to rainfall in the open. *Illinois Acad. Sci. Trans.* 48: 55–61.

Bormann, F.H. 1966. The structure, function, and ecological significance of root grafts in *Pinus strobus* L. *Ecol. Monogr.* 36: 1–26.

Bormann, F.H., G.E. Likens, D.W. Fisher, and R.S. Pierce. 1968. Nutrient loss accelerated by clearcutting of a forest ecosystem, *Symposium on Primary Productivity and Mineral Cycling in Natural Ecosystems*, H.E. Young (ed.). U. Maine Press, Orono, Maine, pp. 187–193.

Bornebusch, C.H. 1930. The fauna of forest soils. *Det Forstlige Forsøgsvaesen i Denmark* 11: 1–224.

Boughey, A.S. 1968. *Ecology of Populations*. MacMillian, Toronto, Canada.

Bowen, G.D. 1973. Mineral nutrition of ectomycorrhizae, Chapter 5 in Marks, C.G. and T.T. Kozlowski. *Ectomycorrhizae*. Academic, New York.

Boyce, J.S. 1961. *Forest Pathology*, 3rd ed. McGraw-Hill, New York.

Bradley, J.G. 1974. When smoke blotted out the sun. *The American West* 11(5): 4–9.

Braun-Blanquet, J. 1951. *Plant Sociology*. McGraw-Hill, New York.

Braun, E.L. 1950. *Deciduous Forests of Eastern North America*. Blakiston, Phila., Pa.

Brewer, W.H. 1894. The woodlands and forest systems of the United States, in *Statistical Atlas of the U.S., Based on the Results of the 9th Census, 1870*. F.A. Walker. J. Bien., Lith., New York.

Bray, J.R. and E. Gorham. 1964. Litter production in forests of the world. *Advan. Ecol. Res.* 2: 101–157.

Brix, H. 1967. An analysis of dry matter production of Douglas-fir seedlings in relation to temperature and light intensity. *Can. J. Bot.* 45: 2063–2072.

Broadfoot, W.M. and H.D. Burke. 1958. Soil-moisture constants and their variations. *USDA For. Serv. Southern For. Expt. Sta. Occ. Pap.* 166.

Broadfoot, W.M. 1967. Shallow water impoundment increases soil moisture and growth of hardwoods. *Soil Sci. Soc. Am. Proc.* 31: 562–564.

Brown, A.A. and K.P. Davis. 1973. *Forest Fire Control and Use,* 2nd ed. McGraw-Hill, New York.

Brown, C.L. and H.E. Sommer. 1975. *An Atlas of Gymnosperms Cultured in vitro: 1924–1974.* Georgia Forest Res. Council, Macon, Ga.

Brown, G.W. and J.T. Krygien. 1970. Effect of clearcutting on stream temperature. *Water Resour. Res.* 6: 1133–1139.

Brown, J.H., Jr. and T.W. Hardy, Jr. 1975. Summer water use by white pine and oak in Rhode Island. *Rhode Island Agr. Expt. Sta. Bull.* 414, Kingston, R.I.

Buck, J.M., R.S. Adams, J. Cone, M.T. Conkle, W.J. Libby, C.J. Eden, and M.J. Knight. 1970. California tree seed zones. *USDA For. Serv. California Division and the Calif. Div. of Forestry.* Sacramento, Calif.

Bunting, B.T. 1965. *The Geography of Soils.* Aldine, Chicago.

Burton, D.H., H.W. Anderson, and L.F. Riley. 1969. Natural regeneration of yellow birch in Canada, in *Proc. Birch Sympos.* U. of New Hampshire, Durham, N.H.

Byram, G.M. and G.M. Jemison. 1943. Solar radiation and forest fuel moisture. *J. Agr. Res.* 67: 149–176.

Byram, G.M. 1948. Terrestrial radiation and its importance in some forestry problems. *J. For.* 46: 653–658.

Byrne, J.G., C.R. Gass, and C.K. Losche. 1965. Relation of forest composition to certain soils in the southern Appalachian Plateau in Youngberg, C.T. Forest-soil relationships in North America, in *2nd North America Forest Soils Conf.,* Oregon State U. Press, Corvallis, Ore. pp. 199–214.

Cajander, A.K. 1926. The theory of forest types. *Acta Forest Fennica* 21: 1–108.

Campbell, R.A. and D.J. Durzan. 1975. Induction of multiple buds and needles in tissue cultures of *Picea glauca. Can. J. Bot.* 53(16): 1652–1657.

———. 1976. The potential for cloning white spruce via tissue culture, in *12th Lake States For. Tree Improve. Conf. Proc. 1975.* USDA Forest Service Gen. Tech. Rep. NC-26, pp. 158–166.

Campbell, R.K. and A.I. Sugano. 1975. Phenology of bud burst in Douglas-fir related to provenance, photoperiod, chilling, and flushing temperature. *Bot. Gaz.* 136(3): 290–298.

Campbell, T.E. 1955. Freeze damages shortleaf pine flowers. *J. For.* 53: 452.

Carmean, W.H. 1954. Site quality of Douglas-fir in S.W. Washington and its relationship to precipitation, elevation, and physical soil properties. *Soil Sci. Soc. Am. Proc.* 18: 330–334.

———. 1956. Suggested modifications of the standard Douglas-fir site curves for certain soils in southwestern Washington. *For. Sci.* 2: 242–250.

———. 1965. Black oak site quality in relation to soil and topography in southeastern Ohio. *Soil Sci. Soc. Am. Proc.* 29: 308–312.

———. 1975. Forest site quality evaluation in the U.S. *Advan. Agronomy* 27: 209–269.

Chappell, D.E. 1962. Value growth of pine pulpwood on the George Walton experimental forest. *USDA For. Serv. Southeastern Exp. Sta. Pap.* 140.

Chisman, H.H. and F.X. Schumacher, 1940. On the tree-area ratio and certain of its applications. *J. For.* 38: 311–317.

Choate, G.A. 1961. Estimating Douglas-fir site quality from aerial photographs. *USDA For. Serv. Pac. N.W. For. and Rng. Expt. Sta. Res. Pap.* 45.

Clements, F.E. 1916. Plant succession: Analysis of the development of vegetation. *Carnegie Inst. Wash. Publ.* 242, Washington, D.C.

Cliff, E.P. 1939. Relationship between elk and mule deer in the Blue Mountains of Oregon, in *Transactions 4th North American Wildlife Conf.*, Wildlife Management Inst., Washington, D.C., pp. 560–569.

Cochran, P.H. 1969. Thermal properties and surface temperature of seedbeds. USDA For. Serv., Pacific Northwest For. and Range Expt. Sta., Portland, Oregon.

Coile, T.S. 1940. Soil changes associated with loblolly pine succession on abandoned agricultural land of the Piedmont plateau. *Duke U. School of Forestry Bull.* 5.

———. 1948. Relation of soil characteristics to site index of loblolly and shortleaf pines in the lower Piedmont region of North Carolina. *Duke U. School of Forestry Bull.* 13.

———. 1952a. Soil and growth of forests. *Advan. Agronomy* IV: 329–398.

———. 1952b. Soil site relations of the southern pines. *Forest Farmer* 10(7): 10, 11, 13; 10(8): 11–12.

Cole, D.W., S.P. Gessel, and S.F. Dice. 1967. Distribution and cycling of N, P, K, and Ca in a second growth Douglas-fir ecosystem, in *Symposium on Primary Prod. and Mineral Cycling in Natural Ecosystems*, H.E. Young (ed.). U. of Maine Press, Orono, Maine.

Conner, R.N. and C.S. Adkisson. 1976. Discriminant function analysis: a possible aid in determining the impact of forest management on woodpecker nesting habitat. *For. Sci.* 22(2): 122–130.

Cook, David I. and David F. Van Haverbeke. 1971. Trees and shrubs for noise abatement. *Nebraska Agr. Exp. Sta. Res. Bull.* 246. Lincoln, Nebr.

Cooper, C.F. 1960. The ecology of fire. *Sci. Am.* 204(4): 150–160.

Cooper, R.W. 1965. Wind movement in pine stands. *Georgia Forest Res. Pap.*, No. 33. Georgia Forest Res. Council, Macon, Ga.

Corbett, E.S. and R.P. Crouse. 1968. Rainfall interception by annual grass and chapparal—losses compared. USDA For. Serv., *Pacific S.W. For. and Range Expt. Sta. Res. Pap. PSW-48.*

Critchfield, W.B. 1957. Geographic variations in *Pinus contorta. Maria Moors Cabot Foundation Publ.*, No. 3. Harvard U., Cambridge, Mass.

———. 1963. Hybridization of the Southern pines in California. 1962, in *Proc. Forest Genetics Workshop.* Southern Forest Tree Improvement Comm. No. 22.

Critchfield, W.B. 1965. Crossability and relationships of the California big-cone pines, in *Joint Proc. 2nd Genetics Workshop. USDA For. Serv. Res. Pap.* NC-6, pp. 36–44.

Curlin, J.W. 1970. Models of the hydrologic cycle, Ch. 18, pp. 268–285, in *Analysis of Temperate Forest Ecosystems,* P.E. Reichle (ed.). Springer-Verlag, New York.

Curtis, J.T. and R.P. McIntosh. 1950. The interrelations of certain analytic and synthetic phytosociological characters. *Ecology* 31: 434–455.

———. 1951. An upland forest continuum in the prairie-forest border region of Wisconsin. *Ecology* 32: 476–496.

Curtis, R.O. and D.L. Reukema. 1970. Crown development and site estimates in a Douglas-fir plantation spacing test. *For. Sci.* 16(3): 287–301.

Curtis, W.R. 1960. Moisture storage by leaf litter. *USDA For. Serv. Lake States For. Expt. Sta. Tech. Note* 577.

Daubenmire, R.I. 1947. *Plants and Environment.* Wiley, New York.

Daubenmire, R. 1961. Vegetative indicators of rate of height growth in ponderosa pine. *For. Sci.* 7: 24–34.

Daniel, T.W. and J. Schmidt. 1972. Lethal and nonlethal effects of the organic horizons forested soils on the germination of seeds from several associate conifer species of the Rocky Mountains. *Can. J. For. Res.* 2: 179–184.

Davey, C.B. 1970. Soil microorganisms, their relationships to nutrient cycling, tree nutrition and forest fertilization. *Short Course on Forest Fertilization.* T.V.A., Muscle Shoals.

De Bell, B.S. 1971. The phytoxic effects of cherry bark oak. *For. Sci.* 17(2): 180–185.

Derr, H.J. and W.F. Mann, Jr. 1959. Guidelines for direct-seeding longleaf pine. *USDA Forest Service Occ. Pap., Southern For. Exp. Sta.,* 171.

Doak, K.D. 1934. Fungi that produce ectotrophic mycorrhizae in afforestation. *J. For.* 32: 22–29.

Dochinger, L.S. and C.E. Selisker, 1970. Air pollution and the chlorotic dwarf disease of eastern white pine. *For. Sci.* 16: 46–55.

Doolittle, W.T. 1958. Site index comparisons for several forest species in the Southern Appalachians. *Soil Sci. Soc. Am. Proc.* 22: 455–458.

Doss, B.D. and W.M. Broadfoot. 1956. Properties of 91 southern soil series. *USDA For. Serv. Southern For. Expt. Sta. Occ. Pap.* 147.

Douglas, J.E. and W.T. Swank. 1972. Streamflow modification through management of eastern forests. *USDA For. Serv. Res. Pap.* SE-94.

————. 1975. Effects of management practices on water quality and quantity: Coweeta Hydrologic Laboratory, N.C. *USDA For. Service Gen. Tech. Rep.* NE-13.

Downs, A.A. and W.E. McQuilkin. 1944. Seed production of Southern Appalachian oak. *J. For.* 42(12): 913–920.

Drury, W.H. and I.C.T. Nisbet. 1973. Succession. *J. Arnold Arboretum* 54(3): 331–368.

Dunning, D. 1928. A tree classification for the selection forests of the Sierra Nevada. *J. Agr. Res.* 36: 775–781.

Durzan, D.J. and R.A. Campbell. 1974. Prospects for the mass production of improved stock of forest trees by cell and tissue culture. *Can. J. For. Res.* 4(2): 151–174.

Duvigneaud, P. and S. Denaeyer-De Smet. 1970. Biological cycling of minerals in temperature deciduous forests, in *Analysis of Temperate Forest Ecosystems*, Ecological Studies 1. D.E. Reichle (ed.). Springer-Verlag, New York.

Eaton, T.H., Jr. and R.E. Chandler, Jr. 1942. The fauna of the forest humus layers in New York. *Cornell Agr. Expt. Sta., Memoir* 247.

Einspahr, D. and A.L. McComb. 1951. Site index of oaks in relation to soil and topography in northeastern Iowa. *J. For.* 49: 719–723.

Eis, S. 1962. Statistical analysis of several methods of estimation of forest habitats and tree growth near Vancouver, B.C. *U. Brit. Columbia Fac. of Forestry, Forestry Bull.* No. 4, Vancouver, British Columbia, Canada.

Esau, K. 1960. *Anatomy of Seed Plants.* Wiley, New York.

Fergus, C.L. 1956. Frost cracks on oak. *Phytopathology* 46: 297.

Fernow, B.E. 1905. Forest terminology. *Forestry Quarterly* 3: 255–268.

Fielding, J.M. 1967. The influence of silvicultural practice on wood properties. *Intl. Rev. For. Res.* 2: 95–126.

Finney, H.R., N. Holoway, and M.R. Heddleson. 1962. The influence of microclimate on the morphology of certain soils of the Allegheny plateau Ohio. *Soil Sci. Soc. Am. Proc.* 26: 287–292.

Forbes, R.S. 1955. *Forestry Handbook.* Ronald Press, New York.

Forcier, L.K. 1973. Reproductive strategies and the co-occurrence of climax tree species. *Sci.* 189: 808–809.

Fowells, H.A. (ed.). 1965. Silvics of forest trees in the United States. *USDA For. Serv. Agr. Hndbk.* 271. U.S. Govt. Printing Office. Wash., D.C. 20402.

Fowells, H.A. and G.H. Schubert. 1956. Seed crops of forest trees in the pine region of California. *USDA Tech. Bull.* 1150.

Fowler, D.P. 1963. Effects of inbreeding in red pine, *Pinus resinosa* Art. Ph.D. diss., Yale University, New Haven, Conn.

Fowler, D.P. 1965. Effects of inbreeding in red pine (*Pinus resinosa* Art) pollination, studies II and III. *Silvae Genetics* 14: 12–23, 37–46.

Fowler, D.P. and T.W. Dwight. 1964. Provenance differences in stratification requirements of white pine. *Can. J. Bot.* 42: 669–673.

Fralish, S.S. 1968. The influence of soil texture on the amount of sugar maple present in the aspen stands of northern Wisconsin. *Bull. Ecol. Soc. Am.* 49(2): 79.

Fraser, D.A. 1956. Ecological studies of trees at Chalk River, Ontario, Can. II, Ecological conditions and radical growth. *Ecology* 37: 777–789.

Fried, M. and H. Broeshart. 1967. *The Soil-Plant System in Relation to Inorganic Nutrition*. Academic, New York.

Fritts, H.H. 1961. An analysis of maximum summer temperature inside and outside the forest. *Ecology* 42:436–440.

Fuentes, J.M. 1971. Interaction between planting site and seed source in loblolly pine. *Forst Res. Tech. Note. No. 32*. Southland Expt. For. Internatl. Paper Co., Bainbridge, Ga.

Gaiser, R.N. 1950. Relation between soil characteristics and site index of loblolly pine in the Coastal Plain Region of Virginia and North Carolina. *J. For.* 48: 271–275.

Gammon, Glenn L. 1969. Specific gravity and wood moisture variation of white pine. *USDA Forest Service. Res. Note* NE-99.

Gant, R.E. and E.E.C. Clebsch. 1975. The allelopathic influences of *Sassafrass albidum* in old-field succession. *Ecology* 56(3): 604–615.

Gates, D.M. 1965. Energy, plants, and ecology. *Ecology* 46(1,2) (also partly reported in *Sci. Am.* 213(6): 76–84).

———. 1965. Heat transfer in plants. *Sci. Am.* 213(6): 76–84.

Gay, L.W. and K.R. Knoerr. 1975. The forest radiation budget. *Duke U. Sch. of For. and Envir. Studies. Bull. 19*. Durham, N.C.

Gemmer, E.W. 1932. Well-fed pines produce more cones. *USDA For. Serv. For. Worker* 8(5): 15.

Gessel, S.P. and A.N. Balci. 1965. Amount and composition of forest floors under Washington coniferous forests, in *Forest-Soil Relationships in North America. Second North American Forest Soils Conf. 1963*. Oregon State Univ., Corvallis, Oregon.

Gevorkiantz, S.R., P.O. Rudolph, and P.J. Zehngraff. 1943. A tree classification for aspen, jack-pine, and second growth red pine. *J. For.* 41: 268–274.

Gilbert, Adrian M. 1965. Stand differentiation ability in northern hardwoods. *USDA Forest Serv. Res. Pap.* NE-37.

Gill, J.D., R.M. Degraaf, and J.W. Thomas. 1974. Forest habitat management for non-game birds in Central Appalachia. *USDA For. Serv. Res. Note,* NE-192.

Gingrich, S.F. 1967. Measuring and evaluating stocking and stand density in upland hardwood stands in the Central States. *For. Sci.* 13(1): 38–52.

Gleason, H.A. 1939. Individualistic concept of the plant association. *Am. Midland Naturalist.* 21: 92–110.

———. 1975. Delving into the history of American ecology. 1952 letter to C.H. Mueller printed in *Bull. Ecol. Soc.* 56(4): 7–10.

Golley, F.B., L. Ryszkowski, and J.T. Sokur. 1974. The role of small mammals in temperate forests, grasslands, and cultivated fields. Chap. 10 in Golley, Petrusewicz, and Ryszkowski, IBP No. 5 *Small Mammals: Their Productivity and Population Dynamics.* Cambridge Univ. Press, Cambridge.

Gosz, J.R., R.T. Holmes, G.E. Likens, and F.H. Borman. 1978. The flow of energy in a forest ecosystem. *Sci. Am.* 238(3): 92–102.

Graber, R.E. 1969. Seed losses to small mammals after fall sowing of pine seed. *USDA Forest Serv. Res. Pap.* NE-135.

Graham, K. 1963. *Concepts of Forest Entomology.* Reinhold, New York.

Graham, Samuel. 1954. Scoring tolerance of forest trees. *Michigan Forestry Note,* No. 4.

Grant, V. 1971. *Plant speciation.* Columbia Univ. Press, New York.

Gregory, R.A. and B.F. Wilson. 1968. A comparison of cambial activity of white spruce in Alaska and New England. *Can. J. Bot.* 46: 733–734.

Grier, C.C. 1975. Wildfire effects on nutrient distribution and leaching in a coniferous ecosystem. *Can. J. For. Res.* 5: 599–607.

Grigal, D.F. and L.F. Ohmann. 1975. Classification, description, and dynamics of upland plant communities within a Minnesota Wilderness area. *Ecol. Monogr.* 45(4): 389–407.

Gullion, G.W. 1972. Improving your forested lands for ruffed grouse. *Minnesota Agr. Exp. Sta., Misc. Jour. Ser., Publ. No. 1439,* 1–34, St. Paul, Minn.

Haig, I.T. 1929. Colloidal content and related soil factors as indicators of site quality. *Yale Univ. School of Forestry Bull.* 24.

Halligan, J.P. 1975. Toxic terpenes from *Artemisia California. Ecology* 56: 999–1003.

Halls, L.K. and J.L. Shuster. 1965. Tree-herbage relations in pine-hardwood forests of Texas. *J. For.* 63: 282–283.

Hamilton, W.J., Jr. and D.B. Cook. 1940. Small mammals and the forest. *J. For.* 38: 468–473.

Hanley, D.P., W.C. Schmidt, and G.M. Blake. 1975. Stand structure and successional status of two spruce-fir forests in southern Utah. *USDA Forest Service Res. Pap.* INT-176.

Hannah, P.R. 1968. Topography and soil relations for white and black oak in southern Indiana. *USDA For. Serv. Res. Pap.* NC-25.

Hansen, E.A. 1976. Determining moisture-nutrient requirements for maximum fiber yield. *USDA For. Serv. Gen. Tech. Rept.* NC-21, pp. 43–46.

Hare, R.C. and G.L. Switzer. 1969. Introgression with shortleaf pine may explain rust resistance in western loblolly pine. *USDA For. Serv. Res. Note* SO—88.

Harris, W.F., P. Sollins, N.T. Edwards, B.E. Dinger, and H.H. Shugart. 1975. Analysis of carbon flow and productivity in a temperate deciduous forest ecosystem, in *Proc. Symp. Productivity of World Ecosystem, Seattle, Wash.* pp. 116–122. Natl. Acad. of Sci., Washington, D.C.

Hart, G. and H.W. Lull. 1963. Some relationships among air, snow, and soil temperatures and soil frost. *USDA For. Serv. Note* NE-3.

Hassinger, J.D., S.A. Liscinsky, and S.P. Shaw. 1975. Wildlife. Chap. 6 in *Clearcutting in Pennsylvania.* State U., School of For. Resources, U. Park, Pa.

Harvey, A.E., M.F. Jurgensen, and M.J. Larsen. 1976. Intensive fiber utilization and prescribed fire: effects on the microbial ecology of forests. *USDA For. Serv. Gen. Tech. Rept.* INT-28.

Hatch, A.B. and K.D. Doak. 1933. Mycorrhizal and other features of the root systems on *Pinus. J. Arnold Arboretum* 14: 85–88.

Heiberg, S.O. and A.L. Leaf. 1960. Potassium fertilization of coniferous plantations in New York. *Trans. 7th Intern. Congr. Soil Sci.* 3: 376–383.

Heidmann, L.S. 1976. Frost heaving of tree seedlings: a literature review of causes and possible control. *USDA For. Serv. Gen. Tech. Rept.* RM-21.

Hellmers, H. and W.P. Sundahe, 1959. Response of *Sequoia sempervirens* (*D.Don*) Endl. and Pseudotsuga menziesii (Mirh.) Franco seedlings to temperature. *Nature Lond.* 184 (4694): 1247–1248.

Hellmers, H. and James Bonner. 1959. Photosynthetic limits of forest tree yield. *Proc. Soc. Am. For.* 59: 32–35.

Henderson, G.S. 1974. The ecosystem approach to forest nutrition and water relations, in *Proc. 3rd North American Forest Biology Workshop, pp. 64–76. College of Forestry and Natural Resources, Fort Collins, Colo.*

Henderson, G.S. and W.F. Harris. 1975. An ecosystem approach to characterization of the nitrogen cycle in a deciduous forest watershed in B. Bernier and C.H. Winget, in *Forest Soils and Land Management* (Proc. 4th North American For. Soils Conf., 1973, Laval U., Quebec U.). Laval Press, Montreal, Canada.

Hepting, C.H. 1971. Diseases of forest and shade trees of United States. *USDA For. Serv. Agr. Handbk.* No. 386.

Hermann, R.K. and D.P. Lavender. 1968. Early growth of Douglas-fir from various altitudes and aspects in southern Oregon. *Silvae Genetica* 17(4): 143–151.

Hickok, K.H. and R.J. Hutnik. 1966. Dry matter production and energy conversion in red pine plantations of different spacings. *Res. Briefs Sch. For. Penna. State U.* 1(2) 48–52.

Hocker, H.W., Jr. 1956. Certain aspects of climate as related to the distribution of loblolly pine. *Ecology* 37(4): 824–833.

———. 1961. Germination of eastern white pine seed and early establishment of white pine seedlings on prepared site. *New Hampshire Agr. Exp. Sta. Tech. Bull.* 103, Durham, N.H.

———. 1962. Stimulating conelet production of eastern white pine. *New Hampshire Agr. Expt. Sta. Tech.* Bull. 107, Durham, N.H.

———. 1969. Estimating white pine seed fall from the number of seed locations in opened cones. *J. For.* 67: 813.

Hodges, J.D., 1967. Patterns of photosynthesis under natural environmental conditions. *Ecology* 48: 234–242.

Hofmann, J.G. 1949. D.F. thesis. Duke U. School of For. (see Coile, 1952a).

Hogg, W.H. 1965. A shelterbelt study: relative shelter, effective winds and maximum efficiency. *Agr. Meteor. (Amsterdam)* 2(5): 307–315.

Holkias, N.A., F.J. Weihmeyer, and A.H. Hendrickson. 1955. Determining water needs for crops from climatic data. *Hilgardia* 24(9): 207–233.

Holman, H. 1964. Forest ecological studies on drained peat land in the Province of Uppland, Sweden. *Studia Forestalia Suecica,* No. 16.

Holmes, J.W. and S.Y. Shim. 1968. Diurnal changes in stem diameter of Canary Island pine (*Pinus canariensis,* C. Smith) caused by soil water stress in varying microclimate. *J. Expt. Bot.* 19: 219–232.

Hooven, E.F. 1956. Field tests of tetramine treated Douglas-fir seed. *Oregon State Board of Forestry Res. Note.* No. 29, Corvallis, Ore.

Hoover, M.D. 1953. Interception of rainfall in a young loblolly pine plantation. *USDA Forest Serv. Southeastern For. Expt. Sta. Pap.* 21.

Horn, H.S. 1974. The ecology of secondary succession. *Ann. Rev. Ecol. and Systematics.* 5: 25–37.

———. 1975. Forest succession. *Sci. Am.* 232(5): 90–98.

Hornbeck, J.W. 1964. The importance of dew in watershed management research. *USDA For. Serv. Res. Note* NE-24.

Horsley, S.B. 1974. Allelopathic inhibition of black cherry seedling growth by fern, grass, goldenrod, and Aster, in *Proc. 3rd North American Forest Biology Workshop.* Colorado State U., Fort Collin, Colo.

Hough, A.F. 1945. Frost pocket and other microclimates in forests of the northern Allegheny Plateau. *Ecology* 26: 235–250.

Huberman, M.A. 1943. Sunscald of eastern white pine, *Pinus strobus L. Ecology* 24: 456–471.

Husch, B. and W.H. Lyford. 1956. White pine growth and soil relationship in southeastern New Hampshire. *New Hampshire Agr. Expt. Sta. Tech. Bull.* 95. Durham, N.H.

Husch, B. 1959. Height growth of white pine in relation to selected environmental factors on four sites in southeastern New Hampshire. *New Hampshire Agr. Expt. Sta. Tech. Bull.* 100. Durham, N.H.

Jenny, Hans. 1941. *Factors of soil formation.* McGraw-Hill, New York.

Johnson, E.A. 1952. Effect of farm woodlot grazing on watershed values in the Southern Appalachian mountains. *J. For.* 50: 109–113.

Johnston, J.P. 1941. Height-growth periods of oak and pine reproduction in the Missouri Ozarks. *J. For.* 39: 67–68.

Jones, J.R. 1969. Review and comparison of site evaluate on methods. *USDA For. Serv. Res. Pap.* RM-51.

Jorgensen, J.R., C.G. Wells, and L.J. Metz. 1975. The nutrient cycle: key to continuous forest production. *J. For.* 73(7): 400–405.

Judson, S. 1968. Erosion of the land, or what's happening to our continents. *Am. Sci.* 56(4): 356–374.

Keen, F.P. 1943. Ponderosa pine tree classes redefined. *J. For.* 41: 249–253.

Kellison, R.C., R.D. Heeren, and S. Jones. 1976. Species selection as related to soils in the Atlantic Coastal Plain, in *Proc. 6th Southern Soils Workshop. U.S.D.A. For. Serv.,* SAF Publ.

Kendeigh, S.C. 1947. Bird population studies in the coniferous forest biome during a spruce budworm outbreak. Biology Bull. 1 Dept. Lands and Forests. Ontario, Canada.

———. 1961. *Animal Ecology.* Prentice-Hall, Englewood Cliffs, N.J.

Kershaw, K.A. 1973. *Quantitative and Dynamic Plant Ecology,* 2nd ed. American Elsevier, New York.

Kimmins, J.P. 1972. The ecology of forestry—the ecological role of man, and forester, in forest ecosystems. *The Forestry Chron.* 48: 301–307.

Kingsley, N.P. 1976. The forest resources of N.H. *USDA For. Serv. Res. Bull.* NE-43.

Kitching, R. 1967. Water use by tree plantations in England. *J. Hydrology* 5(2): 206–213.

Kittredge, J. 1944. Estimation of the amount of foliage of trees and stands. *J. For.* 42(12): 905–912.

———. 1948. *Forest Influences.* McGraw-Hill, New York.

Knauf, T.A. and M.J. Bilan. 1974. Needle variation in loblolly pine from mesic and xeric sources. *For. Sci.* 20(1): 88

Knudsen, L.L. 1950. Interrelations of some soil properties of coastal plain soils. M.F. thesis. Duke U. School of For., Durham, N.C.

———. 1950a Relationship between soil properties and growth of slash price. Ph.D. thesis. Duke U. School of For., Durham, N.C.

Kohler, M.A., T.S. Nordenson, and W.E. Fox. 1955. Evaporation from ponds and lakes. *U.S. Dept. of Commerce, Weather Bureau Res. Pap.* 38.

Komarek, E.V., Sr. 1967. The nature of lightning fires, in *Proc. 7th California Tall Timbers Fire Ecology conf.,* pp. 51–41, Tallahassee, Fla.

Koppen, W. 1923. *Die Klimate der erde.* Wide Gryter, Berlin.

Korschgen, L.J. 1967. Feeding habits and foods. Chap. 7 in Hewitt, O.H. (ed.), *The Wild Turkey and Its Management.* The Wildlife Society, Washington, D.C.

Korstian, C.F. 1927. Factors controlling germination and early survival in oak. *Yale U. Sch. of For. Bull.* 19.

Kotok, E.S. 1965. Timber quality research-another concept. *Forest Prod. J.* 15(10): 459–462.

Kozlowski, T.T. 1964. Shoot growth in woody plants. *Bot. Rev.* 30(3): 335–392.

———. 1971. *Growth and development of trees.* Academic, New York.

———. 1972. *Seed Biology,* vols. 1–3, in *Physiological Ecology Series.* Academic, New York, N.Y.

Kozlowski, T.T. and C.E. Ahlgren. 1974. *Fire and Ecosystems.* Academic, New York.

Kramer, P.J. 1943. Amount and duration of growth of various species of tree seedlings. *Plant Physiology* 19:239–251.

———. 1949. *Plant and soil water relationships.* McGraw-Hill, New York.

———. 1957. Some effects of various combinations of day and night temperatures photoperiod on height growth of loblolly pine seedlings. *For. Sci.* 3(1): 45–55.

———. 1969. *Plant and soil water relationships* (Chap. 9). McGraw-Hill, New York.

Kriebel, H.B. 1958. Geographic differentiation in seed dormancy and juvenile growth rate of Ontario sugar maple, in *Proc. 6th Canadian Comm. Forest Tree Breeding* (2): R7–11, Forestry Res. Div., Canada Dept. of Northern Affairs and Natural Res., Ottawa, Canada.

Krugman, S.L., W.I. Stein, and D.M. Schmitt. 1974. Seed biology, Chap. 1 in C.S. Schopmeyer. *USDA For. Serv. Agr. Handbook* 450.

Lack, D. and L. Lack. 1933. Territory reviewed. *British Birds* 27: 179–199.

Larcher, W. 1975. *Physiological Ecology,* transl. M.A. Biederman-Thorson. Springer-Verlag, New York.

Larsen, C.S. 1937. The employment of species types and individuals in forestry. *Royal Veterinary and Agricultural College Yearbook, 1937.* Copenhagen, Denmark.

Larsen, P.R. 1969. Wood formation and the concept of wood quality. *Yale Sch. For. Bull.* 74.

———. 1973. Auxin gradients and the regulation of cambium activity, in T.T. Kozlowski (ed.), *Tree Physiology Colloquium.* Cooperative Extension Programs, U. of Wisconsin, Madison, Wisc.

Larson, J.S. 1929. Fires and forest succession in the Bitter Root Mountains of northern Idaho. *Ecology* 10: 67–68.

Leak, W.B., D.S. Solomon, S.M. Filip. 1969. A silvicultural guide for northern hardwoods. *USDA For. Serv. Res. Pap.* NE-143.

———. 1970. Some preliminary estimates of energy utilization in evenaged northern hardwoods. *USDA For. Serv. Res. Note* NE-108.

———. 1976. Relation of tolerant species to habitat in the White Mountains of New Hampshire. *USDA For. Serv. Res. Pap.* N–351.

Ledig, F.T. 1974. Photosynthetic capacity: developing a criterion for the early selection of rapidly growing trees. *Yale Sch. of For. and Environ. Studies Bull.*, No 85, pp. 19–39.

Ledig, F.T. and J.H. Fryer. 1971. The serotinous cone habit in *Pinus rigida* as related to selection in introgression. IUFRO Congress, Gainesville, Fla. 1971. Working Group Quantitative Genetics.

Lee, R. and C.R. Sypolt. 1974. Toward a biophysical evaluation of forest site potential. *For. Sci.* 20(2): 145–154.

Leikola, M. 1967. Observations on wind conditions in managed Scots pine stand. *Silva. fenn.* 1(3): 57–72 (*For. Abstr.* 1968(1): 21).

Lemieux, G.J. 1961. An evaluation of Paterson's CVP Index in eastern Canada. *Canada Dept. of For., For. Res. Branch Tech.*, Note No. 112.

Leonard, R.E. 1961. Interception of precipitation by Northern hardwoods. *USDA For. Serv. Northeastern For. Expt. Sta. Pap.* 159.

Libby, W.J. 1958. The backcross hybrid Jeffrey × (Jeffrey × Coultier) pine. *J. For.* 56: 840–842.

Little, S. 1974. Effects of fire on temperate forests: Northeastern United States, in T.T. Kozlowski and C.E. AHlgren. *Fire and Ecosystems.* Academic, New York.

Lloyd, M.G. 1961. The contribution of dew to the summer water budget of Northern Idaho. *Am. Meteor. Soc. Bull.* 42: 572–580.

Loehr, R.C. 1972. Relative contribution of contaminants from non-point sources. Dept. Agr. Eng. Conell U. in Stone E. 1973. The impact of timber harvest on soils and water. *Rept. of the President's Advisory Panel on Timber and the Environment.* U.S. Govt. Printing Office, Washington, D.C.

Lorio, P.L. and J.D. Hodges. 1977. Tree water status affects induced southern pine beetle attack and broad production. *USDA For. Serv. Res. Pap.* SO-135.

Love, L.D. 1952. The Fraser expt. for.—its works and aim. *USDA Rocky Mountain For. and Range Expt. Sta. Pap.* 8.

Lowe, W.J. 1974. Morhological variation in balsam fir related to seed source. Ph.D. thesis, Univ. of New Hampshire, Durham, N.H.

Lull, H.W. 1965. Factors influencing water production from forested watersheds, in *Proc. Municipal Watershed Management Symp.* U. of Mass. Coop. Extension Serv. Publ. 446, Amherst, Mass.

Lunt, H.A. 1932. Profile characteristics of New England soil. *Connecticut Agr. Expt. Sta. Bull.* 342.

Lutz, H.J. 1928. Trends and silvicultural significance of upland forest successions in southern New England. *Yale U. Sch. of For. Bull.* 22.

Lutz, H.J. and R.F. Chandler, Jr. 1946. *Forest Soils.* Wiley, New York.

Luftus, N.S., Jr. 1975. Response of yellow-poplar seedlings to simulated drought. *USDA For. Serv. Res. Note.* SO-194.

Lyford, W.H., Jr. 1941. Mineral composition of freshly fallen white pine and red maple leaves. *New Hampshire Agr. Exp. Sta. Tech. Bull.* 77. Durham, N.H.

Lynch, D.W. 1958. Effects of stocking on site measurements and yield of second growth ponderosa pine in the Island Empire. *USDA For. Serv. Inter-mountain For. and Rng. Expt. Sta. Res. Pap.* 56.

Lyr, H. and G. Hoffman. 1967. Growth rates and growth periodicity of tree roots. *Intl. Rev. For. Res.* 2: 181–206.

Macfadyen, A. 1963. *Soil organisms,* J. Doeksen and J. Vander Drift (eds.). North-Holland, Amsterdam.

Mader, D.L. and D.F. Owen. 1961. Relationships between soil properties and red pine growth in Massachusetts. *Soil Sci. Soc. Proc. Amer.* 25: 62–65.

Mader, D.L. and H.W. Lull. 1968. Depth, weight, and water storage of the forest floor in white pine stands in Mass. *USDA For. Serv. Res. Pap.* NE-109.

Madgwich, H.A.I. 1964. Estimation of surface area of pine needles with special reference to *Pinus resinosa. J. For.* 62: 636.

Maki, T.E. 1968. Drainage and soil moisture control in forest production, in N.E. Linnartz, *The Ecology of Southern Forests* (17th Annual Forestry Symposium). Louisiana State Univ. Press, Baton Rouge, La.

Malac, B.F. 1968. Research in forest fertilization at Union-Camp Corp., in *Proc. Sympo. Forest Fertilization, Theory and Practice.* T.V.A., Knoxville, Tenn.

Manley, S.A.M. 1972. The occurrance of hybrid swarms of red and black spruce in central New Brunswick. *Can. J. For. Res.* 2: 381–391.

Mann, W.F., Jr. and J.M. McGilvray. 1974. Response of slash pine to bedding and phosphorus application in the southeasten flatwoods. *USDA For. Serv. Res. Pap.* SO-99.

Marquis, D.A. 1973. Cherry-maple in *silvicultural systems for major forest types of the U.S. USDA For. Serv. Agr. Hndbk.,* No. 445.

Martin, A.C., H.S. Zim and A.L. Nelson. 1961. *American wildlife and plants.* Dover Publ. Inc., New York.

Marx, D.H. 1973. Mycorrhizae and feeder root diseases, in G.C. Marks and T.T. Kozlowski. *Ectomycorrihizae their ecology and physiology.* Academic, New York.

————. 1976. The role of mycorrhizae in forest production, in *Tappi Conference Papers, Annual Meeting,* 1977. Atlanta, Georgia.

Marx, D.H. and C.B. Davey. 1969. The influence of ectotrophic mycorrhizal fungi on the resistance of pine roots to pathogenic infections IV. Resistance of naturally occurring mycorrhizae to infections by *Phytophthora cinnamoni*. *Phytopathology* 59: 559–565.

Mattson, W.J. and N.D. Addy. 1975. Phytophagous insects as regulators of primary production in forest ecosystems. *Science* 190(4214): 515–522.

McClurkin, D.C. 1953. Soil and climatic factors related to the growth of longleaf pine. *USDA For. Serv. Southern For. Expt. Sta. Occ. Pap.* 132.

McGee, C.E. and L. Della-Bianca. 1967. Diameter distribution in natural yellow-poplar stands. *USDA For. Serv. Res. Pap.* SE-25.

McHattie, L.B. and R.J. McCormack. 1961. Forest microclimate: a topographic study in Ontario. *J. Ecology* 49(2): 301–334.

McIntosh R.P. 1967. The continuum concept of vegetation. *Bot. Rev.* 33(2): 130–187.

Mergen, F. 1958. Natural polyploidy in slash pine. *For. Sci.* 4(4): 283–295.

Merriam, C.H. 1898. Life zones and crop zones of the U.S. *USDA Biological Survey Bull.* 10.

Metz, L.J. 1950. Relationship between soil properties and growth of loblolly pine in the S.E. Coastal plain. Ph.D. thesis. Duke U., Sch. of For., Durham, N.C.

———. 1954. Forest floor in the Piedmont Region of South Carolina. *Soil Sci. Soc. Am. Proc.* 18(3): 335–338.

Metz, L.J. and J.E. Douglas. 1959. Soil moisture depletion under several Piedmont cover types. *USDA For. Serv. Tech. Bull.* No. 1207.

Meyer, B.J. and D.B. Anderson. 1952. *Plant physiology*, 2nd ed. D. Van Nostrand, New York.

Miller, P.R. and J.R. McBride. 1975. Effects of air pollutants on forests, in J.B. Mudd. and T.T. Kozlowski. *Response of Plants to Air Pollution.* Academic, New York.

Miller, W.D. and T.E. Maki. 1957. Planting pines in pocosins. *J. For.* 55: 659–663.

Minckler, Leon. 1961. Silvicultural considerations, in *The Challenges of Forestry*. State U. of New York, College of For., Syracuse, N.Y.

Mitchell, K.J. 1975. Dynamics and simulated yield of Douglas-fir. *For. Sci. Monog.* 17, Suppl. Vol. 21(4).

Mitscherlich, E.A. 1909. Des gesetz des minimums und des gesetz de abnehmenden Bodenerbrags. *Landw. Jahrh.* 38: 537–552.

Moehring, D.M. and C.W. Ralston. 1967. Diameter growth of loblolly pine related to available soil moisture and rate of soil moisture loss. *Proc. Soil Sci. Soc. Amer.* 31(4): 560–562.

Möller, C.M. 1946. Untersuehungen uber Laubmenge, stoffverlust and stoff production des Waldes. *Det. Forst. Forstei Denmark* 17: 1–289.

Möller, C.M., D. Müller, and J. Nielson. 1954. Graphic representation of dry matter production of European beech. *Det. Forstl. For. sogv. Denmark* 21: 327–335.

Morris, J.E. 1967. Racial variation in sand pine seedlings, in *Proc. 9th South. Conf. For. Tree Improv.*, Knoxville, Tenn.

Muller, C.H. 1966. The role of chemical inhibition (allelopathy) in vegetational composition. *Bull. Torrey Bot. Club* 93: 332–351.

Muller, D. 1928. Die Kohlensaureassimilation bei asktischen pflazen und die abhanqiqkeit der assimilation von der temperature. *Planta* 6: 22–39.

Murie, O.J. 1951. *The Elk of North America.* The Stackpole Co., Harrisburg, Penn.

Musselman, R.C., D.T. Lester, and M.S. Adams. 1975. Localized ecotypes of *Thuja Occidentalis* L. in Wisconsin. *Ecology* 56: 647–655.

Myer, W.H. 1938. Yield of even-aged stands of Ponderosa pine. *USDA For. Serv. Tech. Bull.* 630.

Nelson, J.C., B.A. Roach, O.L. Frank, W.W. Ward. 1975. Timber management, in *Clearcutting in Pennsylvania.* Sch. For. Res., Penn State U., State College, Pa.

Nelson, R.M. 1959. Drought estimation in southern forest fire control. *USDA For. Serv. Southeastern For. Expt. Sta. Pap.* No. 99.

Nelson, T.C., T. Lotti, E.V. Brandes, and K.B. Trousdell. 1961. Merchantable cubic foot volume growth in natural loblolly pine stands. *USDA For. Serv. Southeastern For. Expt. Sta. Pap.* No. 127.

Nicholson, S.A. and C.D. Monk. 1974. Plant species diversity in old-field succession in the Georgia Piedmont. *Ecology* 55: 1075–1085.

Nienstaedt, Hans and E.B. Snyder. 1974. Principles of genetic improvement of seed. *USDA For. Serv. Agr. Handbook,* No. 450.

N.C. State, School of Forest Resources 1977. 21st Annual Report North Carolina State Univ. Cooperative Tree improvement and hardwood research programs, pp. 13–14, Raleigh, N.C.

Oblerlander, G.T. 1956. Summer fog precipitation on the San Francisco peninsula. *Ecology* 37(4): 851–852.

Odum, E.P. 1969. The strategy of ecosystem development. *Science* 164: 262–270.

Oosting, H.J. 1950. *The Study of Plant Communities.* Freeman, San Francisco, Calif.

Ovington, J.D. 1957. Dry matter production by *Pinus sylvestris. Ann. Botany* 21: 288–314.

———. 1961. Some aspects of energy flow in *Pinus sylvestrus. Ann. Bot.* 25(97): 12–20.

Ovington, J.D. and H.A.I. Madgwick. 1959. Distribution of organic matter and plant nutrients in a plantation of scots pine. *For. Sci.* 5(4): 344–355.

Panshin, A.J. and Carl De Zeeuw. 1964. *Textbook of Wood Technology*, vol 1. McGraw-Hill, New York.

Paterson, S.S. 1956. The forest area of the world and its potential producivity. *Meddelande fran Gotegorgs Universitets Geografiska Institute, Goteborg*, No. 51.

Patric, J.H. 1966. Rainfall interception by mature coniferous forests of southeasten Alaska. *J. Soil Water Conserv.* 21(6): 229–231.

Paul, Benson H. 1963. The application of silviculture in controlling the specific gravity of wood. *USDA For. Serv. Tech. Bull.*, No. 1288.

Pease, A.S. 1926. *Animal Ecology*. McGraw-Hill, New York.

Penman, H.L. 1948. Natural evaporation from open water, bare soil and grass. *Proc. Royal Soc. (London)*, A193: 120–145.

Perla, R.I. and M. Martinelli, Jr. 1976. *Avalanche Handbook. USDA For. Serv. Agr. Handbk.*, No. 489.

Perry, T.D. and G.W. Baldwin. 1966. Winter breakdown of the photosynthetic apparatus of evergreen species. *For. Sci.* 12(3): 374–384.

Peterson, E.B. 1965. Inhibition of black spruce primary roots by a water-soluble substance in *Kalmia angustifolia*. *For. Sci.* 11(4): 473–479.

Pharis, R.P. and H. Hellmers. 1964. Photosynthesis of conifers during and after sub-freezing temperatures. *Plant Physiol.* 39 (abst. suppl.): xlvi, xlvii.

Philbrook, James S. 1971. A stocking guide for eastern white pine silviculture in southern New Hampshire. MS thesis, University of New Hampshire, Durham, N.H.

Philbrook, J.S., J.P. Barrett, and W.B. Leak. 1973. A stocking guide for eastern white pine. *USDA For. Ser. Res. Note.* NE-168.

Pierce, R.S. and C.W. Martin, C.C. Reeves, G.E. Likens, and I.H. Bormann. 1972. Nutrient loss from clear cuttings in New Hampshire in Watersheds in Transition. *Am. Water Resource Assoc. Proc. Sec.* 14: 285–295.

Pitelka, F.A. 1941. Distribution of birds in relation to major biotic communities. *American Midland Naturalist.* 25(1): 113–137.

Post, B.W. and R.O. Curtis. 1970. Estimation of northern hardwood site index from soils and topography in the Green Mountains of Vermont. *Vermont Agr. Expt. Sta. Bull.* 664. Burlington, Vt.

Qashu, H.K. and D.D. Evans. 1967. Water disposition in a stream channel with riparian vegetation. *Soil Sci. Soc. Am. Proc.* 31(2): 263–269.

Ralston, C.W. 1951. Some factors related to the growth of longleaf pine in the Atlantic Coastal Plain. *J. For.* 49: 408–412.

———. 1964. Evaluation of forest site productivity. *International Review For. Res.* 1: 171–201.

Ralston, C.W. and G.E. Hatchell. 1971. Effects of prescribed burning on physical properties of soil, in *Proc. Prescribed Burning Symposium, 1971.* *USDA For. Serv. Southeastern For. Expt. Sta.*, Asheville, N.C.

Ray, R.G. 1941. Site types and rate of growth. *Canada Dept. of Mines and Res. Dominion For. Serv., Silvicultural Res. Note,* No. 65.

Read, R.A. 1965. Windbreaks for the Central Great Plains. *USDA Agric. Handbook* 250.

Reifsnyder, W.E. and H.W. Lull. 1965. Radiant energy in relation to forests. *USDA Tech. Bull.* No. 1344. U.S. Govt. Printing Office. Wash., D.C.

Reigner, I.C. 1964. Calibrating a watershed by using climatic data. *USDA For. Serv. Res. Pap.* NE-15

Reineke, L.H. 1933. Perfecting a stand-density index for evenaged forests. *J. Agr. Res.* 46: 627–638.

Rennie, P.J. 1963. Methods of assessing forest site capacity. *Commonwealth For. Rev.* 42(4), No. 114: 306–317.

Reynolds, E.R.C. and C.S. Henderson. 1967. Rainfall interception by beech, larch, and Norway spruce. *Forestry* 40(2): 165–184.

Riebold, R.J. 1971. The early history of wildfires and prescribed burning, in *Proc. Prescribed Burning Symposium, 1971. USDA For. Serv. Southeastern For. Expt. Sta.*, Asheville, N.C.

Roe, A.L. 1967. Productivity indicators in western larch forests. *USDA For. Serv. Res. Note* INT-59.

Rogerson, T.L. 1964. Estimating foliage on loblolly pine. *USDA For. Serv. Res. Note* SO-16.

Romberger, J.A. 1969. Apical meristems of trees, why we study them. *USDA Agric. Sci. Rev. Second Quarter.*

Rose, C.W. 1966. *Agricultural Physics.* Pergamon, New York.

Salo, D.J., R.E. Inman, B.J. McGurk, and J. Verhoeff. 1977. *Silvicultural Biomass Farms,* (vol. III) *Land Suitability and Availability.* MTR-7347. Energy Research and Development Adm., Natl. Tech. Info. Serv., U.S. Dept. Commerce.

Sargent, C.S. 1884. *Report of the forest of North America.* U.S. Dept. of Interior, Census Off., Wash., D.C.

Sartz, R.S. and G.R. Trimble, Jr. 1956. Snow storage and melt in a northern hardwood forest. *J. For.* 54(8): 499–502.

Sartz, R.S. 1961. Comparison of bulk density of soil in abandoned land and forestland. *USDA For. Serv. Lake States For. Expt. Sta. Tech. Note* 601.

Schalin, I. 1966. *Alnus incana* in forestry practice. *Metsat. Ackak.* 93: 362–366, 370.

Schaller, F. 1968. *Soil Animals.* The U. of Michigan Press, Ann Arbor, Mich.

Schmidt, W.C., R.C. Shearer, and A.L. Roe. 1976. Ecology and silviculture of western larch. *USDA For. Serv. Tech. Bull.* No. 1520.

Schnur, G.L. 1937. Yield, stand, and volume tables for even-aged upland oak forest. *USDA Tech. Bull.* No. 560.

Schomaker, C.E. 1968. Solar radiation measurements under a spruce and birch canopy during May and June. *For. Sci.* 14(1): 31–38.

Schopmeyer, C.S. (tech. coordinator). 1974. Seeds of woody plants in the U.S. *USDA For. Serv. Agr. Handbook.* No. 450 (supercedes Misc. Pbl. No. 654, 1948).

Schultz, P.P. 1969. What is being learned from intensive management of slash pine. *For. Farmer* 28(6): 8–9, 16.

Schumacher, F.X. 1939. A new growth curve and its application to timber-yield studies. *J. For.* 37(1): 819–820.

Schumacher, R.X. and T.C. Coile. 1960. *Growth and yields of natural stands of the southern pines.* T.S. Coile, In., Durham, N.C.

Shantz, H.L. and R. Zon. 1924. *Atlas of American Agriculture.* USDA, Wash., D.C.

Shaw, S.P. and T.H. Ripley. 1965. Managing the forest for sustained yield of woody browse for deer. *Proc. Soc. Am. Foresters,* 229–233.

Shearer, R.C. 1974. Early establishment of conifers following prescribed broadcast burning in western larch/Douglas-fireforests, in *Tall Timbers Fire Ecology Rept. No. 14.* Fire and Land Management Symposium, Missula, Montana.

Shoulders, E. and T.A. Terry. 1968. Climate at the seed source affects longleaf pine performance in Louisiana plantation. *USDA For. Serv. Res. Note* SO-78.

Shreve, F. 1917. A map of the vegetation of the U.S. *Geog. Rev.* 3: 119–125.

Siccama, T.G. 1974. Vegetation, soil and climate on the Green Mountains of Vermont. *Ecol. Monog.* 44: 325–349.

Silen, Roy R. 1970. The seed source question of ponderosa pine. School of For., Oregon State U., Corvallis, Ore.

Sjolte-Horgensen, J. 1967. The influence of spacing on the growth of coniferous plantations. *Int. Rev. For. Res.* 2: 43–94.

Squillace, A.E. and R.T. Bingham, 1958. Localized ecotype variation in western white pine. *For. Sci.* 4: 20–34.

Smith, A.D. 1940. A discussion of the application of a climatological diagram, the hythergraph, to the distribution of natural vegetation types. *Ecology* 21:184–191.

Smith, D.M. 1962. *Practice of Silviculture,* 7th ed. Wiley, New York.

Smith, J.H. 1964. Root spread can be estimated from crown width of Douglas-fir, lodgepole pine, and other British Columbia tree species. *For. Chron.* 40(4): 456–473.

Smith, W.H. 1970. *Tree Pathology: A Short Introduction.* Academic, New York.

Society of American Foresters. 1950. *Forestry Terminology a glossary of technical terms used in forestry*. Soc. of American Foresters, Wash., D.C.

Society of American Foresters. 1954. *Forest Cover Types of North America*. Soc. of Am., For., Wash., D.C.

Squillace, A.E. 1965. Combining superior growth and timber quality with high gum yield in slash pine, in *Proc. 8th Southern Forest Tree Improv. Conf.*, pp. 73–76, Georgia Forestry Res. Council, Macon, Ga.

Stanley, O.B. 1938. Indicator significance of lesser vegetation in the Yale forest near Keene, N.H. *Ecology* 19: 188–207.

Steinbrenner, E.C. 1968. Progress and needs in tree nutrition research in the Northwest, in *Forest Fertilization, Theory and Practice, Proc. Sym. on Forest Fert*. I.V.A., Knoxville, Tenn.

Steinbrenner, E.C., S.W. Suffield, and R.K. Campbell. 1960. Increased cone production of young Douglas-fir following nitrogen and phosphorous fertilization. *J. For.* 58(2): 105–110.

Stephens, E.R. and F.R. Burleson. 1967. Analysis of the atmosphere for light hydrocarbon. *J. Air Pollut. Contn. Assn.* 17: 147–153.

Stoate, T.N. 1950. Nutrition of the pine. *Bull. For. Bur. Australia*. 30.

Stoddard, H.L. 1931. *The Bobwhite Quail. Its Habits, Preservation, and Increase*. Charles Scribner's Sons, New York.

Stoeckeler, J.H. 1948. The growth of quaking aspen as affected by soil properties and fire. *J. For.* 46: 727–737.

Stoehr, H.A. 1946. M.F. thesis. Duke U., Sch. of For.

Stone, E.C. 1957. Dew as an ecological factor. II, the effect of artificial dew on the survival of *Pinus ponderosa* and associated spp. *Ecology* 38: 414–442.

———. Variation in the root-growth capacity of ponderosa pine transplants, in R.K. Hermann (ed.) *Regeneration of Ponderosa Pine*. School of Forestry. Oregon State Univ., Corvallis, Ore.

Stone, E.L., R.R. Morrow, and D.S. Welch. 1954. A malady of red pine on poorly drained sites. *J. For.* 52: 104–114.

Stone, E.L. and R.F. Fisher. 1969. An effect of conifers on available soil nitrogen. *Plant and Soil* 30(1): 134–138.

Stone, E.L., Jr. 1971. Effects of prescribed burning on long-term productivity of coastal plain soils, in *Prescribed Burning Symposium Proceedings Southeastern Forest Expt. Sta. USDA Forest Service*, Asheville, N.C.

Stout, B.B. 1956. Studies of root systems of deciduous trees. *Black Rock For. Bull*. No. 15. Harvard U., Cambridge, Mass.

Stransky, J.J. and D.R. Wilson. 1964. Terminal elongation of loblolly and shortleaf pine seedlings under soil moisture stress. *Soil Sci. Soc. Am. Proc.* 28: 439–440.

Stuart, G.W. and W.E. Sopper. 1968. A study of rainfall interception by an oak hickory forest cover on a watershed in central Pennsylvania. *Res. Briefs Sch. For. Resour. Pa. Sta. Univ.* 3(1): 14–16.

Swank, W.T. and H.T. Schreuder. 1974. Comparison of three methods of estimating surface area and biomass for a forest of young white pine. *For. Sci.* 20(1): 91–100.

Swift, L.W., Jr. and J.B. Messer. 1971. Forest cuttings raise temperatures of small streams in the southern Appalachians. *J. Soil and Water Conserv.* 26: 111–116.

Swift, L.W., Jr. 1976. Algorithm for solar radiation on mountain slopes. *Water Resources Res.* 12(1): 108–112.

Switzer, G.L., L.E. Nelson, and W.H. Smith. 1968. The mineral cycle in forest stands, in *Forest Fertilization, Theory and Practice. Proc. Sym. on For. Fert.* T.V.A., Knoxville, Tenn.

Tadaki, Y. 1966. Some discussions on the leaf biomass of forest stands and trees. *Bull. Govt. For. Expt. Sta.* No. 184. Tokyo, Japan.

Talbot, P.H.B. 1971. *Principles of Fungal Taxonomy.* St. Martin's Press, New York.

Taylor, R.I. 1939. The application of tree classification—making lodgepole pine for selection cutting. *J. For.* 37: 772–782.

T.V.A. 1968. *Forest Fertilization, Theory and Practice. Proc. of Symp. on Forest Soil Fert., 1967.* Tenn. Valley Authority, Muscle Shoals, Ala.

Thornthwaite, C.W. 1931. The climates of North America according to a new classification. *Geog. Rev.* 21: 633–655.

———. An approach toward a rational classification of climate. *Geol. Rev.* 38: 55–94.

Thornthwaite, C.W. and F.K. Hare. 1955. Climate classification in forestry. *Unasylva* 9: 51–59.

Toumey, James W. and C.F. Korstian. 1947. *Foundations of silviculture,* 2nd ed. Wiley, New York.

Transeau, E. 1905. Forest centers of eastern N. America. *Am. Naturalist* 39: 875–889.

Trappe, J.M. 1962. Fungus associates of ectomycorrhizae. *Bot. Rev.* 28: 538.

Treshow, M. 1970. *Environment and Plant Response.* McGraw-Hill, New York.

Trimble, G.R., Jr., 1969. Diameter growth of individual hardwood trees. *USDA For. Serv. Res. Pap.* NE-145.

Trimble, G.R., Jr. and H.W. Lull. 1956. The role of forest humus in watershed management in New England. *USDA For. Serv. Northeastern For. Expt. Sta. Pap.* 85.

Trimble, G.R., Jr. and S. Weitzman. 1956. Site index studies of upland oaks in the northern Appalachians. *For. Sci.* 2: 162–173.

Trimble, G.R., Jr., R.S. Sartz, and R.S. Pierce. 1958. How type of soil frost affects infiltration. *J. Soil and Water Conserv.* 13(2): 81–82.

Trimble, George R., Jr., James H. Patric. John D. Gill, George H. Moeller, and James N. Kochenderfer. 1974. Some options for managing forest land in the central Appalachians. *USDA For. Serv. Gen. Tech. Rep.* NE-12.

Trippensee, R.E. 1948. *Wildlife Management—Upland Game and General Principles.* McGraw-Hill, New York.

Trousdell, K.B. and M.D. Hoover. 1955. A change in ground water level after clear-cutting of loblolly pine in the Coastal Plain. *J. For.* 53: 495–498.

Tryon, E.H. and R.P. True. 1968. Radial increment response of Appalachian hardwood species to a spring freeze. *J. For.* 66: 488–591.

Tubbs, C.H. 1976. Effect of sugar maple exudate on seedlings of northern conifer species. *USDA For. Serv. Res. Note* NC-213.

Urie, D.H. 1959. Pattern of soil moisture depletion varies between red pine and oak stands in Michigan. *USDA For. Serv. Lake States For. Expt. Sta. Tech. Note* 564.

Ursic, S.J. and F.E. Dendy. 1965. Sediment yields from small watersheds under various land uses and forest covers. Federal Interagency Sediment Conf. Proc. *USDA Misc. Publ.* 970, pp. 47–52.

USDA. 1955. *Soil. Year Book of Agriculture.* U.S. Govt. Printing Office, Wash., D.C.

USDA Forest Service. 1958. Timber resources for America's future. *USDA Forest Service, Forest Resource Doc. Nos. 14.* U.S. Govt. Printing Office, Wash., D.C.

———. 1971a. Wildlife Management Handbook, Southern Region. *USDA Forest Service* FSH 2609.

———. 1971b. *Prescribed burning symposium.* Asheville, N.C.: Southeastern Forest Expt. Station.

Van Bavel, C.H.M. 1966. Potential evaporation: the combination concept of its experimental verification. *Water Resources Res.* 2(3): 455–467.

Viro, P.J. 1974. Effects of forest fire on soil, in T.T. Kozlowski and C.E. Ahlgren. *Fire and Ecosystems.* Academic, New York.

Vogelmann, H.W., T. Siccama, D. Leedy, and D.C. Ovitt. 1968. Precipitation from fog moisture in the Green Mountains of Vermont. *Ecology* 49(6): 1205–1207.

Waggoner, P.E. and R.H. Shaw. 1952. Temperature of potato and tomato leaves. *Plant Physiology* 27: 710–724.

Waggoner, P.E. 1975. Micrometeorological models, in J.L. Monteith (ed.). *Vegetation and the Atmosphere.* Academic, New York.

Waksman, S.A. 1952. *Soil Microbiology.* John Wiley, New York.

Wallace, G.J. 1963. *An Introduction to Ornithology,* 2nd ed. Macmillan, New York.

Ward, W.W. 1964. Live crown ratio and stand density in young, even-aged red oak stands. *For. Sci.* 19(1): 56.

Waring, R.H. 1970. Matching species to site, in R.K. Hermann (ed.), *Regeneration of Ponderosa Pine.* School of Forestry, Oregon State Univ. Corvallis, Ore.

Weaver, J.E. and R.E. Clements. 1938. *Plant Ecology.* McGraw-Hill, New York.

Webb, W.L. 1973. Timber and Wildlife, Appendix N. *Report of the President's Advisor, Panel on Timber and the Environment.* U.S. Govt. Printing Office, Wash., D.C.

Weitzman, S. and R.R. Roy. 1959. Snow behavior in forests of northern Minnesota and its management implications. *USDA For. Serv. Lake States For. Expt. Sta. Pap.* No. 69.

Wells, C.G. and D.M. Crutchfield. 1969. Foliar analyses for predicting loblolly pine response to phosphorus fertilization on wet sites. *USDA For. Serv. Res. Note* SE-128. Ashville, N.C.

Wells, C.G. 1971. Effects of prescribed burning on soil chemical properties and nutrient availability, in *USDA For. Serv. Prescribed Burning Symposium Proceedings.* Southeastern Forest Expt. Sta., Asheville, N.C.

Wells, O.O. 1969. Results of the southwide pine seed source study through 1968–1969, in *Proc. of the 10th Southern Conf. on Forest Tree Improv.,* pp. 117–129, Eastern Tree Seed Laboratory, Macon, Ga.

Wells, O.O. and P.C. Wakeley. 1966. Geographic variation in survival, growth and fusiform rust infection of planted loblolly pine. *For. Sci. Monogr.* 11.

————. 1970. Variation in shortleaf pine from several geographic sources. *For. Sci.* 16(4): 415–423.

West, A.J. 1959. Snow evaporation and condensation. *Proc. 1959 Western Show Conf.,* pp. 66–74, U.S. Army Corps of Engineers, Portland, Ore.

White, D.P. and A.L. Leaf. 1956. Forest fertilization, a bibliography with abstracts, on the use of fertilizers, and soil amendments in forestry. *World Forest Series Bull.* 2 (Tech. Publ. 81). State Univ. College of Forestry, Syracuse, N.Y.

Whitford, H.N. 1901. Genetic development of the forests of northern Michigan. *Bot. Gaz.* 31: 289–325.

Whittaker, R.H. 1956. Vegetation of the Great Smoky Mountains. *Ecol. Monog.* 21: 1–80.

————. 1967. Gradient analysis of vegetation. *Biol. Rev.* 49: 207–264.

Whittaker, R.H. and P.P. Feeny. 1971. Alleochemics: Chemical interaction between Species. *Science* 171: 757–770.

Wilde, S.A. 1958. *Forest Soils.* The Ronald Press Co., New York.

————. 1976. *Woodlands of Wisconsin. Science, art and letters of growing timber.* Wisc. Agr. Extension Serv., Univ. of Wisc., Madison, Wisc.

Wilde, S.A., E.C. Teinbrenner, R.S. Pierce, R.C. Dosen, and D.J. Pronin. 1953. Influence of a forest cover on the state of the ground water table. *Soil Sci. Soc. Am. Proc.* 17: 65–67.

Williams, C.B., Jr. and C.T. Dyrness. 1967. Some characteristics of forest floors and soils under the fir/hemlock stands in the Cascade range. *USDA For. Serv. Sta. Note* PNW-37.

Williston, H.L. 1966. Forest floor in loblolly pine plantations as related to stand characteristics. *USDA For. Serv.* SO-26.

Wilson, F.G. 1951. Control of growing stock in evenaged stands of conifers. *J. For.* 49: 692–695.

Winton, L.L. 1968. Plantlets from aspen tissue cultures. *Science* 160: 1234–1235.

Witkamp, M. 1966. Decomposition of leaf litter in relation to environment microflora, and microbial respiration. *Ecology* 47(2): 194–201.

———. 1971. Soils as components of ecosystems. *Ann. Rev. Ecol. and Systematics.* 2: 85–110.

Woodbury, A.M. 1947. Distribution of pigmy conifers in Utah and northwestern Arizona. *Ecology* 28(2): 113–126.

Woodruff, N.P. and W.W. Zingg. 1953. Wind tunnel studies of shelterbeef models. *J. For.* 51: 173–178.

Wright, H.E., Jr. 1974. Landscape development, forest fires, and wilderness management. *Science* 186 (4163): 487–495.

Wright, J.W. 1976. *Introduction to forest genetics.* Academic, New York.

Yocom, H.A. and E.R. Lawson. 1977. Tree percent from naturally regenerated shortleaf pine. *South I. App. For.* 1(2): 10–11.

Youngberg, C. and C. Davey. 1970. Soils and forest growth, in *Proc. 3rd. North American Forest Soils Conf.* Oregon State Univ. Press, Corvallis, Ore.

Youngberg, C.T. and L. Hu. 1972. Root nodules on mountain mahogany. *For. Sci.* 18(3): 211–212.

Youngman, A.L. 1967. *An ecotypic differentiation approach to the study of isolated populations of Pinus taeda in south central Texas.* Abstr. of thesis, in Dissert. Abstr. 27B (9), (3006), (O.R.S.).

Zahner, R. 1954. Estimating loblolly pine sites in the Gulf Coastal Plain. *J. For.* 52: 448–449.

———. Evaluating summer moisture deficiencies. *USDA For. Serv. Southern For. Exp. Sta. Occ. Pap.* 150.

———. Field procedures for soil-site classification of pine land in south Arkansas and north Louisiana. *USDA For. Serv. Southern For. Expt. Sta. Occ. Pap.* 155.

Zahner, R. and A.R. Stage. 1966. A procedure for calculating daily moisture stress and its utility in regressions of tree growth on weather. *Ecology* 47(1): 64–74.

Zavitkovski, J. 1976. Biomass studies in intensively managed forest stands, in *Intensive Plantation Culture*. USDA Forest Service Gen. Tech. Rept. NC-21.

Zimmerman, M.H. 1964. Effect of low temperature on ascent of sap in trees. *Plant Physio.* 39(4): 568–572.

Zimmermann, M.H. and C.L. Brown. 1971. *Trees, Structure and Function*. Springer-Verlag, New York.

Zobel, Bruce, 1961. Inheritance of wood properties in conifers. *Silvae Genetica* 10(3): 63–70.

Zobel, B.J. 1971. The genetic improvement of southern pines. *Sci. Am.* 225(5): 94–103.

Zobel, B.J. and E.H. Aldin, Jr. 1962. Effect of bole straightness on compression wood of loblolly pine. *N.C. State University at Raleigh, Sch. of For. Res. Tech. Rept.* No. 15.

Zon, R. 1941. Climate and the nation's forests, in *1941 Yearbook of Agriculture Climate and Man*. U.S. Govt. Printing Office, Wash., D.C.

Appendix I

COMMON NAMES	SCIENTIFIC NAMES
Ailanthus	*Ailanthus altissima* (Mill.) Swingle
Alder, speckled	*Alnus rugosa* (Du Roi) Spreng.
Alder, red	*Alnus rubra* Bong.
American hornbean (see Beech)	
Apple	*Malus pumila* Mill.
Ash, American mountain	*Sorbus americana* Marsh.
black	*Fraxinus nigra* Marsh.
green or red	*Fraxinus pennsylvanica* Marsh.
oregon	*Fraxinus latifolia* Benth.
white	*Fraxinus americana* L.

Common and Scientific Names of Tree Species Mentioned in the Text[a]

COMMON NAMES	SCIENTIFIC NAMES
Aspen, largetooth or bigtooth	*Populus grandidentata* Michx.
quaking or trembling	*Populus tremuloides* Michx.
Baldcypress (see Cypress)	
Basswood, American, or American linden	*Tilia americana* L.
Beech, American	*Fagus grandifolia* Ehrh.
blue, or American hornbeam	*Carpinus caroliniana* Walt. (*Carpinus betulus virginiana* Marsh.)
Bigcone Douglas-fir (see Fir)	
Birch, gray or field	*Betula populifolia* Marsh.
river	*Betula nigra* L.
sweet or black	*Betula lenta* L.
white or paper	*Betula papyrifera* Marsh.
yellow	*Betula alleghaniensis* Britton (*B. lutea* Michx.)
European white birch	*Betula pendula* Roth.
Blackgum (see Gum)	
Bluebeech (see Beech)	
Boxelder (see Elder)	
Buckeye, yellow	*Aesculus octandra* Marsh.
Butternut	*Juglans cinerea* L.
Catalpa, northern or hardy	*Catalpa speciosa* Warder
southern	*Catalpa bignonioides* Walt.
Ceanothus, feltleaf	*Ceanothus arboreus* Greene
Cedar, Atlantic white	*Chamaecyparis thyoides* (L.) B.S.P.
eastern red	*Juniperus virginiana* L.
incense cedar	*Libocedrus decurrens* Torr.
rocky mountain juniper	*Juniperous scopulorum* Sarg.
Port-Orford-cedar	*Chamaecyparis lawsoniana* (A. Murr.) Parl.
yellow cedar, Alaska	*Chamaecyparis nootkatensis* (D. Don) Spach
northern white, or eastern arborvitae	*Thuja occidentalis* L.
western red	*Thuja plicata* Don
Cherry, black	*Prunus serotina* Ehrh.
chokecherry	*Prunus virginiana* L.
pin	*Prunus pensylvanica* L. f.
Chestnut, American	*Castenea dentata* (Marsh.) Borkh.
Chinkapin, golden	*Castanopsis chrysophylla* (Bougl.) A.
Coffeetree, Kentucky	*Gymnocladus dioica* (L.) K. Koch
Cottonwood (see Poplar)	
Cucumber tree	*Maganolia acuminata* L.

COMMON NAMES	SCIENTIFIC NAMES
Currant, black	*Ribes nigrum* L.
Cypress, bald	*Taxodium distichum* (L.) Rich.
Dogwood, flowering	*Cornus florida* L.
Douglas-fir (see Fir)	
Elder, American	*Sambucus canadensis* L.
Box	*Acer negundo* L.
Elm, American or white	*Ulmus americana* L.
rock or cork	*Ulmus thomasii* Sarg.
slippery	*Ulmus rubra* Muhl.
Fir, alpine or subalpine	*Abies lasiocarpa* (Hook.) Nutt.
bigcone Douglas-fir	*Pseudotsuga macrocarpa* (Vasey) Mayr
balsam	*Abies balsamea* (L) Mill.
California red fir	*Abies magnifica* A. Murr.
Douglas	*Pseudotsuga menziesii* (Mirb.) Franco
grand or lowland white	*Abies grandis* (Dougl.) Lindl.
noble	*Abies procera* Rehd. (*A. nobilis* (Dougl.) Lindl.)
Pacific silver or silver	*Abies amabilis* (Dougl.) Forbes
white	*Abies concolor* (Gord. and Glend.) Lindl.
Gum, black, or black tupelo	*Nyssa sylvatica* Marsh.
red or sweet	*Liquidambar styraciflua* L.
tupelo, or water tupelo	*Nyssa aquatica* L.
Hackberry	*Celtis occidentalis* L.
Hawthorn	*Crataegus* spp.
Hazel, California	*Corylus cornuta* var *californica* (A.DC.) Sharp
beaked	*Corylus cornuta* var *cornuta* Marsh.
Hemlock, eastern	*Tsuga canadensis* (L.) Carr.
mountain	*Tsuga mertensiana* (Raf.) Sarg.
western	*Tsuga heterophylla* (Raf.) Sarg.
Hickory, bitternut	*Carya cordiformis* Wangenh.) K. Koch
mockernut	*Carya tomentosa* Nutt.
pignut	*Carya glabra* (Mill.) Sweet
shagbark	*Carya ovata* (Mill.) K. Koch
Holly, American	*Ilex opaca* Ait.
Honeylocust (see Locust)	
Hophornbeam, eastern	*Ostrya virginiana* (Mill.) K. Koch
Hornbeam (see Beech)	
Horsechestnut (see Buckeye)	
Hawthorn	*Crataegus* spp.

COMMON NAMES	SCIENTIFIC NAMES
Juniper (see Cedar)	
Kentucky coffeetree (see Coffee tree)	
Larch, American, or Tamarack	*Larix laricina* (Du Roi) K. Koch
European	*Larix decidua* Mill. (L. europaea DC.)
Japanese	*Larix leptolepis* Murr.
subalpine	*Larix lyallii* Parl.
western	*Larix occidentalis* Nutt.
Laurel, California	*Umbellularia Californica* (Hook and Arn.)
Linden (see Basswood)	
Locust, black or yellow	*Robinia pseudoacacia* L.
honey	*Gleditsia triacanthos* L.
Madrone, Pacific	*Arbutus menziesii* Pursh
Magnolia, southern	*Magnolia grandiflora* L.
Mangrove	*Rhizophora mangle* L.
black	*Avicennia nitida* Jacq.
Maple, red (including trident)	*Acer rubrum* L.
silver	*Acer saccharinum* L.
striped	*Acer pensylvanicum* L.
sugar	*Acer saccharum* Marsh.
vine	*Acer circinatum* Pursh
Oak, bear	*Quercus ilicifolia* Wangenh.
black	*Quercus velutina* Lam.
blackjack	*Quercus marilandica* Muenchh.
bur	*Quercus macrocarpa* Michx.
California white	*Quercus lobata* Née
Caynon live	*Quercus chrysolepis* Liebm
chestnut	*Quercus prinus* L.
Gambel	*Quercus gambelii* Nutt.
live	*Quercus virginiana* Mill.
northern pin	*Quercus ellipsoidalis* E.J. Hill
northern red or eastern red	*Quercus rubra* L. (Q. *borealis* var. *maxima* (Marsh.) Ashe)
Oregon white	*Quercus garryana* Dougl.
overcup	*Quercus lyrata* Walt.
pin	*Quercus palustris* Muenchh.
post	*Quercus stellata* Wangenh.
scarlet	*Quercus coccinea* Muenchh.
southern red	*Quercus falcata* Michx.
swamp white	*Quercus bicolor* Willd.
turkey	*Quercus laevis* Walt. (Q. *catesbaei* Walt.)
white	*Quercus alba* L.
willow	*Quercus phellos* L.

COMMON NAMES	SCIENTIFIC NAMES
Osage orange	*Maclura pomifera* (Raf.) Schneid
Pecan	*Carya illinoensis* (Wangenh.) K. Koch
Persimmon, common	*Diospyros virginiana* L.
Pine, Austrian	*Pinus nigra* Arnold
bishop	*Pinus muricata* D. Don.
bristlecone	*Pinus aristata* Englem.
Coulter	*Pinus coulteri* D. Don.
Digger	*Pinus sabiniana* Dougl.
eastern white	*Pinus strobus* L.
foxtail	*Pinus balfouriana* Grev. and Balf.
jack	*Pinus banksiana* Lamb.
Jeffrey	*Pinus jeffreyi* Grev. and Balf.
knobcone	*Pinus attenuata* Lemm.
limber pine	*Pinus flexilis* James
loblolly	*Pinus taeda* L.
lodgepole	*Pinus contorta* Dougl.
longleaf	*Pinus palustris* Mill.
Monterey	*Pinus radiata* D. Don
pinyon	*Pinus edulis* Engelm.
pitch	*Pinus rigida* Mill.
pond	*Pinus serotina* Michx.
ponderosa or western yellow	*Pinus ponderosa* Laws.
(Rocky Mountain form)	*Pinus ponderosa* var. *scopulorum* Engelm.
red or Norway	*Pinue resinosa* Ait.
sand	*Pinus clausa* (Chapm.) Vasey
Scotch	*Pinus sylvestris* L.
shortleaf	*Pinus echinata* Mill.
slash	*Pinus elliotti* Engelm.
spruce	*Pinus glabra* Walt.
sugar	*Pinus lambertiana* Dougl.
table-mountain	*Pinus pungens* Lamb.
Virginia or scrub	*Pinus virginiana* Mill.
western white	*Pinus monticola* Dougl.
whitebark	*Pinus albicaulis* Engelm.
Poplar, balsam	*Populus balsamifera* L.
California, or black cottonwood	*Populus trichocarpa* Torr. and Gray
Eastern, or eastern cottonwood	*Populus deltoides* Bartr.
yellow, or tuliptree	*Liriodendron tulipifera* L.
Redbud, eastern	*Cercis canadensis* L.
Redcedar (see Cedar)	
Redwood	*Sequoia sempervirens* (D. Don) Endl.
Sassafras	*Sassafras albidum* Nutt.) Nees

COMMON NAMES	SCIENTIFIC NAMES
Serviceberry	*Amelanchier* spp.
Sequoia, giant	*Sequoia gigantea* (Lindl.) Decne.
Sourwood	*Oxydendron arboreum* (L.) DC.
Spruce, bigcone (see bigcone Douglas-fir)	
Spruce, black	*Picea mariana* (Mill.) B.S.P.
blue	*Picea pungens* Engelm.
Engelmann	*Picea engelmannii* Parry
Norway	*Picea abies* L.
red	*Picea rubens* Sarg.
Sitka	*Picea sitchensis* (Bong.) Carr.
white	*Picea glauca* Moench) Vosa
Sumac, staghorn	*Rhus typhina* L.
winged or shining	*Rhus copallina* L.
Sycamore, America	*Platanus occidentalis* L.
Tamarack *(see Larch)*	
Tanoak	*Lithocarpus densiflorus* (Hook and Arn.) Rehd.
Torreya, California	*Torreya californica* Torr.
Tuliptree (see Poplar)	
Tupelo (see Gum)	
Viburnum	*Viburnum* spp.
Walnut, black	*Juglans nigra* L.
Willow, arctic	*Salix glauca* L.
black	*Salix nigra* Marsh.
Witch-hazel	*Hamamelis virginiana* L.
Yellow-poplar (see Poplar)	
Yew, Pacific	*Taxus brevifolia* Nutt.

[a]Names of trees are based on E.L. Little. 1953. *Check List of Native and Naturalized Trees of the United States. Agriculture Handbook No. 41, U.S. Forest Service, Wash., D.C. Names of other species are from various sources.*

Appendix II.

COMMON NAMES	SCIENTIFIC NAMES
	MARSUPIAL
Opossum	*Didelphis marsupialis*
	INSECTIVORE
Mole, hairytail	*Parascalops breweri*
eastern	*Scalopus aquaticus*

Common and Scientific Names of Wildlife and Bird Species Mentioned in the Text[a]

COMMON NAMES	SCIENTIFIC NAMES
Shrew, masked	*Sorex cinereus*
smoky	*Sorex fumeus*
dusky	*Sorex obscurus*
shortail	*Blarina brevicauda*

HARES AND RABBITS

Hare, snowshoe (varying)	*Lepus americanus*
Cottontail, eastern	*Sylvilagus floridanus*

RODENTS

Marmot, yellowbelly	*Marmota flaviventris*
Squirrel, ground	
golden-mantled	*Citellus lateralis*
chipmunk, eastern	*Tamias striatus*
chipmunk least (western)	*Eutamias minimus*
Squirrel, tree	
gray, eastern	*Sciurus carolinensis*
western	*Sciurus griseus*
fox	*Sciurus niger*
tassel-eared	*Sciurus aberti*
red	*Tamiasciurus hudsonicus*
chickaree	*Tamiasciurus douglasi*
Squirrel, flying	
northern	*Glaucomys sabrinus*
southern	*Galucomys volans*
Rat, bushy-tailed wood	*Neotoma cinerea*
Beaver	*Castor canadensis*
mountain (Aplodontia)	*Aplodontia rufa*
Mouse, white-footed deer	*Peromyscus maniculatus*
white footed	*Peromyscus leucopus*
woodland, jumping	*Napaeozapus insignis*
Voles	
boreal red-backed	*Clethrionomys gapperi*
pine	*Pitymys pinetorum*
Porcupine	*Erethizon dorsatum*

CARNIVORES

Wolf, gray	*Canis lupus*
Fox, red	*Vulpes fulva*
gray	*Urocyon cinereoargenteus*
Bear, black	*Ursus americanus*
grizzly	*Ursus horribilis*
Raccoon	*Procyon lotor*
Martin	*Martes americana*
fisher	*Martes pennanti*
Weasel, shorttail	*Mustela erminea*
least	*Mustela rixosa*
longtail	*Mustela frenata*

COMMON NAMES	SCIENTIFIC NAMES
Wolverine	*Gulo luscus*
Skunk, striped	*Mephitis mephitis*
spotted	*Spilogate putorius*
Mountain lion	*Felis concolor*
Lynx	*Lynx canadensis*
bobcat	*Lynx rufus*

EVENTOED UNGULATES

Antelope (American pronghorn)	*Antilocapra americana*
Deer, mule	*Odocoileus hemionus*
white-tailed	*Odocoileus virginianus*
Elk (Wapiti), Roosevelt	*Cervus canadensis Rosseveltis merriam*
Rocky Mountain	*C.R. Nelsoni Bailey*
Moose	*Alces alces*
Mountain sheep (Bighorn)	*Ovis canadensis*

BIRDS

Duck, wood	*Aix sponsa*
Golden eagle	*Aquila chrysaëtos canadensis*
bald	*Haliceetus leucocephalus*
Hawk, goshawk	*Accipiter gentilis atricapillas*
sharp-shinned	*Accipiter striatus velox*
Cooper's	*Accipiter cooperii*
red-tailed	*Buteo jamaicensis*
red-shouldered	*Buteo lineatus*
broad-winged	*Buteo platypterus platypterus*
pigeon	*Falco columbarius columbarius*
hawk sparrow	*Falco sparveruis*
Grouse, spruce	*Canachites canadensis*
blue	*Dendragapus obscurus*
ruffed	*Bonasa umbellus*
prairie chicken	*Tympamuchus cupido*
sharp-tailed	*Pedioecetes phasianellus*
Quail, bob-white	*Colinus virginianus*
Turkey, Mexican	*Meleagris gallopavo gallopavo*
eastern	*Meleagris gallopavo silvestrii*
Florida	*Meleagris gallopavo osceóla*
S. Rocky mountain	*Meleagris gallopavo merriami*
Woodcock	*Philohela minor*
Dove, mourning	*Zenaidura macroura*
Whip-poor-will	*Caprimulgus vociferus*
Owl, great horned	*Bubo virginianus*
barred	*Strix varis*
saw-whet	*Aegolius acadia acadica*
screech	*Otus asio*
pigmy	*Glaucidium gnoma*

COMMON NAMES	SCIENTIFIC NAMES
Swift, chimney	*Chaetura pelagica*
Woodpecker, flicker	*Colaptes auratus*
black-backed three-toed	*Picoides arcticus*
pileated	*Hylatomus pileatus*
red-headed	*Melanerpes erythrocephalus*
	erythrocephalus
yellow-bellied sapsucker	*Sphyrapicus varius varius*
Williamson's sapsucker	*Sphyrapicus thyroideus*
hairy	*Dendrocopus villosus*
downy	*Dendrocopus pubescens*
white-headed	*Dendrocopus albolarvatus*
red-cockaded	*Dendrocopus borealis*
red-bellied	*Centurus carolinus*
Flycatcher, Acadian	*Empidonax virescens*
Traill's	*Empidonax traillii*
western	*Empidonax fifficilis*
Hammond's	*Empidonax hammondii*
olivesided	*Nuttallornis borealis*
Pewee, eastern wood	*Contopus virens*
western wood	*Contopus sordidulus*
Swallow, tree	*Iriodoprocne bicolor*
Jay, Canada (grey)	*Perisoreus canadensis*
blue	*Cyanocitta cristata*
Stellar's	*Cyanocitta stelleri*
Raven	*Corvus corex*
crow	*Corvus ossifragus*
Clark's nutcracker	*Nucifraga columbiana*
Titmouse, tuffed	*Parus bicolor*
chickadee, black capped	*Parus atricapillus*
mountain	*Parus gambeli*
Nuthatch, white-breasted	*Sitta carolinensis*
red-breasted	*Sitta canadensis*
pigmy	*Sitta pygmaea*
Creeper, brown	*Certhia familiaris*
Wren, Carolina	*Thyrothorus ludovicianus*
winter	*Thoglodytes troglodytes*
house	*Thoglodytes aëdon*
Catbird	*Dumetella carolinensis*
Thrasher, brown	*Toxostoma rufum rufum*
Robin	*Turdus migratorius*
Townsend's solitaire	*Myadestes townsendi*
Thrush, wood	*Hylocichla mustelina*
hermit	*Hylocichla guttata faxoni*
olive-backed	*Hylocichla ustulata*
Swainson's	*Hylocichla ustulata*

COMMON NAMES	SCIENTIFIC NAMES
veery	*Hylocichla fuscescens*
varied	*Ixoreus naevius*
Bluebird, eastern	*Sialia sialis*
mountain	*Sialia currucoides*
Kinglet, golden-crowned	*Regulus satrapa satrapa*
ruby-crowned	*Regulus calendula calendula*
Starling	*Sturnus vulgaris vulgaris*
Vireo, red-eyed	*Vireo olivaceus*
solitary	*Vireo solitarius*
yellow-throated	*Vireo flavifrons*
Warbler, black and white	*Mniotilta varia*
Wilson's	*Wilsonia pusilla pusilla*
Backman's	*Vermivora bachmanii*
Nashville	*Vermivora ruficapilla ruficapilla*
yellow-throat	*Gedthlypis trichas*
chestnut-sided	*Dendroica pensylvanica*
cerulean	*Dendroica cerulea*
black-throated blue	*Dendroica coerulescens*
Audubon's	*Dendroica auduboni*
black-throated green	*Dendroica virens*
Townsend's	*Dendroica townsendi*
Hermit	*Dendroica occidentalis*
yellow	*Dendroica petechia*
Kirtland's	*Dendroica Kirtlandii*
Chat, yellow-breasted	*Icteria virens virens*
Ovenbird	*Seiurus aurocopillus*
water-thrush, northern	*Seiurus noveboracensis*
Louisiana	*Seiurus motacilla*
Redstart, American	*Septophaga ruticilla*
Lark, meadow	*Sturnello magna*
Oriole, Baltimore	*Icterus galbula*
Tanager, scarlet	*Piranga olivacea*
western	*Piranga ludoviciana*
Cardinal	*Richmondena cardinalis*
Grosbeak, rose-breasted	*Pheucticus ludoviciarus*
pine	*Pinicola enucleator leucura*
evening	*Hesperiphona vespertina vespertina*
Pine siskin	*Spinus pinus pinus*
Goldfinch	*Spinus tristis tristis*
Indigo bunting	*Passerina cyanea*
Purple finch	*Carpodacus purpureus purpureus*
Cassin's	*Carpodacus cassinii*
Red crossbill	*Loxia curvirostra*
Towhee	*Pipilo erythrophthalmus*

COMMON NAMES	SCIENTIFIC NAMES
Sparrow, chipping	*Spizella passerina passerina*
song	*Melospize melodia*
Lincoln's	*Melospize lincolnii*
white-throated	*Zonotrichia albicollis*
fox	*Passerella iliaca iliaca*
Junco, Oregon	*Junco oreganus*
grey-headed	*Junco caniceps*
slate-colored	*Junco hyemalis*

[a]Names based on those used by H.H. Collins. 1959. *Complete Field Guide to American Wildlife*. Harper, New York; R.T. Peterson. 1947. *A Field Guide to the Birds*. Houghton Mifflin, Boston; R.T. Peterson. *A Field Guide to Western Birds*. Houghton Mifflin. Boston. W.H. Burt and R.P. Grossenheider. 1976. *A Field Guide to the Mammals*, 3d ed. Houghton Mifflin, Boston.

Glossary

ABIOTIC: The nonliving elements (factors) of the environment, that is, soil, climate, physiography.

ABUNDANCE: Ratio of number of individuals to the number of quadrats in sample, that is, mean number of individuals of a species per quadrat on which the species occurs. Abundance: frequency ratio $= \dfrac{100D}{f^2}$, where

$$D = \text{density} \quad = \frac{\text{Total number of individuals}}{\text{Total number of quadrats}}$$

$$F = \text{frequency} = \frac{\text{Number of occupied quadrats}}{\text{Total number of quadrats}} \times 100$$

AGE, STAND: The average age of the trees that compose a stand. In practice, applied to even-aged stands by obtaining the average age of representative dominants. See EVEN-AGED and UNEVEN-AGED.

AGE, TREE: The number of years elapsed since the germination of the seed, or the budding of the sprout or root sucker.

ALL-AGED: Applied to a stand in which theoretically trees of all ages up to and including those of the felling age are found. See EVEN-AGED and UN-EVEN-AGED.

ALLELE: One of a pair of genes located at the same locus in homologous chromosomes and controlling the same character in a diploid individual.

ALLELOPATHY: The suppression of germination, growth, or the limiting of the occurrence of plants as the result of the release of chemical inhibitors by some plants.

ANATOMY: The study of the intimate structure and form of plants; the study of the formation and arrangement of the parts of plants; the study of minute structures, histology.

ARTHROPODS: Members of a group of small soil animals including crustaceans, mites, spiders, ants, and insects.

ASEXUAL REPRODUCTION: Reproduction without fertilization which includes various forms of vegetative reproduction, layering, stump sprouts, root sprouts, also includes grafting and reproduction by cuttings. See SPROUT and LAYER.

ASCOMYCETES: A group of fungi producing a saclike ascus in which ascospores are borne. This group contains some of the most destructive fungi, but few cause wood decay. Examples: chestnut blight, nectria canker, larch canker, needle cast, and blights of conifers.

ASPECT: The direction toward which a slope faces.

ASSOCIATION: A unit of vegetation essentially uniform in general appearance, ecological structure, and floristic composition; an ecological unit smaller than a plant formation.

ATMOMETER: An instrument for measuring evaporation, esp. the Livingston porous cup style.

AUTECOLOGY: The relations of individual plants to their habitats, the relations and adaptation of individual species to their environments.

AVAILABLE SOIL WATER: The amount of water retained in a soil between field capacity and the permanent wilting percentage. This should not be confused with amount of water available for good growth which can be substantially less. Available water is that portion of soil capillary water that contributes to growth and survival of plants.

AUXIN: A growth regulator that promotes cell elongation. See HORMONE.

BARK: Tissues of a woody stem outside the vascular system. Inner bark: the physiologically active layer of tissue between the cambium and the lastformed periderm. See Phloem. Outbark: The layer of dead tissue, of a dry corky nature, outside the last-formed periderm.

BASIDIOMYCETES: A group of fungi containing the wood and root rots, and the rusts, that forms a club-like structure on which spores are borne after nuclear fusion and meiosis. Some fungi in the group are responsible for the formation of root mycorrhiza.

BIOMASS: The quantity of biological matter present on a unit area. The total may be separated into plant and animal mass, which may be further divided into the mass of standing crop of the tree portion of a stand, that is, total biomass above ground standing crop separated into foliage, branch, stem, and flowers.

BIOMES: Biotic communities, vegetative formations that represent communities similar in climate, soil, and vegetative life form. Includes all successional and climax communities. Principal terrestrial biomes are temperate deciduous forests, conifers forest, woodland, chaparral, tundra, grassland, tropical savanna, and broad-leaved evergreen. See FORMATION.

BIOTIC FACTORS: The relation of organisms to each other from an ecological view point (syn. biota, the flora and fauna of a region or ecosystem).

BOLE: The trunk of a tree. It may extend to the top of the tree as in some conifers, or it may be lost in the ramification of the crown, as in deciduous species (syn. trunk, stem).

BRANCH: The extension of a bud that contains apical and lateral buds and a lateral meristem. See STEM.

BROWSE: Leaves, small twigs, and shoots of shrubs, seedling, and sapling trees, and vines available for forage for livestock and wildlife.

CAMBIUM: A thin layer of longitudinally dividing cells between the xylem and phloem which gives rise to secondary growth. The lateral meristem.

CANKER: A definite, relatively localized, necrotic lesion primarily of the bark and cambium.

CAPACITY, CARRYING: In wildlife management, the optimum density of game that a given environment or range is capable of sustaining permanently.

CAPILLARY WATER: Water that fills the smaller pores less than 0.05 mm in diameter and that by adhesion to the soil particles and cohesion of the water molecules themselves can resist the force of gravity and remain suspended in the soil. This water constitutes the major source of water for tree growth except in soils having a high water table.

CATION-EXCHANGE CAPACITY: Total capacity of soil colloids for holding cations. The order of activity of the most important ions in diminishing order is H^+ Ba^+ Ca^{++} Mg^{++} K^+ NH_4^+ Na^+. The atomic weight of Ca is 40, its valence

2, for one milliequivalent of Ca = (40/2) × 0.001 = 0.020 gm. A soil with a capacity for 5 m.e./100 gm. will hold in replaceable form 0.020 × 5 = 0.1 gm of Ca/100 gm of soil. A ½ foot of soil over one acre weighing 2,000,000 pounds would absorb 200 pounds of Ca.

CARNIVORE: An organism that consumes mostly flesh. See OMNIVORE and HERBIVORE.

CLAY: Soil particle fraction below 0.002 mm in size.

CLEARCUTTING: An area on which the entire timber stand has been cut. Removal of the entire stand in one cut. Reproduction obtained with or without planting or artificial seeding.

CLIMATE: Meteorological phenomena of an area characterized by the patterns of temperature, precipitation, wind and other weather over a period of time.

CLIMAX: The final stage of a succession that continues to occupy an area as long as climate and soil conditions remain unchanged: climatic climax a climax in which climate is the controlling factor. A plant community that has reached a relatively stable condition in which it is able to reproduce itself indefinitely under existing conditions: a mature plant community.

CLIMOGRAPH: An expression of the climatic conditions of a particular area by a graph of which one coordinate is the mean monthly temperature and the other the mean monthly precipitation (syn. hytherograph).

CLINE: A geographic gradient usually assumed to be genetically controlled and resulting from adaptation to change in temperature or moisture, or both.

CLONE: All plants reproduced asexually from a common ancestor (ortet) and having identical genotypes, named clones, are given non-Latin names preceded by the abbreviation "cl."

CODOMINANT, TREES: Trees with crowns forming the general level of the crown cover and receiving full light from above, but comparatively little from the sides, usually with medium-sized crowns more or less crowded on the sides.

COMBINING ABILITY: The relative ability of an organism to transmit genetic superiority to its offspring.

COMBINING ABILITY, GENERAL: The relative ability of an organism to transmit genetic superiority to its offspring when crosed with other organisms in general. High combining ability usually implies the presence of genes with additive effects.

COMBINING ABILITY, SPECIFIC: The relative ability of an organism to transmit genetic superiority to its offspring when crossed with specific other individuals. High specific combining ability usually implies the presence of dominance, overdominance, or epistasis.

COMMUNITY: A group of plants growing together (syn. stand), or all of the plants and animals of an area (syn. ecosystem).

COMPENSATION POINT: The light intensity or temperature at which the rates of photosynthesis and respiration are equal.

COMPLETE FLOWER: A flower containing sepals, petals, stamens, and at least one pistil.

COMPOSITION: The relative proportions of the various species included in the total cover on a given area. Applied separately for forest, brush, and range plants.

COMPRESSION WOOD: Abnormal wood formed on the lower side of branches and inclined boles of conifer trees. See TENSION WOOD.

CONIFER: A tree belonging to Division Coniferophyta, usually evergreen with cones and needle-shaped leaves, and producing wood known commercially as "soft wood."

CONSORTING: A species of trees that, although it may be found in pure stands, is most often found as a major segment in a mixed stand, that is, one of the more abundant species.

CONTINUUM: The occurrence of tree species in a region in a continuously shifting series of combinations with a definite sequence or pattern, the resultant of a limited floristic complement acting on, and acted upon by, a limited range of abiotic factors. Such a gradient of communities is called a vegetative continuum.

COPPICE: A newly cut-over area regenerated primarily by sprouts. A method of renewing the forest in which reproduction is by sprouts.

COTYLEDON: The seed leaf, a leaf-like organ within the seed, a leaf-like structure folded within the seed in which food for the new plant is usually stored. The number of cotyledons in a seed is the basis for the primary divisions of the seed plants into monocotyledons and dicotyledons.

COVER TYPE: See FOREST COVER TYPE.

CROSSOVER: Breakage at the same locus on homologous chromosomes and the subsequent exchange between homologous chromosomes at the metaphase stage of meiosis.

CROWN: The branches and foliage of a tree; the upper portion of a tree. The leaves as foliage are an outgrowth of the vascular system, and are mainly concerned with photosynthesis. The branches join the stem or other branches. See Deliquescent and Excurrent.

CROWN CLASSIFICATION: See DOMINANT TREE, CODOMINENT TREE, INTERMEDIATE TREE, OVERTOPPED TREE.

CROWN FIRE: A fire that runs through the tops of living trees or brush.

CULTIVAR: Cultivated variety or strain.

DECIDUOUS: Broad-leaved plants that drop their leaves at the end of each growing season. See HARDWOOD.

DELIQUESCENT (DECURRENT): To ramify into fine divisions, as the veins of a leaf or the trunk or branches of a tree. Having a large number of branches or branching so that the stem is lost in the branches, as maple or oak stems.

DENSITY, STAND: Density of stocking expressed in number of trees, basal area, volume, or other criteria on an acre or hectare basis. See STOCKING.

DETRITUS: Fresh dead or partially decomposed organic matter.

DEUTEROMYCETES: A large miscellaneous, artificial group of fungi in which sexual reproduction does not occur or has not been found. Contains most of the wilts and some damping-off fungi.

DEW POINT: The temperature at which the relative humidity becomes 100%, that is, the air is water saturated. See RELATIVE HUMIDITY.

DIAMETER BREAST HEIGHT: (dbh): The diameter of a tree at 4.5 feet (1.3 m) above average ground level. Additional abbreviations o.b. and i.b. designate whether the diameter refers to measurements outside or inside the bark.

DIFFUSE-POROUS WOOD: Wood in which the pores are of fairly uniform or of only gradually changing size and distribution throughout the growth ring. See RING-POROUS WOOD and POROUS WOOD.

DIOECIOUS: Producing male and female flowers on different plants.

DIPLOID: (a) having $2n$ or two sets of homologous chromosomes. (b) Pertaining to the chromosome number of vegetative rather than gametic tissue. (c) The result of mitotic cell division. See HAPLOID.

DISEASE (NONINFECTOUS): An injurious unbalancing of normal functions as a result of injury from atmospheric phenomenon or mechanically by animals, (including man), and of sufficient duration or intensity to cause disturbance or cessation of vital activity.

DISEASE (PATHOGEN): An injurous unbalancing of normal functions as a result of attacking by bacteria, virus, fungus, or other plant, and of sufficient duration or intensity to cause disturbance or cessation of vital activity.

DOMINANCE: The relative basal area of a species to the total basal area of all species in a statum. The species having the highest relative basal area is considered the dominant species (syn. predominant).

DOMINANT GENE: A gene that prevents its allele from having a phenotypic effect. See RECESSIVE GENE.

DOMINANT TREES: Trees with crowns extending above the general level of the crown cover and receiving full light from above and partly from side; larger than the average trees in the stand, and with crowns well developed but possibly somewhat crowded on the sides.

DORMANCY, SEED: State where the seed embryo is incapable of growth; a state when the metabolic processes are slowed. Especially applies to respiration (syn. after-ripening).

DORMANCY, VEGETATIVE: A physical state when the metabolic processes are slowed, with deciduous plants the period without leaves.

DUNNING TREE CLASSIFICATION: A system of classifying trees into seven classes according to age (stages or maturity), dominance (crown class), crown development, and vigor; originally designed for use in the selection forests of the Sierra Nevada mountains.

EARLY WOOD: The less dense, larger-celled, first formed part of a growth layer along the trunk and branches of a tree (syn. spring wood).

ECOTONE: A transition between two or more diverse communities, such as that between the grassland and forest biomes. It is a tension zone containing overlapping communities of each community; there may be species present in the ecotone not found in either of the bordering communities.

ECOTYPE: A subdivision of a species resulting from the selective action of a particular environment and showing adaptation to that environment. The word ecotype carries connotations of difference and of adaptation, whereas the word race carries a connotation of difference, but not necessarily of adaptation. Ecotypes may be geographic, climatic, elevational, or soil.

ECOSYSTEM: The entire system of life and its environmental and geographical factors that influence all life, including the plants, animals, and the environmental factors.

EMBRYO: That portion of a seed resulting from the union of female and male gametes and developing into a mature plant.

ECOLOGY: The science that deals with the relation of plants and animals to their environment and to the site factors that operate in controlling their distribution and growth.

ECTOMYCORRHIZA: Subsisting on the surface, that is, ectomycorrhizal fungi live on the surface of tree roots, attached to the epidermal cell layer, and penetrating intercellular space of cortical cells. See ENDOMYCORRHIZA.

EMMIGRATION: Movement out of an area.

ENDOGENOUS: The development within (Endogenesis): (1) forming new tissue within the plant; (2) originating from internal tissues; (3) forming inside another organ of a plant. See EXOGENOUS.

ENDOSPERM: Food storage tissue contained in the seed and surrounding the embryo. The endosperm is small or aborted in most hardwood tree seeds. Endosperm has $3n$ chromosomes ($2n$ from the female and $1n$ from the male) in angiosperms and $1n$ chromosomes (all from female) in gymnosperms.

ENDOMYCORRHIZA: Subsisting within the interior of the plant cell, that is, endomycorrhizal fungi penetrate the interior of tree cortical root cells. See ECTOMYCORRHIZA.

ENZYME: An organic catalyst that is produced by living cells which bring about specific (usually intracellular) transformations. Each enzyme catalyzes only one reaction or type of reaction within a very limited range.

EPICORMIC BRANCH: A shoot arising from a dormant bud on a bole. See SPROUT.

EPISTASIS: The interaction of non-alleles within an individual. The dominance of nonalleles.

EVEN-AGED: Applied to a stand in which relatively small age differences exist between individual trees. The maximum difference in age permitted in an even-aged stand is usually 10 to 20 years, though where the stand will not be harvested until it is 100 or 200 years, larger differences of up to 25% of the rotation age may be allowed. See UNEVENAGED.

EXCURRENT: Stem of a tree that extends from base to tip without dividing, as spruce or hemlock stems.

EXOGENOUS: (1) Produced on the outside of another body. (2) Produced externally, as spores on the tips of hyphae. (3) Growing by outer additions of annual layers, as the wood in dicotyledons. See ENDOGENOUS.

FACE-OF-TREE: In log-grading and sawing, one-fourth of the circumference of a log for its entire length.

FAMILY: The offspring of a single tree after open pollination or of a single tree after controlled pollination. *Half-sib*, offspring of a single (usually female) tree and having different parents of the other sex. *Full-sib*-offspring of a single pair of trees, usually resulting from controlled pollination.

FERTILIZATION: The fusion of two gametes to form a zygote. See GAMETE and ZYGOTE. The application of organic or inorganic fertilizers to a forest stand.

FIELD CAPACITY: Condition exists when free drainage of water diminishes and water movement slows, and moisture content does not change appreciably between measurements. In laboratory analysis, equilibrium is obtained at approximately −0.1 to −0.3 bars.

FLEDGE (FLEDGLING): To acquire the feathers necessary for flight, for example, a young bird just fledged.

FOOD CHAIN: Designates the feeding of one group of organisms upon another. At the base are the autotrophic plants (green plants) and along the way are various heterophopic organisms (plants and animals). These latter may be plants (fungi) or animals as plant eaters (herbivous); plant and animal eaters (omonivous); animal feeders (carnivores). See TROPHIC.

FOREST COVER TYPE: A descriptive term used to group stands of similar character as regards composition and development due to certain ecological factors, by which they may be differentiated from other groups of stands. The term suggests repetition of the same character under similar conditions. A type is *temporary* if its character is due to passing influence such as logging; *permanent* if no appreciable change is expected and the character is due to ecological factors alone; *climax* if it is the ultimate stage of a succession of temporary types. A *cover* type is a forest type now occupying the ground, no implication being conveyed as to whether it is temporary, permanent, or climax.

FORMATION: A fully developed climax community representative of a regional climate, includes the developmental stages within its borders, and includes a number of *Associations*. See BIOMES.

FREQUENCY: The measure of the percentage of a species occurrence in a number of equal sized sample areas or quadrats within a community. It is a statistical concept concerned with the uniformity of species distribution within a community.

FROST HEAVING: Lifting of seedlings or small plants above their normal soil level as a result of expansion accompanying ice formation in frozen soil.

FROST POCKET: A basin, or an area immediately above a constriction in a valley, where cold air drainage is prevented with the resulting accumulation of cold air to the extent that it exerts a profound influence upon plant life. An area where cold air is prevented from further movement.

GAMETE: A haploid generation male pollen cell or a female egg cell capable of developing into a zygote (embryo) after fusion with a germ cell of the opposite sex.

GENOTYPE: (a) The entire genetic constitution of an organism, expressed or latent. (b) The genetic constitution of an individual with respect to a few genes under consideration.

GRADIENT ANALYSIS: Abundance and importance value of plant species along an environmental gradient, for example, increase or decrease in the number of individuals of a species or race that occurs along a temperature gradient.

GRASS: A member of the Division Anthophyta; Class *Monocotyledonae*.

GRAVITATIONAL WATER: Water that fills the noncapillary pores of the soil larger than 0.05 mm in diameter; which drain readily under the influence of gravity; which are filled with air unless the soil has a high watertable.

GREGARIOUS: Species occurring in groups within a community and which may form pure stands of species in the natural state.

GROUND FIRE: A fire that not only consumes all the organic materials of the forest floor, but also burns into the underlying soil itself, for example, a peat fire. (Usually combined with but not to be confused with a surface fire.)

GROUND WATER: Subsurface water occupying interstices in a zone of saturation. See WATER TABLE.

GUARD CELL: One of two epidermal cells in a plant leaf or needle that encloses a stoma. The stoma when open permit CO_2 and O_2 to enter the leaf interior and H_2O to evaporate.

HABITAT: The immediate environment occupied by an organism, that is, in forestry, habitat usually refers to the animal habitat. See SITE.

HAPLOID: Having one set ($1n$) of chromosomes, as in a gamete, and as a result of meiosis. See DIPLOID.

HARD PAN: An indurated (hardened) or cemented soil horizon. The soil may have any texture and is compacted or cemented by iron oxide, organic material, silica, calcium carbonate, or other substances.

HARDWOOD   (deciduous broad leaf): Generally one of the botanical group of trees that have broad leaves, in contrast to the conifers. Member of the Division Anthophyta; Class *Dicotyledonae*.

HEARTWOOD: The inner core of a woody stem, wholly composed of nonliving cells and usually differentiated from the outer enveloping layer (sapwood) by its darker color. See SAPWOOD.

HERB: A flowering plant in which the stem does not become woody or persistent, but dies annually, or, after flowering, dies down to the ground; it includes both forbs and grasses. May be perennial. See SHRUB, TREE, and GRASS.

HERBIVORE: An organism that consumes living plants or their parts. See OMNIVORE and CARNIVORE.

HERITABILITY: That portion of the total variation in a trait due to genetic factors. In a broad sense the portion of phenotypic variation due to all genetic factors. In a narrow sense that portion of total phenotypic variation due only to the additive genetic effects and most indicative of the superiority that can be transmitted by seed.

HETEROSIS: Hybrid vigor. See COMBINING ABILITY, SPECIFIC.

HETEROZYGOUS: Possessing different alleles at a locus, originating from a union of gametes of different genotypes.

HOMOZYGOUS: Possessing the same alleles at a locus, originating from a union of gametes of the same genotype.

HORMONE, PLANT: An internal secretion (such as auxins) formed in the actively growing parts of plants that diffuse to other tissues and regulate and influence development of the other tissues.

HUMUS: The lower part of the litter layer consisting principally of amorphous organic matter, located immediately above the A horizon. A complex colloidal mixture.

HYBRID: Offspring of organisms of dissimilar genotype, often but not confined to a cross between different species, for example, crosses among races, varieties, subspecies. See INTERSPECIFIC HYBRID, INTRASPECIFIC HYBRID, INTROGRESSION.

HYDRIC: Tending to be wet. Hydrophytic plants grow in water or in very wet soils. See MESIC and XERIC.

HYDROLOGIC CYCLE: The movement of water in nature through a complete cycle, commencing as atmospheric water vapor, passing into liquid and solid form as precipitation, then into the ground surface or as run-off or ground water to larger bodies of water from which places, by transpiration and evaporation, water returns as vapor to the atmosphere.

HYPHA: The simple or branched threadlike filament that compose the weblike mycelium of fungi that develops by apical growth and usually becomes transversely septate as it develops.

HYPOCOTYL: The short stem of an embryo seed plant, the portion of the axis of the embryo seedling between the attachment of the cotyledons and the radicle.

IMMIGRATION: Movement into an area.

IMPERFECT FLOWER: A flower that has stamens or pistils but not both.

IMPORTANCE VALUE: Obtained by adding together relative frequency (frequency of species as a percentage of total frequency values of all species), relative density (number of individuals of species as a percentage of total number of plants), and relative dominance (basal area for species as percentage of total basal area).

INBREEDING: The breeding of closely related individuals; if carried through enough generations the families can become homozygous, that is, have the same genotypes within their chromosomes.

INTERMEDIATE TREES: Trees shorter than those in the two upper classes (dominant, codominant), but with crowns either below or extending into the crown cover formed by codominant and dominant trees, receiving a little direct light from above, but none from the sides; usually with small crowns considerably crowded on the sides.

INTERNODE: A region of the stem between two successive nodes. See NODE.

INTER-SPECIFIC HYBRID: Cross between individuals of different species. Taxonomically identified by listing both species separated by an "x."

INTRA-SPECIFIC HYBRID: Cross between individuals within the same species, but of different genotypes, for example, cross between suspected races or ecotypes.

INTROGRESSION, HYBRIDIZATION: Long-continued interspecific hybridization leading to an infiltration of genes from one species into another.

INVERSION: See TEMPERATURE INVERSION.

ISOENZYME: Analysis of plant proteins by substituting known enzymes to a plant substrate to determine whether a reaction will occur. To determine presence of various enzymes within plant tissue.

KEEN TREE CLASSIFICATION: A system of classifying ponderosa pine into 16 classes according to four age groups (stages of maturity) and four crown vigor groups (determined by crown size, dominance, and thrift).

LAKE STATES TREE CLASSIFICATION: A system of classifying trees in even-aged stands according to position in the stand, crown density, soundness, form, and utility (pulpwood, sawtimber, etc.).

LATE WOOD: The denser, smaller-celled, later-formed part of a growth layer along the trunk and branches of a tree (syn. summerwood).

LATERAL ROOT: A root that is confined to a horizontal growth pattern in the soil following the surface soil horizons or if developed at a lower level continues a lateral growth. See TAP ROOTS.

LAW OF THE MINIMUM: Liebig's law of the minimum: "When a process is conditioned as to its rapidity by a number of separate factors, the rate of the process is limited by the pace of the slowest factor." Mitscherlich's restatement: "The increase of crop produced by unit increment of the lacking factor is proportional to decrement from the maximum."

LAYER: To reproduce by the rooting of a branch that is first connected with the parent tree.

LEAF DIMORPHSISM: Difference between early leaves—leaves produced from previous years buds—and late leaves—leaves produced by current season buds; late season leaves are usually larger. Present in sweet gum, yellow-poplar, aspen, cottonwood, birch.

LEAF GAP: An area in the vascular region of a stem where parenchyma instead of vascular tissue differentiates; located immediately above a *leaf trace* where a vascular bundle connects the vascular system of a leaf with that of the stem. The *leaf bud* develops into a stem with leaves.

LENTICLE: A small breathing pore in the bark of trees and shrubs, a corky aerating organ that permits gases to diffusion between the plant and the atmosphere.

LINKAGE: The association of characters from one generation to the next due to the fact that the genes controlling the characters are located on the same chromosome. Linkage group. The genes located on a single chromosome or the characters controlled by such genes.

LIVE-CROWN RATIO: Proportion of total tree height consisting of living crown. Expressed as proportion, that is, $\dfrac{\text{Length of live crown}}{\text{Total tree height}}$

LIVESTOCK: Domestic animals kept primarily for farm, ranch, or market purposes, including beef and dairy cattle, hogs, sheep, goats, and horses.

LONG-SHOOTS: Bear lateral buds some of which develop into long-shoots; usually develop from a terminal bud.

MAST: Nuts and seeds accumulated on the forest floor and often serving as food for livestock and wildlife, esp. nuts of oak and beech, and seeds of certain pines as longleaf and pinyon.

MEIOSIS: A form of cell division in which chromosome number is reduced from $2n$ to $1n$. The form of cell division giving rise to gametophytes from diploid tissue.

MERISTEM: Any terminal (apical) or lateral growth zone of a tree, that is, the cambium zone, buds, and root tips of living trees.

MESIC: Moderately moist where dry periods are not of long duration. Mesophytes are plants which grow in soils that contain moderate amounts of moisture. See HYDRIC and XERIC.

MESOPHYLL: The parenchyma tissue between the upper and lower epidermis of a leaf; the cells usually contain chloroplasts. See PALISADE MESOPHYLL.

MEGASPOREGAMETOPHYTE: The few-celled haploid generation portion of a seed plant arising from a meiotic division and giving rise through meitosis to the female gametes. Female inflorescence.

MICROPHYLE: A canal into the nucleus formed by an outgrowth of the integuments. It remains a minute pore in the testa of the seed through which water passes prior to germination, and from which the radicle emerges.

MICROSPOREGAMETOPHYTE: The few-celled haploid generation portion of a seed plant arising from a meiotic division and giving rise through meiosis to the male gametes. Male inflorescence.

MINERAL CYCLE: The movement of a mineral ion in nature through a complete cycle-uptake, retention, restitution, and leaching within the soil-plant system.

MINERAL ION: Electrically charged atoms or groups of atoms formed by disassociation of a molecule, and the form by which most mineral elements enter plant roots. Ions move from outside into the plant root cells as a result of ionic diffusion (syn. essential element). See CATION-EXCHANGE CAPACITY.

MINERAL SOIL HORIZON: Soil horizon containing less than 30% organic matter if the mineral contains more than 50% clay or less than 20% organic matter if the mineral fraction has no clay. Intermediate clay content requires proportional contents of organic matter. Usually constituted by the A, B, and C master horizons.

MITOSIS: The form of cell division in which chromosome number is not reduced. Each daughter cell receiving exactly the same number of chromosomes—$2n$ in vegetative cells or can involve $1n$ gametophytic cells.

MIXED STAND: A stand in which less than 80% of the trees in the main canopy are of a single species.

MONOECIOUS: Bearing separate male and female flowers on the same tree, or complete flowers.

MOR HUMUS: A type of forest humus layer of unincorporate organic material, usually matted or compacted or both, distinctly delimited from the mineral soil, unless the latter has been blackened by washing in organic matter (syn. raw humus). See MULL HUMUS.

MORPHOLOGY: The study of form and structure of plants or other biological organisms. The study of the external shape in contrast to function. Morphological species: A species based upon change in morphology in relation to varied environments.

MULL HUMUS: A type of forest humus layer consisting of organic and mineral matter so mixed that the transition to the underlying mineral layer is not sharp. See MOR HUMUS.

MUTATION: A sudden change in genotype. Usually a "gene" mutation is meant, but the term is sometimes used in a broader sense to include changes due to polyploidy, chromosome deletions, chromosome inversions, or chromosome duplication.

MYCELIUM: A collective term for the vegetative hyphae of a fungus.

MYCORRHIZAE: A symbiotic association between nonpathogenic or weakly pathogenic fungi and living cortical cells of a plant root. Mycorrhizae form on unthickened roots of trees and other plants. See ENDOMYCORRHIZAE and ECTOMYCORRHIZAE.

NAVAL STORES: Products derived from the oleoresins of certain trees, mainly terpentine and resins, and obtained primarily from longleaf and slash pines.

NICHE: A term used to describe the status of a plant or animal in its community, that is, its biotic, trophic, and abiotic relationships. All the components of the environment with which an organism or population interacts.

NITROGEN CYCLE: The circulation of nitrogen in nature. Dead organic matter is converted to ammonium compounds during decay in the soil. These are converted to nitrates, through nitrification as a result of bacterial action, which can be utilized again by plants. *Nitrosomonas* oxidize ammonium ions to nitrites, and *Nitrobacter* oxidize nitrites to nitrates.

NODE: The region of the stem where one or more leaves are attached. See INTERNODE.

NODULE: Enlargements on the roots of certain plants within which are found masses of nitrogen-fixing bacteria on legumenous plants, and actinomycetes on non-legumes.

NON-POROUS WOOD: Wood devoid of pores or vessels. Featured by all conifers and a few hardwood species. See POROUS WOOD.

OCCASIONAL: A species found occasionally in a given plant community, but is not always a regular member of that community, nor does it occur in large numbers.

OMNIVORE: An organism whose diet is broad, including both plant and animal food. See CARNIVORE and HERBIVORE.

ORGANIC SOIL HORIZON: A surface horizon of a soil formed from organic litter derived from plants and animals, and designated as the 0 horizons. These horizons do not include horizons formed by illuviation of organic material into mineral material, nor do they include horizons high in organic matter formed by a decomposing root mat below the surface of a mineral material, usually designated as 0_1 and 0_2, however, L and F can be used to designate the 0_1 and H the 0_2 layer.

ORGANIC SOIL: Soils having organic soil materials that extend from the surface to: (a) a depth within 10 cm or less of a lithic or paralithic contact, provided the thickness of the organic soil materials is more than twice that of the mineral soil above the contact; or (b) any depth if the organic material rests on fragmental material and the interstices are filled with organic materials; or (c) have organic materials that have an upper boundary within 40 cm of the surface; (i) having a bulk density 0.1 gm/cubic centimeters; (ii) the organic soil is saturated with water 6 mos. of the year.

ORTET: The original ancestor of a vegetatively propagated clone. See CLONE.

OVERSTORY (overwood): That portion of the trees in a forest stand forming the upper crown cover. See STRATIFIED and UNDERSTORY.

OVERTOPPED TREES: Trees with crowns entirely below the general level of the crown cover receiving no direct light either from above or from the sides (syn. suppressed).

PALISADE MESOPHYLL: A leaf tissue composed of slightly elongated cells containing chloroplasts; located just beneath the upper leaf epidermis, and above the spongy mesophyll in broad leaved plants, and some conifers.

PARCH BLIGHT: A condition in conifers where the foliage turns brown in late winter or early spring as a result of dessication due to low-temperature injury (syn. winter-kill, physiological drought).

PERCHED WATER-TABLE: Presence of water in a soil horizon as a result of poor internal drainage, or due to topographic position, or impermeable soil layer. Occurs during season of high precipitation or late winter snowmelt.

PERMANENT WILTING PERCENTAGE: It is the soil moisture content at which plants remain permanently wilted unless water is added to the soil. Soil water potential at wilting can vary from -5 to -200 bars. Because of the shape of the water potential-water content drying curve, large changes in water potential at higher tensions accompany minor decreases in water content, so permanent water for plant growth is approximately 15 bars.

PHENOTYPE: The visible characters of a plant. The product of a genotype and its environment.

PHLOEM: Inner bark. The principle tissue concerned with the translocation of elaborated food produced in the leaves, or other areas, downward in the branches, stem, and roots. Primary phloem external to secondary phloem.

PHOTOSYNTHESIS: The manufacture of food, mainly sugar, from carbon dioxide and water in the presence of chloroplasts utilizing light energy and releasing oxygen.

$$6\ CO_2 + 12\ H_2O \xrightarrow[\text{light energy}]{\text{chloroplasts}} C_6H_{12}O_6 + 6O_2 + 6H_2O$$

Carbon dioxide furnishes the O_2 produced as a result of photosynthesis.

PHYCOMYCETES: Algae fungi bearing nonsepate, branching filaments not organized into compact bodies of definite form. Containing several damping-off fungi.

PHYSIOGRAPHIC LOCATION: The location of a stand with respect to slope aspect, slope position, and slope inclination, and the slope conformation relative to the overall terrain.

PHYSIOLOGY: The study of the vital functions of a plant or animal as a living organism, for example, metabolic activity or vital function.

PLUMULE: The terminal bud of an embryo in seed plants. It is a rudimentary shoot.

POLE: Begins at end of sapling stage and ends when trees reach 8–9 inches dbh (or whenever sawtimber is measured); during this stage height growth predominates and economic bole length is attained. See SEEDLING, SAPLING, and SAWTIMBER.

POLYPLOID: Having more than twice the basic number (n) of chromosomes of the ancestral species in its vegetative cells. A cell, tissue, or organism having three sets ($3n$) is triploid, four sets ($4n$) tetraploid, and so forth. See DIPLOID, HAPLOID.

POROUS WOOD: Wood with pores or vessels. Featured by nearly all hardwood species. See NON-POROUS WOOD, DIFFUSE POROUS WOOD, RING POROUS WOOD, and SEMI-RING POROUS WOOD.

PRECLIMAX: The state of vegetation preceding the development of a climax, a more xeric community, that is, usually drier and/or hotter than environment of contiguous climax.

POST CLIMAX: The state of vegetation more mesic than the developed climax usually cooler and/or moister than contiguous climax.

POSTNATAL: Subsequent to birth, relating to an infant immediately after birth.

PRIMARY SUCCESSION: A plant succession beginning on land that has not borne vegetation in its present form, for example, from bare soil.

PROVENANCE: The ultimate natural origin of a tree or a group of trees (syn. geographic origin). Trees having a common center of origin.

PURE STAND: A stand in which at least 80% of the trees in the main crown canopy are of a single species.

RACE: A genetic subdivision of a species having distinctive characteristics when grown in a particular environment. May be used to identify an ecotype, or a provenance. Differences among races may or may not be adaptive, especially when used to identify a portion of a cline. A taxonomic variety or subspecies and so recognized.

RADICLE: The embryonic root of seed-plants.

RAMET: An individual member of a clone, descended through vegetative propagation from an ortet. See ORTET.

RECESSIVE, GENE: A gene without phenotypic effect when present in the heterozygous state. See DOMINANT GENE.

RECOMBINATION: Obtaining new combination of genes through independent segregation of chromosomes at meiosis and through cross overs.

RED BELT: Winter drying confined to altitudinal zones; it approximately follows the topographic contours. See PARCH BLIGHT.

RELATIVE HUMIDITY: The percent saturation of an air mass. 100% minus relative humidity equals the saturation deficit in percent.

RELATIVE TOLERANCE: See TOLERANCE, RELATIVE.

RESINS: A class of inflammable, amorphous, vegetable substances secreted by certain plants and trees, and characterizing the wood of many coniferous species. They are oxidation or polymerization products of the terpenes, and consist of mixtures of aromatic acids and esters. Produced from oleoresins.

RESPIRATION: A series of complex oxidation-reduction reactions whereby living cells obtain energy through the breakdown of organic material and in which some intermediate materials can be utilized for syntheses of plant products. One form of aerobic respiration is

$$C_6 H_{12}O_6 + 6O_2 \xrightarrow{\text{enzymes}} 6CO_2 + 6H_{20} + 674 \text{ kg-cal energy}$$

RING-POROUS WOOD: Wood in which the pores of one part of a growth ring are in distinct contrast in size or number (or both) to those of the other part. See SEMI-RING POROUS WOOD and POROUS WOOD.

ROSIN: The solid residue obtained after distilling off the turpentine from oleoresin.

ROOT: The lower portion of the axis of a higher plant. It is usually branching and does not bear leaves or buds. It anchors the plant in the soil and by means of root-hairs and growing root tips absorbs water and dissolved minerals.

ROTATION: The period of years required to establish and grow timber crops to a specified condition of maturity.

SAND: Soil particle fraction between 0.05 and 2.0 mm.

SAPLING (thicket stage):   Begins with end of seedling stage and ends when trees reach 4 inches dbh; crowns are well elevated and usually many of lower branches have died. See SEEDLING, POLE and SAWTIMBER.

SAPWOOD: Living, physiologically active wood of pale color; includes the more recent annual layers of xylem that are active in translocation of water and minerals. See HEARTWOOD.

SAWTIMBER: Begins at end of pole stage when height growth falls off and begins the period of maximum diameter growth. Terminates when trees become over-mature and die, or are cut. See SEEDLING, POLE and SAWTIMBER.

SECONDARY SUCCESSION: A succession initiated by an abiotic or biotic agent or agents and after the ground has been cleared of its original vegetation.

SEED TREE CUTTING: Removal of the mature timber in one cut, except for a small number of seed trees left singly or in small groups to provide a source of seed for the next tree crop.

SEEDLING: Youngest trees from time of germination until they reach breast height (4.5 feet), usually accompanied with death of lower branches. See SAPLING, POLE AND SAWTIMBER.

SELECTION, NATURAL: Elimination of trees on the basis of their inability to survive in a particular niche, or the ability of a tree to produce seed and to reproduce in a niche.

SELECTION CUTTING: Removal of mature timber, usually the oldest or largest trees, either as single scattered trees or in small groups at relatively short intervals, commonly 5 to 20 years, repeated indefinitely, by means of which the continuous establishment of natural reproduction is maintained.

SEMI-RING-POROUS WOOD: Wood in which the pore diameter and number diminishes gradually from inner to outer portions of the growth ring. See POROUS WOOD and RING-POROUS WOOD.

SEXUAL REPRODUCTION: Exchange of genetic material between male and female gametes through fusion of the pollen tube with the ovule releasing the male gamete into the egg cell resulting in a zygote.

SHELTERWOOD CUTTING: Removal of the mature timber in a series of cuttings, which extend over a period of years equal usually to not more than one-quarter and often not more than one-tenth of the time required to grow the crop, by means of which the establishment of natural reproduction under the partial shade of seed trees is encouraged.

SHORT-SHOOTS: Lateral shoot that do not produce branches, but produce buds and leaves in successive years. May be auxin controlled, but not proven. Short-shoots bear buds that either are completely inhibited or do not develop, or which develop into flowers (syn. spur-shoots).

SHRUB: A woody perennial plant differing from a perennial herb by its persistent and woody stem, and from a tree by its low stature and habit of branching from the base. See TREE and HERB.

SILT: Soil particle fraction between 0.002 and 0.05 mm.

SILVICULTURE: The art of producing and tending a forest; the application of the knowledge of silvics in the treatment of a forest; the theory and practice of controlling forest establishment, composition, and growth.

SILVICS: The history and general characteristics of forest trees and stands, with particular reference to environmental factors as they affect tree selection, growth, and reproduction, and the growth and development of forest stands.

SINKER ROOT: A root, other than a tap root, that grows straight downward in the soil.

SITE: An area considered as to its ecological factors with reference to capacity to produce forests or other vegetation; the combination of biotic, climatic, and soil conditions of an area. A specific location in the forest.

SITE INDEX: An expression of forest site quality based on the height of the dominant stand at an arbitrarily chosen age.

SITE QUALITY: The relative ability of a site to produce tree growth, that is, site index.

SITE TYPE (indicator plant): The presence of a plant or group of plants that serves to denote what some other species might do if grown in the same place, or to demonstrate that there had been some particular or material change in site, cover, soil, water, or use condition (syn. habitat type).

SOIL: The top layer of the earth's surface, composed of finely divided disintegrated rock containing more or less organic material that is penetrated by the roots of plants and includes the surface soil (A horizon), the subsoil (B horizon), and the upper portion of the substratum (C horizon), and in a forested situation having a layer of organic debris on the surface, $(O_1 - O_2)$, and having in addition to its abiotic component, a biotic element as well.

SOIL BULK DENSITY: Ratio of dry-weight of a soil mass to its volume (syn. volume weight ratio).

SOIL PARTICLE VOLUME: Equals $\dfrac{\text{soil bulk density}}{2.65}$. See SOIL PORE VOLUME.

SOIL PORE VOLUME: Equals $1 - \dfrac{\text{soil bulk density}}{2.65}$, where 2.65 equals the average density of the soil forming minerals.

SOIL TEXTURE: The relative proportion of the various size groups of individual soil particles: fine fraction is silt- and clay-sized particles < 0.05 mm; coarse fraction is the stone, gravel, and sand > 0.05 mm. See SAND, SILT, and CLAY.

SOLUM: The A and B horizons of the soil. See SOIL.

SPROUT: A shoot from a dormant bud at the base of a tree or from an exposed root or stump (syn. root sucker, stump sprout). A stump with one or more sprouts is called a stool.

STAG-HEADED: A tree dead at the top as a result of injury, disease, or deficient moisture or nutrients.

STAND: An aggregation of trees or other growth occupying a specific site and sufficiently uniform in species composition, age arrangement, and density as to be distinguishable from the forest or other growth on adjoining areas.

STAND DENSITY: See DENSITY.

STEM (bole): The main ascending axis of a tree or plant contains the primary and secondary vascular systems and the lateral meristenatic tissue, the cambium, and may or may not have an apical meristen depending on crown form. See BRANCH, EXCURRENT, and DELIQUESCENT.

STOCKING: An indication of the number of trees in a stand, compared to the desirable number for best growth and management, such as well stocked, overstocked, partially stocked, or understocked.

STOMATE (stoma): An opening surrounded by guard cells that opens into an internal air cavity below the epidermus of a leaf. A breathing pore in the epidermis of a leaf.

STRATIFICATION: The burying of dormant seed in a cool moist medium to overcome dormancy, or for artificial storage. See DORMANCY, SEED.

STRATIFIED: Applied to crown cover arranged in layers corresponding to species or age classes, or to size of plant as tree layer, shrub layer, herb layer. See UNDERSTORY and OVERSTORY.

STREAM RUNOFF (water yield): The total outflow of a drainage basin through surface channels, or subsurface channels and surface channels if there are exposed subsurface aquifers.

SUBCLIMAX: Succession stage immediately preceding the climax. This condition may be the result of extremely slow development to climax, or to a continuing disturbance such as fire. Sometimes designated as permanent.

SUBSPECIES: A subdivision of a species, morphologically and genetically distinct from other subspecies.

SUBSTRATE: The material upon which an enzyme or fermenting agent acts; the material in or upon which a fungus grows or to which it is attached, the matrix.

SUBSTRATA: See SUBSTRATE.

SUCCESSION: The progressive development of the vegetation toward its highest ecological expression, the climax. The replacement of one plant community by another. See PRIMARY SUCCESSION and SECONDARY SUCCESSION.

SURFACE FIRE: A fire that runs over the forest floor and burns only the surface litter, the loose debris, and the smaller vegetation.

SYNECOLOGY: The study of plant communities, their taxonomy, life history, and relationship to other biotic communities.

SYNGAMEON: The sum total of species or semispecies linked by frequent or occasional hybridization in nature; a hybridizing group of species.

TAPROOT: The main root of a tree that strikes downward with or without heavy branching until it either reaches an impenetrable layer or one so lacking in oxygen or moisture that further downward growth is impossible. See LATERAL ROOT and SINKER ROOT.

TEMPERATURE INVERSION: In meteorology, inversion of the vertical gradient of temperature. The temperature of the air is ordinarily observed to become lower with increasing elevation (altitude), but occasionally the reverse is the case, and when the temperature increases with elevation there is said to be an inversion.

TENSION WOOD: Abnormal wood formed on the upper side of branches and inclined boles of hardwood trees. See COMPRESSION WOOD.

TERRITORY: An isolated area defended by one individual of a species or by a breeding pair against intruders of the same species and in which the owner of the territory makes itself conspicuous.

TOLERANCE, ABSOLUTE: The physiological limits of a species adaptability, that is, tolerance to heat and cold; excess moisture and moisture deficiency; response to nutrients; snow and ice; and so forth. See TOLERANCE, RELATIVE.

TOLERANCE, RELATIVE: The capacity of a tree to reproduce and grow in the shade of and in competition with other trees (syn. shade tolerance, light tolerance). See TOLERANCE, ABSOLUTE.

TOPOGRAPHIC INDEX: Pattern of slope profile or configuration profile of a contour. May be judged to be convex, turning outward; concave, turning inward; straight. Convex forms represent poorer growing conditions, concave are best, and straight are intermediate.

TRANSPIRATION: The loss of water vapor from the plants through the stomates.

TREE: A woody plant having one well-defined stem and a more or less definitely formed crown and roots, usually attaining a height of at least 8 feet. See SHRUB and HERB.

TREE CLASSIFICATION: A designation of all trees in a forest that are alike in certain specific characteristics, such as vigor (See CROWN CLASSIFICATION), insect susceptibility (See DUNNING CLASSIFICATION and KEEN CLASSIFICATION), or value, according to such attributes as crown class, age class, size, form, diameter, log grades and clear length (See LAKE STATES CLASSIFICATION).

TREE HEIGHT: The length of tree stem from the stump to a point in the top of specified diameter, or to the leader tip, or to the maximum extent of the crown.

TREE VOLUME: The amount of wood in a tree or stand according to some unit of measurement, for example, cubic foot, board foot.

TROPHIC: Pertaining to nutrition, esp. the life pyramid among animals; the exchange of energy at each level; plant and heterotrophic plants or animals, these latter are divided into herbivores, omnivores, carnivores. See FOOD CHAIN.

UNDERSTORY: That portion of the trees in a forest stand below the overstory. See OVERSTORY and STRATIFIED.

UNEVEN-AGED: Presence of more than two distinct (at least 3) age classes and a range of size classes (seedling, sapling, pole and sawtimber) is present. See ALL-AGED and EVEN-AGED.

VARIETY: A taxonomic subdivision of a species based on minor characteristics and often an exclusive geographic range, or an assemblage of cultivated individuals of useful and reproducible character. See RACE and CULTIVAR.

VEGETATIVE REPRODUCTION: See ASEXUAL, REPRODUCTION.

WATERSHED: The whole surface drainage area that contributes water to a lake. The total area above a given point on a stream that contributes water to the flow at that point (syn. drainage basin, catchment basin, river basin).

WATER TABLE: Presence of gravitational water in a soil as a result of an impermeable layer (See HARDPAN) preventing free drainage or because the topography limits ground water penetration (syn. perched water table).

WEED: Any herbaceous or woody intruder in a forest; a herbaceous nongrass-like plant occurring on the range; or any species that invades cultivated and fallow fields and pastures. See HERB.

WILTING POINT: The amount of water contained in a soil, measured on a dry-weight basis, at which plants wilt. Permanent wilting occurs when a plant can be placed in a saturated atmosphere and in which it does not recover. The lower limit of capillary water potential. See FIELD CAPACITY; PERMANENT WILTING PERCENTAGE.

WILDLIFE: The wild fauna.

WINDFALL: A tree uprooted or broken off by wind. An area on which trees have been thrown by the wind.

XERIC: Dry tending to be droughty. Xerophytes are plants that grow in soils with scanty water supply. See HYDRIC and MESIC.

XYLEM: The portion of the vascular system that consists of tracheal tissue and wood parenchyma; woody tissue, the wood of the vascular system. The xylem conducts fluids upward from the roots, through the stem and into the branches and leaves in plants and provides mechanical support (syn. wood).

YIELD TABLE: A table showing the progressive change in a stand's development at periodic intervals covering the range of age of a species on given sites. It may include information on average diameter and height, basal area, number of trees, volumes, and other essential data. An empirical yield table is prepared for actual stand conditions; a normal yield table is prepared for fully stocked stand conditions.

ZYGOTE: The diploid cell resulting from fertilization of one gamete by another.

GENERAL REFERENCES

Ford-Robertson, F.C. 1971. *Terminology of Forest Science, Technology, Practice and Products.* Soc. of Am. Foresters, Wash., D.C., p. 202.

Schwarz, C.F., E.C. Thor, and G.H. Elsner, 1976, Wildland planning glossary. *USDA Forest Service Gen. and Tech. Rept.* PSW-13.

Snyder, E.B. 1972. Glossary for forest tree improvement workers. *USDA Forest Service Southern Forest Expt. Sta.*

Swartz, D. 1971. *Collegiate Dictionary of Botany.* Ronald Press, New York.

Munns, E.N. (ed.). 1950. *Forestry Terminology.* Soc. of Am. Foresters, Wash., D.C.

Index

Ruffed grouse, 348, 354
Runoff and vegetation change, 384

Sand pine, variation within, 21
Saprophytic fungi, 321
Sapsuckers (insects), 328
Sawflies, 328
Scale insects, 328
Schumacher and Coile, stocking equation, 108
Scots pine, mineral contents, 264
Seasonal height growth, 88, 90
Secondary succession and site change, 152
Sedimentary rocks, 259
Seed:
    seedbed, 151. *See also* Germination
    seedcoat pathogens, 161
    seed dispersal, 49-51, 150
    seed dormancy, 51-54
    seed loss, 104
    seed production, fertilizer effect, 267
        seed-year frequency, 55, 149
        trees, 57, 351
    seed weight, 150
    seed zones, 14, 15
Seedling establishment and growth, 103-104
Selection, artificial, 26, 33-39
    natural, 11-13, 28, 30
Sexual reproduction, *see* Reproduction
Shantz and Zon, 164
Shelterbelt, use of trees for, 393
Shoot elongation, 84-88, 90, 94
Shortleaf pine:
    forest floor under, 271
    littleleaf disease, 148
    site quality, 310
    soil moisture and growth, 251
Shortleaf pine x loblolly pine hybrid, 14, 23
Shrew, 285
Sibling species, 169
Site, forest, 5
Site change as result of succession, 156
Site index:
    effect on number of trees, 129-131
    harmonized, 112
    measurement, 111
    polymorphic, 112, 398
    species, comparison, 113-114
    stand and yield, 116
Site quality, 144-146, 184-185, 298
Site types, 172, 180-182, 298
    eastern white pine, 181-182
    northern hardwood, 172, 180-181
Size, individual trees, 129-131, 319-320
Slash pine:
    climate and distribution, 235
    fertilizer and growth, 265-266
    fusiform rust, 324, 331
    gum yield, 25-26, 34
    planting off-site, 324

site quality, 311
Sleet, 232
Slope, 291, 295, 300-301
Smog, 227. *See also* Air movement; Inversions
Snow:
    accumulation in stands, 368-370
    melt, 368
    pack, 377
    as precipitation, 232
    and succession, 148
Soil, defined, 241-243
Soil air, 243-248
Soil animals:
    macroorganisms:
        arthropods, 283, 284
        mollusks, 282, 283
        vetebrates, 284, 285
    microorganisms:
        acarina, 282, 283, 284
        nematodes, 282
        protozoa, 282
        rotifers, 282
Soil biota, 242, 274-278
Soil depth and site quality:
    Douglas-fir, 302
    loblolly pine, 302, 305, 306
    longleaf pine, 310, 311
    northern hardwoods, 306-307
    oak, 300-301
    shortleaf pine, 310
Soil drainage, classes, 304, 313-314
    and site quality:
        eastern white pine, 305
        loblolly pine, 305, 306
        longleaf pine, 311
        pond pine, 308
Soil and fire, 360
Soil formation and physiographic location, 294
Soil minerals:
    elements, 259-260
    fertilizers, 265-267
    fire, 361
    retention, 262-266, 270-271
    symbiosis, 274, 277
    tree growth, 186
    uptake, 260-261
    see also Mineral cycling
Soil moisture:
    air temperature, 376
    evapotranspiration, 251-255
    photosynthesis, 250
    physiographic location, 293
    retention, 244-246, 252-253
    tree growth, 250, 311-312
Soil nitrogen cycle, 263, 274
Soil organic matter:
    annual accumulation, 268-271
    fire, 362, 363
    humus, 272